INTELLIGENT ENVIRONMENTS 2021

Ambient Intelligence and Smart Environments

The Ambient Intelligence and Smart Environments (AISE) book series presents the latest research results in the theory and practice, analysis and design, implementation, application and experience of *Ambient Intelligence* (AmI) and *Smart Environments* (SmE).

Coordinating Series Editor:
Juan Carlos Augusto

Series Editors:
Emile Aarts, Hamid Aghajan, Michael Berger, Marc Bohlen, Vic Callaghan, Diane Cook, Sajal Das, Anind Dey, Sylvain Giroux, Pertti Huuskonen, Jadwiga Indulska, Achilles Kameas, Peter Mikulecký, Andrés Muñoz Ortega, Albert Ali Salah, Daniel Shapiro, Vincent Tam, Toshiyo Tamura, Michael Weber

Volume 29

Recently published in this series

ISSN 1875-4163 (print)
ISSN 1875-4171 (online)

Intelligent Environments 2021

Workshop Proceedings of the 17th International Conference on
Intelligent Environments

Edited by

Engie Bashir

Middlesex University Dubai, UAE

and

Mitja Luštrek

Jožef Stefan Institute, Slovenia

Press

Amsterdam • Berlin • Washington, DC

ISBN 978-1-64368-186-3 (print)
ISBN 978-1-64368-187-0 (online)
Library of Congress Control Number: 2021939643
doi: 10.3233/AISE29

Publisher
IOS Press BV
Nieuwe Hemweg 6B
1013 BG Amsterdam
Netherlands
fax: +31 20 687 0019
e-mail: order@iospress.nl

For book sales in the USA and Canada:
IOS Press, Inc.
6751 Tepper Drive
Clifton, VA 20124
USA
Tel.: +1 703 830 6300
Fax: +1 703 830 2300
sales@iospress.com

LEGAL NOTICE

The publisher is not responsible for the use which might be made of the following information.

Preface

Intelligent Environments (IEs) combine physical spaces with ICT and pervasive technology to increase the users' awareness of their surroundings, empower them to carry out their tasks, enrich their experience, and enhance their ability to manage such environments. IEs' growing community, from academia to practitioners, is working on bringing IEs to life. Their work is driven by innovative ideas and technological progress, which makes it possible to place an increasing number of sensors and computing devices in IE in an affordable and energy-efficient manner.

The 17th International Conference on Intelligent Environments is taking place in the challenging times of COVID-19. ICT researchers and engineers have mostly been fortunate in that the pandemic has not significantly impacted their ability to work. Due to social distancing measures, a lot of human activity has moved online and to virtual environments, which is actually stimulating for ICT development. On one hand, this makes conferences such as ours more accessible as it lowers the cost and removes the need to travel. On the other hand, it diminishes the experience. An important aspect of conferences is engaging in informal discussions and sharing ideas in corridors, over a coffee or meal. Despite some efforts, this does not (yet) translate well to virtual environments. This was particularly felt by the workshops at this year's conference, which are unfortunately fewer than in previous years. Nevertheless, we are pleased to include the proceedings of the following workshops, which emphasize multi-disciplinary and transversal aspects of IEs, as well as cutting-edge topics:

- 10th International Workshop on the Reliability of Intelligent Environments (WoRIE'21)
- 3rd International Workshop on Intelligent Environments and Buildings (IEB'21)
- 1st International Workshop on Self-Learning in Intelligent Environments (SeLIE'21)
- 1st International Workshop on Artificial Intelligence and Machine Learning for Emerging Topics (ALLEGET'21)

The proceedings contain contributions reflecting the latest research development in IEs and related areas, focusing on pushing the boundaries of the current state of the art and contributing to an establishment of IEs in the real world. We would like to thank all the contributing authors, as well as the members of the organizing committees and program committees of the workshops for their highly valuable work, which contributed to the success of the Intelligent Environments 2021 event. We are grateful to our technical sponsors, the IEEE Systems Man & Cybernetics Society, IOS Press and the Association for the Advancement of Artificial Intelligence.

The Editors
April 2021

Committees

Workshop Chairs

Engie Bashir, Middlesex University Dubai, UAE
Mitja Luštrek, Jožef Stefan Institute, Slovenia

10th International Workshop on the Reliability of Intelligent Environments (WoRIE'21)

Organizing Committee

Miguel J. Hornos, University of Granada, Spain
Juan Carlos Augusto, Middlesex University, United Kingdom
Carlos Rodriguez-Dominguez, University of Granada, Spain
Aditya Santokhee, Middlesex University, Mauritius Branch

Programme Committee

Oana Andrei, University of Glasgow, United Kingdom
Obinna Anya, Google, USA
Serge Autexier, DFKI Bremen, Germany
Gourinath Banda, Indian Institute of Technology Indore, India
Everardo Barcenas, National University of Mexico (UNAM), Mexico
Saddek Bensalem, Université Grenoble Alpes, France
Stefano Chessa, Università di Pisa, Italy
Eun-Sun Cho, Chungnam National University, Korea
Antonio Coronato, Institute for High Performance Computing and Networking, Italy
Marilia Curado, University of Coimbra, Portugal
Jesús Favela, CICESE, Mexico
Lori Flynn, CERT, USA
Yann-Gaël Guéhéneuc, Concordia University, Canada
Juan A. Holgado-Terriza, University of Granada, Spain
Bahman Javadi, Western Sydney University, Australia
Saru Kumari, Chaudhary Charan Singh University, India
Thibaut Le Guilly, Cryptogarage Inc., Japan
Jun Luo, Nanyang Technological University, Singapore
Paulo Maciel, Federal University of Pernambuco, Brazil
Pedro Merino, University of Málaga, Spain
Daniela Micucci, Università degli Studi di Milano-Bicocca, Italy
Alice Miller, University of Glasgow, United Kingdom
George C. Polyzos, AUEB, Greece
Mario Quinde Li Say Tan, Universidad de Piura, Peru
Stefano Schivo, Open University, The Netherlands
Robert C. Seacord, NCC Group, USA
Alexei Sharpanskykh, Vrije Universiteit Amsterdam, The Netherlands

Sotirios Terzis, University of Strathclyde, United Kingdom
Ayşegül Uçar, Firat University, Turkey
Hirozumi Yamaguchi, Osaka University, Japan
Kevin Yuen, Hong Kong Polytechnic University, Hong Kong

3rd International Workshop on Intelligent Environments and Buildings (IEB'21)

Organizing Committee

Mohammed Bakkali, International University of Rabat, Morocco and University College London, United Kingdom

Programme Committee

Mounir Ghogho, International University of Rabat, Morocco
Mohcine Bakhat, University of Vigo, Spain
Kaveh Dianati, University College London, United Kingdom
Huanfa Chen, University College London, United Kingdom
Natalia Zdanowska, University College London, United Kingdom
Andreu Moià-Pol, Universitat de les Illes Balears Palma, Spain
Bartomeu Alorda, Universitat de les Illes Balears Palma, Spain
Walter Gaj Tripiano, GreenVille Group, Italy

1st International Workshop on Self-Learning in Intelligent Environments (SeLIE'21)

Organizing Committee

Antonio Coronato, CNR-ICAR, Italy
Giovanna Di Marzo Serugendo, University of Geneva, Switzerland
Mohamed Bakhouya, International University of Rabat, Morocco
Ayşegül Uçar, Firat University Elazig, Turkey

Programme Committee

Vincenzo De Florio, Vrije Universiteit Brussel, Belgium
Claudia Di Napoli, CNR-ICAR, Italy
Marie-Pierre Gleizes, University Paul Sabatier, France
Salima Hassas, University Claude Bernard, France
Jose Luis Fernandez-Marquez, University of Geneva, Switzerland
Georgios Meditskos. CERTH, Greece
Muddasar Naeem, University La Parthenope, Italy
Giovanni Paragliola, CNR-ICAR, Italy

WWW & Social Media Committee

Angelo Esposito, CNR-ICAR (Italy)

1st International Workshop on Artificial Intelligence and Machine Learning for Emerging Topics (ALLEGET'21)

Organizing Committee

Andrés Bueno-Crespo, Catholic University of Murcia, Spain
Raquel Martínez-España, Catholic University of Murcia, Spain
Fernando Terroso, Catholic University of Murcia, Spain
Andrés Muñoz, Catholic University of Murcia, Spain

Programme Committee

José Manuel Cadenas Figueredo, University of Murcia, Spain
María del Carmen Garrido Carrera, University of Murcia, Spain
Germán Rodríguez Bermúdez, University Centre of Defence at the Spanish Air Force Academy, Spain
Carlos Periñan Pascual, Polytechnic University of Valencia, Spain
Sofia Ouhbi, United Arab Emirates University, UAE
José María Cecilia Canales, Polytechnic University of Valencia, Spain
José García Rodríguez, University of Alicante, Spain
Albert García García, University of Alicante, Spain
Andrés Muñoz Ortega, Catholic University of Murcia, Spain
Francesco Leotta, Sapienza University of Roma, Italy
Magdalena Cantabella Sabater, Catholic University of Murcia, Spain
Ramón Andrés Díaz Valladares, University of Montemorelos, Mexico
Antonio Llanes Castro, Catholic University of Murcia, Spain
Ginés David Guerrero, National Laboratory for High Performance Computing, Chile
Jesús Antonio Soto Espinosa, Catholic University of Murcia, Spain
Baldomero Imbernón Tudela, Catholic University of Murcia, Spain
Mario Hernandez Hernandez, Autonomous University of Guerrero, Mexico
Seop Na, Software Convergence Education Institute, Chosun University, Korea
Saeid Nahavandi, Institute for Intelligent Systems Research and Innovation, IISRI, Deakin University, Australia
Abdul M Mouazen, Ghent University, Belgium
Paulo Moura Oliveira, UTAD University Vila-Real, Portugal
José Boaventura Cunha, UTAD University Vila-Real, Portugal
Josenalde Oliveira, Agricultural School of Jundiaí- Federal University of Rio Grande do Norte, Brazil
Raúl Parada, Centre Tecnològic Telecomunicacions Catalunya, Spain
Nuria Vela, Catholic University of Murcia, Spain
Gabriel Pérez Lucas, University of Murcia, Spain
José Fenoll Serrano, Instituto Murciano de Investigación y Desarrollo Agrario y Alimentario (IMIDA), Spain
Isabel Garrido Martín, Instituto Murciano de Investigación y Desarrollo Agrario y Alimentario (IMIDA), Spain
Jaehwa Park, Chung-Ang University, Korea
Joan Melià Seguí, Universitat Oberta de Catalunya, Spain
Carlos Monzo Sánchez, Universitat Oberta de Catalunya, Spain
Ioannis Chatzigiannakis, Sapienza University of Rome, Italy
Andrea Vitaletti, Sapienza University of Rome, Italy

Contents

1st International Workshop on Self-Learning in Intelligent Environments (SeLIE'21)

1st International Workshop on Artificial Intelligence and Machine Learning for Emerging Topics (ALLEGET'21)

10th International Workshop on the Reliability of Intelligent Environments (WoRIE'21)

Intelligent Environments 2021
E. Bashir and M. Luštrek (Eds.)
© *2021 The authors and IOS Press.*
This article is published online with Open Access by IOS Press and distributed under the terms
of the Creative Commons Attribution Non-Commercial License 4.0 (CC BY-NC 4.0).
doi:10.3233/AISE210072

Introduction to the Proceedings of WoRIE'21

Carlos RODRÍGUEZ-DOMÍNGUEZ[a], Aditya SANTOKHEE[b],
Miguel J. HORNOS[a,1] and Juan C. AUGUSTO[c]

[a] *Software Engineering Department, University of Granada, Granada, Spain*
[b] *Department of Computing, Middlesex University Mauritius, Mauritius*
[c] *Department of Computer Science, Middlesex University, London, United Kingdom*

This section of the proceedings is a compilation of the accepted contributions to be presented at the 10th International Workshop on the Reliability of Intelligent Environments (WoRIE 2021), which should have been held within the 17th International Conference on Intelligent Environments (IE 2021) in Dubai, UAE, on June 21-24, 2021. However, due to the on-going pandemic caused by the COVID-19 outbreak, the whole conference is being held completely online for a second consecutive year, as happened with IE'20, planned to be held in Madrid (Spain). We believe that it is necessary for greater effort and collaboration between researchers and practitioners to provide a more holistic and unified methodology to develop higher quality IEs. Thus, this event aims to serve as a forum whereby researchers, academics and professionals engaged in the development of Intelligent Environments (IEs) can discuss about how to make them more reliable, safer, and securer, as well as increasing user confidence in them.

We are grateful to Dr. Stefano Chessa for accepting our invitation to give the keynote speech of this edition. His talk will focus on reliability of an e-health prototype operating within the context of IEs. The selected contributions cover wide ranging topic areas representing a well-balanced distribution of theoretical and practical efforts: security issues of cloud-based IoT system, discussion on functionalities offered by recent versions of a computational tool to construct and evaluate mathematical models which estimate system behaviours, findings of the integration between a fuzzy controller and a connected digital home, development of multi-purpose conversational systems using a modular architecture, and an assessment of uncertainty in whole building simulation models. With all this, we hope to have a successful and fruitful workshop, where all the audience actively participate and discuss on the topics addressed. We also expect that readers enjoy the selected papers included in these proceedings.

On a final note, we wish to express our sincere thanks to the authors of the submitted papers for their very interesting and high-quality contributions; WoRIE'21 Program Committee members, for their excellent work and invaluable support during the review process; and IE'21 Workshops Chairs, for their help and support. All of them have made possible to successfully organize the present edition of this workshop, which is being consolidated year after year.

---

[1] Corresponding author, Software Engineering Department, University of Granada, E.T.S. de Ingenierías Informática y de Telecomunicación, 18071 Granada, Spain; E-mail: mhornos@ugr.es.

Intelligent Environments 2021
E. Bashir and M. Luštrek (Eds.)
© 2021 The authors and IOS Press.
This article is published online with Open Access by IOS Press and distributed under the terms
of the Creative Commons Attribution Non-Commercial License 4.0 (CC BY-NC 4.0).
doi:10.3233/AISE210073

Reliability Aspects in E-Health Pilot Experimentations

Stefano CHESSA [a]
[a] *Department of Computer Science, University of Pisa, Italy*

Abstract. In many experimental research projects on e-health that make use of smart technologies (like Internet of Things and Artificial Intelligence), it is common (and good) practice to organize experimental pilot studies aimed at validating the proposed protocols and solutions. Such an experimentation however, is at the boundary of different disciplines, involve humans and it typically addresses a wide range of stakeholders (including medical specialists and patients). Furthermore, it is also dependent on several user-related aspects like acceptance, usability, ethics and on the specific common practices of the applicative domain.

It is clear that this scenario poses several challenges that go beyond the typical challenges of experimentations that just focus on technology. The collection of requirements, for example, needs to address a variety of stakeholders, and in some cases the interaction with some of them must be mediated. The development of the required technology itself may be constrained by ethical requirements, and the planning of the pilot should also address the validation of the medical practice in the use of the developed technology and not just the validation of the technology itself, a fact that makes the technological design blurred with the development of the medical practice. Finally, the pilot themselves are exposed to human factors due to the involvement of end users. All these aspects may, in the end, impact on the reliability of the entire pilot.

Taking inspiration from a recently concluded project comprising a relatively long experimental pilot that involved technologies developed on-purpose, humans and medical protocols, this keynote discusses the reliability issues arose during the pilots and their impact on the final evaluation of the project results.

Keywords. e-health, IoT, artificial intelligence, experimental pilots, human factors

Intelligent Environments 2021
E. Bashir and M. Luštrek (Eds.)
doi:10.3233/AISE210074

Perception of Security Issues in the Development of Cloud-IoT Systems by a Novice Programmer

Fulvio CORNO [a], Luigi DE RUSSIS [a] and Luca MANNELLA [a,1]

[a] *Politecnico di Torino, Corso Duca degli Abruzzi 24, 10129 Torino, Italy*

Abstract. It is very hard (or ineffective) to take an old system and add to it security features like plug-ins. Therefore, a computer system is much more reliable designed with the approach of security-by-design. Nowadays, there are several tools, middlewares, and platforms designed with this concept in mind, but they must be appropriately used to guarantee a suitable level of reliability and safety. A security-by-design approach is fundamental when creating a distributed application in the IoT field, composed of sensors, actuators, and cloud services. The IoT usually requires handling different programming languages and technologies in which a developer might not be very expert.

Through a use case, we analyzed the security of some IoT components of Amazon Web Services (AWS) from a novice programmer's point of view. Even if such a platform could be secure by itself, a novice programmer could do something wrong and leave some possible attack points to a malicious user.

To this end, we also surveyed a small pool of novice IoT programmers from a consulting engineering company. Even if we discovered that AWS seems quite robust, we noticed that some common security concepts are often not clear or applied, leaving the door open to possible issues.

Keywords. AWS, cloud, cybersecurity, IoT, novice programmers, security

1. Introduction

Designing a system is one of the more critical parts of the development life cycle since this phase's decisions could be part of the product's final release. For this reason, it is essential to design new systems having (at least) the main security concepts in mind. In this way, it will be less probable to expose possible attack points when the final product is ready for the users.

Unfortunately, this is not always so easy, especially for a programmer that is new to a particular technology. It is challenging to avoid "by design" security issues if the programmer does not feel comfortable with a specific technology. The Information and Communications Technology world is changing continuously: several technologies, products, frameworks, and software libraries appear and die every year. So, in a diversi-

---

[1]Corresponding Author: Luca Mannella, Politecnico di Torino, Corso Duca degli Abruzzi 24, 10129 Torino, Italy; E-mail: luca.mannella@polito.it.

fied programming field like the Internet of Things (IoT), even an experienced programmer could be considered a novice in a new application domain from a certain perspective.

In this work, we will not talk about someone that is currently learning how to program, but we are going to focus our analysis on *novice IoT programmers*, those software developers that had never developed a full-working and ready-for-production IoT system. To understand better novice IoT programmers' security perception, we analyzed a use case related to a teaching activity recently conducted by our research group. This use case is focused on *Amazon Web Services* (AWS), which currently is one of the most complete, widespread, and reliable cloud computing platforms available on the market [1]. AWS respects *confidentiality, integrity, and availability* (CIA) of the customer's data [2], and it provides services on demand: the customers pay only for resources that effectively consume. According to all these characteristics, it is quite probable that a novice programmer starts from AWS for developing a cloud IoT application with a robust and easy back-end. Even if such a platform could be secure by itself [2], a lack of experience could bring the developers to leave some possible attack points to a malicious user. Therefore, in case something goes wrong, the programmer can not wholly blame Amazon.

For all these reasons, we decided to conduct a survey on a small group of novice IoT programmers — taken from a consulting engineering company in Italy — to understand how much a lack of knowledge could be potentially harmful to the final product of this kind of developer.

In this paper, we are going to have an analysis of the related literature in Section 2. In Section 3, we will present the use case taken into account in our research activity, focusing on the possible reliability issues in Section 3.1. Afterward, in Section 4, we will conduct an AWS analysis focused on the previously highlighted attack points. Meanwhile, in Section 5, we will examine the perception of these issues by our group of novice IoT programmers. To conclude, we will discuss in Section 6 what is the actual severity of the previously cited attack points considering both the AWS design and the developers' perceptions. In the end, in Section 7, we will provide some considerations related to this topic, and we will propose some insights for future works.

2. Related Work

The study of the way of thinking of novice programmers is not a new field of research; in 1989, Soloway and Spohrer published an entire book [3] to understand better the main issues of this class of programmers. More recently, in 2005, Lahtien et al. [4] conducted a quite extensive survey — counting 559 students and 34 teachers from 5 different countries — focused on the most common issues in learning how to programming by a university student. In this study, scholars pointed out that novice programmers' biggest problem should not be to understand a computer science course's concepts but, instead, learning how to apply them. This issue could be one of the reasons why, even if the attendees of our survey are aware of some security concepts, they did not apply them while developing their prototypes.

Even in our research group, this is not the first research activity related to novice programmers. We started studying *novice IoT programmers* of 2 different editions of the "Ambient Intelligence" course (2014 and 2015), held in Politecnico di Torino. In a previous study [5], we highlighted some of the most painful points for this class of developers.

In particular, we discovered that novice programmers perceived as extremely difficult the tasks regarding: integrating various subsystems, interaction with proprietary third-party services, and the configuration of mobile, web, or hybrid applications. Moreover, we are still studying and working on possible solutions to help programmers develop better IoT solutions like a computational notebook focused on IoT technologies [6].

However, it is not so common to find studies related to the *security perception* of a novice programmer. Usually, researchers analyzed this kind of perception from a non-technical point of view. For example, Varga et al. [7] published a work related to the cyber-threat perception of actors belonging to the Swedish financial sector.

Nowadays, the benefits of cloud computing are clear both for technicians and for the general public. However, every technology can expose security threats. Scholars have studied possible security issues of cloud computing since the very beginning of its world-wide diffusion [8][9]. On the one hand, according to the security section of their web-site[2], Amazon puts much effort into keeping AWS a reliable service. In their white papers, Amazon ensures that the IT infrastructure that AWS provides to its customers is designed and managed in alignment with security best practices and several IT security standards [10]. This commitment also seems to be confirmed by the work of other scholars [11]. On the other hand, one of the key points stressed inside the AWS policy is the shared responsibility model [12]. Since AWS and its customers share control over the IT environment, security is not entirely a duty of Amazon, but it is a responsibility shared with its customers. When it comes to managing security and compliance in the AWS Cloud, each party has distinct responsibilities. For this reason, the programmers must not underestimate their role in keeping their application developed through AWS secure.

Meanwhile, regarding IoT systems, the recent literature shows that IoT systems seem to be in a dangerous situation. According to the research activities of Kumar et al. [13], IoT systems are becoming more and more important in people's houses; approximately 40% of houses all around the world have at least an IoT device, and about 70% in North America. In their work, Kumar et al. analyzed an extensive data set of IoT devices (83M) in a considerable amount of real houses (15.5M). They discovered that a surprising number of devices still support File Transfer Protocol (FTP) and Telnet — both considered not secure nowadays. Furthermore, on a specific day, they perform an in-depth analysis of the data coming from the users who are actively using *Avast Wi-Fi inspector*[3]). They find out that 62% of the scanning houses contained at least one known vulnerability. In another recent study [14], Kafle et al. demonstrate a lateral privilege escalation attack on a cloud IoT environment (Google Nest[4]). In their study, they explain as, quite often, the insecurity of the platform is related to errors of third-party programmers. Even if Nest is trying to keep its platform secure through a review process, they do not execute this review on "new" applications (i.e., applications downloaded by less than 50 users) that are more likely applications developed by novice programmers.

Even if other scholars are studying countermeasures to the weaknesses of IoT systems (e.g., through sharp Intrusion Detection Systems [15]), we decided to conduct this work to understand better what is the real *security perception* of novice IoT programmers. In particular, we are interested in getting relevant insights about the possibility of helping them design more reliable IoT systems from the very beginning.

---

[2] https://aws.amazon.com/security/, last visited on February 26th, 2021.
[3] https://www.avast.com/internet-security, last visited on February 28th, 2021.
[4] https://store.google.com/category/google_nest, last visited on February 28th, 2021.

3. Use Case

During the end of 2020, our research group was teaching a professional training course for a consulting engineering company in Turin, Italy. The course's main goal was to teach a small group of programmers how to develop a Cloud-IoT-based application from scratch. This course started by introducing the IoT world, explaining possible advantages, disadvantages, and challenges. It also explained one of the most used protocols in this field: Message Queuing Telemetry Transport (MQTT). After this part, the programmers were introduced to cloud computing technologies starting from a general perspective. Subsequently, they were introduced to Amazon Web Services (AWS), focusing on cloud computing for IoT. In the end, the software developers learned some additional concepts related to how to develop a web server using the Representational State Transfer (REST) approach.

The course was organized in 4 non-consecutive days (one day per week), with theoretical sections in the mornings and practical laboratories in the afternoons. To conclude the course, each attendee has to develop a full-working prototype with all the components shown in Figure 1. This prototype had to be presented in a final lecture.

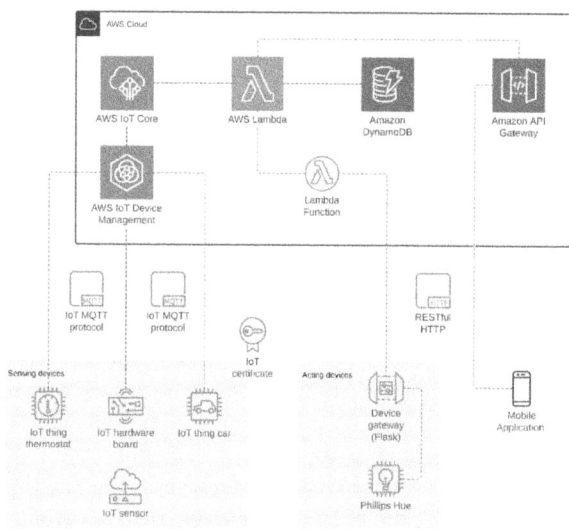

Figure 1. The architectural schema of the use case architecture.

The main components of the project could be divided into four different categories:

- **Sensing devices**: devices used to retrieve some data from the physical world;
- **Acting devices**: devices used to act on the physical world (e.g., a smart lamp);
- **AWS Cloud**: a cloud back-end server able to manipulate and store all the data necessary for the application;
- **Front-end devices**: used to interact with the back-end (e.g., a *mobile application*).

3.1. Main Architecture Attack Points

Considering the proposed the architecture (Figure 1), we defined the main attack points according to our knowledge and to the state of the art of security issues in the IoT field (e.g., [13] [14] [15] [16]). At the end of our analysis, we identified these five main attack points in the use case architecture:

1. the data flow between a *sensor* and the *AWS back-end*;
2. the data flow between *AWS back-end* and a *physical actuator* (e.g., a smart lamp);
3. the data flow between the *AWS API Gateway* and the *user's device*;
4. the developed code stored inside *AWS back-end* (e.g., AWS Lambda component);
5. the *back-end database* offered by Amazon (i.e., Dynamo DB).

For instance, a possible attack that could occur on all the *data flows* could be a Man-In-The-Middle (MITM) attack[5] or a Replay attack[6]. Furthermore, a possible issue, if the database is not well protected and the data stored are not encrypted, could be a steal of information — that could be directly used by the malicious user or sold. Finally, our analysis's last attack point is the code on the AWS back-end, for example, on the Lambda component. This component is probably the most critical point. If the code on the back-end of the application is compromised, the attacker will become able to do whatever she wants with the developed system.

4. AWS Analysis

In this section, starting from the architecture shown in Figure 3, and from the points presented in Section 3.1, we have identified how a malicious user can conduct these attacks, and we verified what countermeasures are used by AWS components.

4.1. Data Flow Protection

Three of the attack points highlighted in Section 3.1 are related to data flows. During transmission, data could be potentially eavesdropped, tampered with, and forged. The most common way to counter these dangers is by using cryptography on the transmitted data. According to AWS documentation [10], users must establish all the connections with the AWS back-end using *HTTPS*. This protocol is an extension of traditional HTTP using the Transport Layer Security (TLS) protocol to encrypt the data flow. For this reason, HTTPS is also called *HTTP over TLS*. TLS is a widespread protocol based on asymmetric cryptography. At the current time, TLS is considered a reliable and secure protocol, so the Amazon approach is coherent with the state-of-the-art. Furthermore, if a user requires additional security, AWS offers the possibility to create an Amazon Virtual Private Cloud (VPC) and use IPsec to contact the VPC.

---

[5] An attack in which a malicious user intercepts and modifies the data meanwhile they are being transmitted.
[6] An attack in which a valid data pack is stored and sent by a malicious user in a different moment.

4.2. AWS Back-End Protection

To have access to AWS Back-end services (like AWS Lambda) users must be authenticated. AWS provides two different types of user account: *root* user and *Identity and Access Management* (IAM) user. The root user is created by a customer when she registers for the first time to AWS. It is the account with the highest privileges. The root account can create several IAM accounts and assign them the necessary privileges (e.g., the right of writing on a particular database or the right to use a specific service). Even in this case, registration and login data flows are protected using HTTPS. Indeed, assuming that AWS implemented the registration and login processes correctly, the main possible failure points are two: an *inadequate privileges policy* and *bad passwords*.

An example of a good privileges policy is the *Principle of Least Privilege* (POLP). This principle's idea is that any user, program, or process should have only the *bare minimum* privileges necessary to perform its function. Furthermore, it is a good habit not to share an account between multiple users. Each user should have their accounts. Even if Amazon strongly recommends using the root account only for creating IAM accounts (and for other very few tasks), this best practice is not enforced at all. From our perspective, AWS should force its customers to create at least one IAM account. This approach will help new users to understand this security concept better.

Instead, the password policy enforced by AWS for root and IAM accounts is quite robust. Currently, it forces the users to create passwords with a minimum length of 8 symbols and at least two of the following three characteristics:

- including both lowercase and uppercase characters:
- including a number;
- including a non-alphanumeric symbol.

Unfortunately, we noticed that AWS seems not to have taken any particular countermeasures for dictionary attacks (for example, a password like "Amaz0nWS" is correctly accepted when registering a new account). On the other hand, for the IAM accounts, there is the possibility of let AWS auto-generates the password. The generated passwords have a length of 16 characters and include all the characteristics previously cited.

4.3. AWS Back-End Database Protection

Amazon provides different database solutions (both relational and NoSQL), and it offers database-related services as data-warehouse. In particular, in our use case, the developers have to use DynamoDB, a NoSQL database service that provides fast and predictable performance with seamless scalability. Amazon DynamoDB is accessible via TSL-encrypted endpoints.

By default, the DynamoDB service encrypts all data at rest to enhance data security. A customer can use the default encryption, the AWS-owned Customer Master Key (CMK), or the AWS-managed CMK to encrypt all data. DynamoDB also offers support to switch encryption keys between the AWS-owned CMK and AWS-managed CMK.

To avoid the risk of losing data, a user can set up automatic backups using a particular template in AWS Data Pipeline that was created just for copying DynamoDB tables. AWS offers the possibility to decide between full or incremental backups in the same or in a different region.

To control who can use the DynamoDB resources and API, the user has to set up permissions in the IAM service. Through IAM policy, a user can also specify a fine-grained access control policy (e.g., to allow or deny access to specific rows or columns). Additionally, each request to the database must contain a valid HMAC-SHA256 signature. HMAC (Hash-based Message Authentication Code) is an authentication code to be sent together with the request, generated using SHA256 (Secure Hash Algorithm 256), a cryptographic hash function belonging to the SHA-2 family. Even in this case, this could be considered a state-of-the-art approach. The AWS Software Developer Kits automatically sign user's requests; however, each user can write their HTTP requests proving the signature in the header of the requests.

5. Developer Perspective on AWS Security

The purpose of this section is to understand if the lack of knowledge of a novice programmer could be compensated or not by the AWS countermeasures cited in Section 4. To do this, we prepared a survey starting from the use case presented in Section 3. In this section, we will present the survey structure and the answers provided by our participants.

5.1. Structure of the Survey

At the very end of the course, after the final review of the projects — when the students are already aware if they successfully passed the exam or not — we asked the attendees to fulfill a survey. Thanks to these survey results, we have an initial idea of the perception of the security issues during the development of a Cloud-IoT system from a novice programmer's point of view, focusing on AWS.

We divided the survey into three distinct sections: "background and individual studying", "possible attack points", and "countermeasures and best practices".

In the first section, we asked the participants how much they think to be cybersecurity experts, how much they think that security is important in IoT systems, and who is in charge of the security of something developed on AWS infrastructure. Then, we showed Figure 1 as a reminder of their systems' architecture. We demanded the attendees to think about the possible security issues of the architecture and how they managed these issues during the development (if they did not manage the issues, we also asked why).

In the section "possible attack points", we introduced what we already explained in Section 3.1. We required the students to sort the attack points from the easiest to the hardest to attack and according to the possible severity if an attack succeeded. Furthermore, the programmers had to specify the worst possible consequence if a malicious user attacks the most critical point and, to conclude, how many of those points they considered while developing their projects.

In the last section, we wanted to understand if they tried to take some countermeasures (e.g., robust passwords, different accounts with different privileges, encryption algorithms, etc.) and if they are aware that AWS provides some components to help them manage IoT security (e.g., AWS IoT Device Defender[7]).

---

[7]https://aws.amazon.com/iot-device-defender/, last visited on February 27th, 2021.

5.2. Survey Results

In total, the course had 9 *novice IoT programmers*; we were able to collect answers from 6 of them (so our survey had an answer rate of 67%). All the attendees were male.

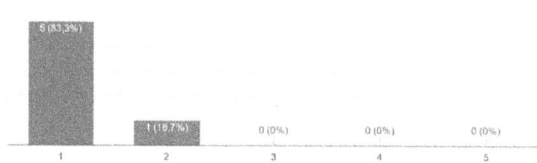

Figure 2. Novice programmers' competence perception about cybersecurity.

From one side, our programmers feel very inexperienced about cybersecurity; on a scale from 1 to 5, 5/6 participants answer 1, while the other select 2. On the other side, they think that cybersecurity is quite important, and it depends on the severity of the implemented software solution. All the novice programmers think that the implemented architecture could include a security issue, but no one said to have done something to mitigate the highlighted problems. At the question related to whom is in charge of the security of what is developed on AWS, 4/6 participants answer "both developer and AWS". Instead, 2/6 consider the responsibility *entirely of the developer.*

The novice programmers consider the AWS database the most secure component among the presented attack points. In contrast, the data flows between the AWS back-end and the sensors/actuators are considered the less secure points. The most critical point in case of a successful attack is the data flow from the back-end to the actuators, followed by the developed Lambda functions on the back-end and the data flow from sensors to the back-end. 5 participants consider as the worst possible consequence a cyber-physical attack: a security breach in cyberspace that impacts the physical environment (e.g., activating or deactivating a machine). They are mainly afraid that an attacker can take control of the system to damage a machine or a person. The second biggest concern seems to be data loss. The novice programmers consider potentially attackable at most 2 of the 5 attack points (3/6 answers). At the same time, 2 participants did not think to anyone of the presented attack points during the development phase.

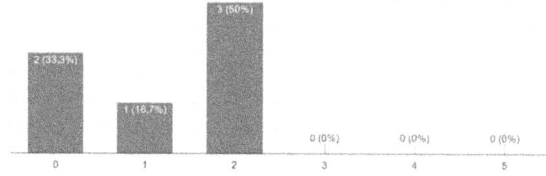

Figure 3. The number of attack points considered attackable during the development phase.

All the participants created a strong password (probably because AWS enforces it), but only 1 developer have used the service for creating an IAM account, while all the others use only the root account. Furthermore, only 2 participants specify that they did not create IAM accounts but are aware that they should had. Talking about data flows cryptography, 4 novice programmers did not check if the platform uses TLS for encryption or not. Only 1 developer verifies that TLS is used on the data flow from the sensors to the AWS back-end. Finally, regarding the data stored inside DynamoDB, 5/6 participants did not think if they were encrypted or not. To conclude, nobody thinks about using an additional tool provided by Amazon to improve his solution's security. Only one participant said to have heard about *IoT Device Defender*, but he decided not to use it for the time constraint.

6. Discussion

It is interesting to notice that even if our group of programmers thinks that cybersecurity is quite critical, they do not feel to be experts in this field. Perhaps, it is a consequence of this perception if they did not pay too much attention to security requirements during their solution development. Furthermore, nobody thought to use an additional tool to compensate for their lack of knowledge and enhance their solution's robustness (e.g., using AWS IoT Device Defender). All the participants thought that the architecture could include security issues, but no one acted to mitigate the problem. This is a little bit strange considering that for two respondents, the solution's security is entirely the responsibility of the developer (meanwhile, for all the others is equally divided between the developer and AWS). We could try to explain this lack of action by considering that this project is just the outcome of a training course, and cybersecurity aspects were not the focus of our course.

The AWS database is considered the most secure point in the architecture. This perception could be quite dangerous considering that most programmers do not take care of database encryption. Even if AWS takes care of data at rest, programmers should remember to properly manage critical information (e.g., users' passwords). Our use case does not include users' registration, but properly hashing users' passwords is crucial in a more complex application. According to our participants, the less secure points are the data flows from and to the AWS back-end. They are mainly apprehensive about cyber-physical attacks. Nevertheless, 4 programmers did not think about the security of the communication channels. Even in this case, Amazon does not allow connection not protected with TLS, so, fortunately, a lack of competence does not create a potential attack point.

We noticed that thanks to AWS policy, all the participants had created a strong password. On the contrary, almost no one had used AWS IAM service to create separate accounts with different privileges. Using a unique account for all the possible operations could create many problems if a malicious user owns the password. From our perspective, this issue could be mitigated if AWS enforce users to create at least an IAM account. A tutorial phase dedicated to this particular issue could help novice programmers understand why it is important to use a good privilege policy like Least Privilege's principle.

7. Conclusions

From this preliminary analysis, we can conclude that AWS seems to be a good choice for implementing a secure Cloud-IoT solution even for a novice IoT programmer. The platform implements concepts at the state-of-the-art, and it has many useful tools to start developing. AWS is quite able to compensate for the lack of knowledge of the novice developers involved in the study. Our main suggestion is related to the accounts' policies. We believe that forcing users to create separate accounts through the AWS IAM service could be a good starting point to improve the novice programmers' awareness of good privileges policies.

We strongly believe that security-by-design is the best way to improve the reliability of every system. To achieve this goal, the primary security principles should be inside programmers' minds since the beginning of the development cycle. Thanks to this study, we understood that novice IoT developers tend to consider their system's security as an important feature but to manage it as an additional component.

In our future works, we will have a more focused survey on a larger sample of novice IoT programmers to understand better this phenomenon's dimension. Later on, we would like to provide some best practices and tools to help such novice programmers develop much more reliable IoT systems.

Acknowledgment

We want to acknowledge the employees who voluntarily decided to participate in the survey to conduct this research activity.

References

[1] Bala R, Gill B, Smith D, Wright D, Ji K. Magic Quadrant for Cloud Infrastructure and Platform Services. Gartner Inc.; 2020. Available from: https://www.gartner.com/doc/reprints?id=1-1ZDZDMTF& ct=200703&st=sb.

[2] Inc AWS. Introduction to AWS Security. 410 Terry Avenue North Seattle, WA 98109 United States: Amazon Web Services Inc.; 2020. Available from: https://d1.awsstatic.com/whitepapers/ Security/Intro_to_AWS_Security.pdf.

[3] Soloway E, Spohrer JC. Studying the novice programmer. 365 Broadway, Hillsdale, New Jersey 07642 United States: Lawrence Erlbaum Associates, Inc.; 1989.

[4] Lahtinen E, Ala-Mutka K, Järvinen HM. A study of the difficulties of novice programmers. Acm sigcse bulletin. 2005;37(3):14–18.

[5] Corno F, De Russis L, Sáenz JP. Pain Points for Novice Programmers of Ambient Intelligence Systems: An Exploratory Study. In: 2017 IEEE 41st Annual Computer Software and Applications Conference (COMPSAC). vol. 1. IEEE; 2017. p. 250–255.

[6] Corno F, De Russis L, Sáenz JP. Towards Computational Notebooks for IoT Development. In: Extended Abstracts of the 2019 CHI Conference on Human Factors in Computing Systems; 2019. p. 1–6.

[7] Varga S, Brynielsson J, Franke U. Cyber-threat perception and risk management in the Swedish financial sector. Computers & Security. 2021;p. 102239. Available from: https://www.sciencedirect.com/ science/article/pii/S0167404821000638.

[8] Carroll M, Van Der Merwe A, Kotze P. Secure cloud computing: Benefits, risks and controls. In: 2011 Information Security for South Africa. IEEE; 2011. p. 1–9.

[9] Behl A, Behl K. An analysis of cloud computing security issues. In: 2012 World Congress on Information and Communication Technologies. IEEE; 2012. p. 109–114.

[10] Inc AWS. Amazon Web Services: Overview of Security Processes. 410 Terry Avenue North Seattle, WA 98109 United States: Amazon Web Services Inc.; 2020. Available from: https://d0.awsstatic. com/whitepapers/aws-security-whitepaper.pdf.

[11] Narula S, Jain A, et al. Cloud Computing Security: Amazon Web Service. In: 2015 Fifth International Conference on Advanced Computing & Communication Technologies. IEEE; 2015. p. 501–505.

[12]   Taggart M, Roach B, Woods P. Amazon Web Services: Risk and Compliance. 410 Terry Avenue North Seattle, WA 98109 USA: Amazon Web Services Inc.; 2020. Available from: `https://d1.awsstatic.com/whitepapers/compliance/AWS_Risk_and_Compliance_Whitepaper.pdf`.

[13]   Kumar D, Shen K, Case B, Garg D, Alperovich G, Kuznetsov D, et al. All things considered: an analysis of IoT devices on home networks. In: 28th USENIX Security Symposium (USENIX Security 19); 2019. p. 1169–1185.

[14]   Kafle K, Moran K, Manandhar S, Nadkarni A, Poshyvanyk D. Security in Centralized Data Store-Based Home Automation Platforms: A Systematic Analysis of Nest and Hue. ACM Transactions on Cyber-Physical Systems. 2020 Dec;5(1):1–27.

[15]   Anthi E, Williams L, Słowińska M, Theodorakopoulos G, Burnap P. A supervised intrusion detection system for smart home IoT devices. IEEE Internet of Things Journal. 2019;6(5):9042–9053.

[16]   Huang DY, Apthorpe N, Li F, Acar G, Feamster N. Iot inspector: Crowdsourcing labeled network traffic from smart home devices at scale. Proceedings of the ACM on Interactive, Mobile, Wearable and Ubiquitous Technologies. 2020;4(2):1–21.

Intelligent Environments 2021
E. Bashir and M. Luštrek (Eds.)
© *2021 The authors and IOS Press.*
This article is published online with Open Access by IOS Press and distributed under the terms
of the Creative Commons Attribution Non-Commercial License 4.0 (CC BY-NC 4.0).
doi:10.3233/AISE210075

The Mercury Environment: A Modeling Tool for Performance and Dependability Evaluation

Thiago PINHEIRO [a,b], Danilo OLIVEIRA [b,c], Rubens MATOS [b,d], Bruno SILVA [b],
Paulo PEREIRA [a,b], Carlos MELO [a,b], Felipe OLIVEIRA [a,b], Eduardo TAVARES [a,b],
Jamilson DANTAS [a,b], and Paulo MACIEL [a,b,1]

[a] *Centro de Informática, Universidade Federal de Pernambuco, Brazil*
[b] *MoDCS research group, Brazil*
[c] *HighQSoft GmbH, Germany*
[d] *Instituto Federal de Educação, Ciência e Tecnologia de Sergipe, Brazil*

Abstract. It is important to be able to judge the performance or dependability metrics of a system and often we do so by using abstract models even when the system is in the conceptual phase. Evaluating a system by performing measurements can have a high temporal and/or financial cost, which may not be feasible. Mathematical models can provide estimates about system behavior and we need tools supporting different types of formalisms in order to compute desired metrics. The Mercury tool enables a range of models to be created and evaluated for supporting performance and dependability evaluations, such as reliability block diagrams (RBDs), dynamic RBDs (DRBDs), fault trees (FTs), stochastic Petri nets (SPNs), continuous and discrete-time Markov chains (CTMCs and DTMCs), as well as energy flow models (EFMs). In this paper, we introduce recent enhancements to Mercury, namely new SPN simulators, support to prioritized timed transitions, sensitivity analysis evaluation, several improvements to the usability of the tool, and support to DTMC and FT formalisms.

Keywords. models, Mercury, DTMC, fault tree, simulation

1. Introduction

There are many tools for supporting performance and/or dependability analysis of computer systems. Some tools have limitations regarding the evaluation of non-exponential models and they support only one formalism or a small set of them, or even a limited number of evaluations. Considering this, the MoDCS[2] research group[3] started the development of the Mercury tool in 2008, having a vision to deliver to the community a tool for supporting performance and dependability evaluations overcoming the limitations found in other tools.

---

[1]Corresponding Author: Head of MoDCS Research Group and Full Professor at Centro de Informática, Av. Jorn. Aníbal Fernandes, s/n - Cidade Universitária, Recife - PE, 50740-560, Brazil; E-mail: prmm@cin.ufpe.br.
[2]Modeling of Distributed and Concurrent Systems.
[3]http://www.modcs.org/.

The Mercury tool has been developed to enable the creation and evaluation of stochastic models, such as: stochastic Petri nets (SPNs) [1], continuous and discrete-time Markov chains (CTMCs and DTMCs) [2], reliability block diagrams (RBDs) [3] and dynamic RBDs (DRBDs) [4], fault trees (FTs) [5], and energy flow models (EFMs) [6]. The tool has been widely adopted in numerous research projects, which had the results published in peer-reviewed journals[4] and conferences[5]. Mercury supports a considerable number of formalisms and empowers a large range of evaluations for each of them, so it can help the academy and industry to make predictions in several application fields. Among other distinguished features of this tool are: the possibility of evaluating non-exponential models through SPN and RBD simulations, supporting more than twenty-five probability distributions; a scripting language; a random variate generator (RVG); computation of reliability importance indices; moment matching [7] of empirical data, as well as sensitivity analysis evaluation of CTMC, SPN, and RBD models.

The MoDCS team is frequently improving the tool by adding new features, updating existing functionalities, and fixing bugs, and new versions are usually released every six months. In this paper, we introduce recent enhancements to Mercury up to version 5.0.2 that were not covered by the previous papers presenting the tool [8,9,10], namely: support to DTMC and FT formalisms; new SPN simulators; prioritized timed transitions; sensitivity analysis evaluation of RBDs; and several improvements to the usability of the tool.

This work is organized as follows. Section 2 presents a comparison between Mercury and similar tools. Section 3 provides an overview of the tool. Section 4 presents the recent enhancements. Section 5 presents a case study as an example to demonstrate the feasibility of using Mercury for supporting infrastructure planning, and finally Section 6 draws the final remarks.

2. Related Tools

This section presents a general comparison of formalisms supported by Mercury and other similar tools. There are many modeling tools available worldwide, each one with its respective pros and cons, as well as with a set of modeling formalisms associated, which varies according to their representation power and complexity. Among the most cited tools, we can mention ReliaSoft BlockSim[6], Relex[7], SHARPE [11], TimeNet [12], Snoopy [13], SPNP[8], and GreatSPN[9]. BlockSim provides the way to evaluate some dependability attributes (reliability, availability, and maintainability) with RBDs and FTs models. The main limitation of BlockSim is that it cannot evaluate components dependencies with its current modeling formalisms. To evaluate more complex scenarios, it is necessary to employ tools that can handle space-state models, such as Relex and SHARPE. Other tools like TimeNET and Snoopy provide powerful Petri Net (PN) modeling and evaluation mechanisms but are limited to PN extensions. On the other hand,

---

[4]Publications in journals: `https://www.modcs.org/?page_id=521`.
[5]Publications in conferences: `https://www.modcs.org/?page_id=525`.
[6]BlockSim: `https://www.reliasoft.com/BlockSim`.
[7]Relex: `http://www.relex.com`.
[8]SPNP: `https://www.informatik.uni-hamburg.de/TGI/PetriNets/tools/db/spnp.html`.
[9]GreatSPN: `http://www.di.unito.it/~greatspn/index.html`.

SPNP and GreatSPN support generalized stochastic Petri nets (GSPNs). Table 1 presents a general comparison of Mercury and these tools.

Table 1. Modeling tools comparison (DRBD only through Script Language).

Formalism Tool	RBD	DRBD	FT	CTMC	DTMC	SPN	EFM
BlockSim	✓		✓				
Relex	✓		✓	✓		✓	
SHARPE	✓		✓	✓		✓	
TimeNet						✓	
Snoopy						✓	
SPNP						✓	
GreatSPN						✓	
Mercury	✓	✓	✓	✓	✓	✓	✓

3. Overview and Software Architecture

This section presents the main features available on Mercury for all supported modeling formalisms. Currently, Mercury supports six formalisms via its GUI editor (see Figure 1) and an overview is depicted in Figure 2.

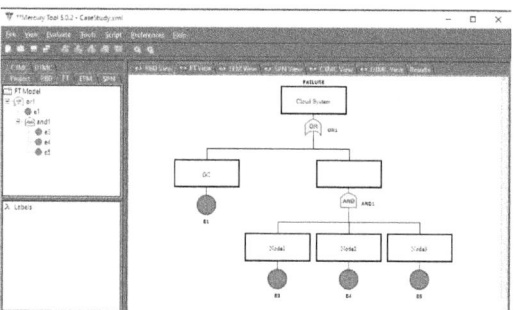

Figure 1. Mercury GUI.

The SPN editor and evaluator allow modeling and evaluating GSPNs. Mercury implements numerical analysis and simulation techniques to support performance and dependability evaluations for SPN models including steady-state and time-dependent metrics. Steady-state metrics are obtained by performing stationary evaluations and time-dependent metrics are obtained by performing transient evaluations. Regarding transient evaluation, the tool provides a simulator to evaluate the mean time to absorption (MTTA) of absorbing models with non-exponential transitions. For stationary analysis, two solution methods are available: Grassmann-Taksar-Heyman (GTH) [14] and Gauss-Seidel [2]. For transient analysis, two solution methods are available: Uniformization (also known as Jensen's method) and Runge-Kutta (4th order) [2]. Additionally, the SPN evaluator allows the execution of experiments, which evaluate the impact of varying a parameter of the model on a chosen metric. The SPN editor has also a feature called *token*

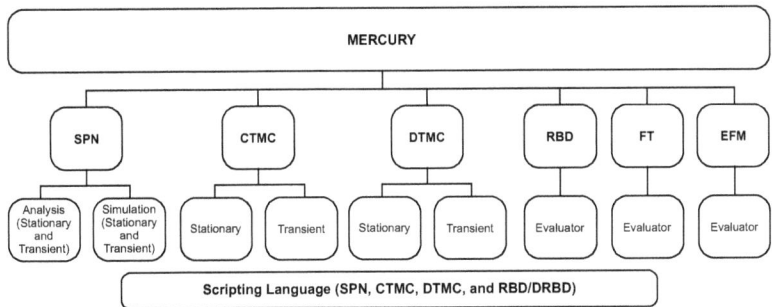

Figure 2. Mercury - Formalisms.

game to assist the validation of models. By using this feature, users can simulate graphically the firing of transitions, step by step. This mechanism allows one to change the state of the model under evaluation considering the current marking state. For example, users can simulate equipment failures as well as the corresponding consequences on the availability of a system. This feature can be useful to test the behavior of transitions that have guard expressions or priorities assigned to them, making it possible to test whether the logical rules applied to the model are properly implemented. Another important feature regarding SPN evaluation is the structural analysis. This allows the user to evaluate the structural properties of the model through analysis of place invariants, siphons, and traps [15].

The CTMC editor and evaluator allow modeling and evaluating CTMCs. Regarding stationary analysis, the GTH and Gauss-Siedel numerical techniques are supported. Regarding transient analysis, the Uniformization and Runge-Kutta numerical techniques are supported. Probabilities of absorption and MTTA may be computed when evaluating absorbing CTMCs. Experiments are also supported. Algebraic expressions using symbolic parameters can be used to define transition rates between states. Also, greek letters can be used to compose the name of parameters. In addition to states and transitions, users can define reward rates assigned to states, which enables the modeling of *Markov reward models*. Mercury also supports sensitivity analysis on CTMCs, making it possible to evaluate the sensitivity of state probabilities with respect to each input parameter entered.

The DTMC editor and evaluator allow modeling and evaluating DTMCs. The same numerical techniques available for CTMCs are also available for DTMCs. When performing stationary analysis, it is possible to compute sojourn times and recurrence time [2]. For absorbing models, Mercury also enables computing mean time to absorption and probabilities of absorption. Expressions including state probabilities may compose metrics. Unlike CTMCs, transitions between states are based on probability in DTMCs. Transition probabilities can be defined by means of algebraic expressions using symbolic parameters. Using greek letters for parameter labels and experiments are also supported.

The RBD editor and evaluator allow modeling and evaluating availability and reliability using block diagrams. The tool supports three types of block configurations: series, parallel, and K-out-of-N (KooN). RBDs provide closed-form equations making it possible to obtain the results more quickly than using other methods, such as SPN simulation. However, there are many situations (e.g., dependency among components) in which modeling using RBD is more difficult than adopting SPN. Mercury provides two methods for computing dependability metrics: SFM (a method based on the structural function) and SDP (sum of disjoint products). The SDP method, based on Boolean algebra, computes metrics considering minimal cuts and minimum paths. Mercury supports the evaluation of the following metrics: mean time to failure (MTTF), mean time to repair (MTTR), steady-state availability, instantaneous availability, reliability, unreliability, uptime, and downtime. In addition, when evaluating time-dependent metrics, multiple points on time may be computed. Experiments are also supported. Additionally, the RBD evaluator allows the calculation of component importance measures, bounds for dependability analysis, structural and logical functions as well as sensitivity analysis. Component importance measures indicate the impact of a particular component with respect to the overall reliability or availability. Thus, the most important component (i.e., that one with the highest importance) should be improved in order to obtain an increase in reliability or availability. Evaluation of dependability bounds [16] is a method to calculate dependability metrics when the model is very large. By applying such a method, approximations of the chosen metric can be obtained more quickly than solving all the closed equations involved. Structural and logical functions are alternative ways of representing the system mathematically, in which the former adopts algebraic expressions and the latter adopts boolean expressions. Mercury also provides a feature to reduce the complexity of RBD models.

The Fault Tree editor and evaluator allow modeling and evaluating availability and reliability using fault trees. FTs and RBDs differ from each other in their purposes. RBDs are success-oriented while FTs are a failure-oriented modeling approach. Using the Mercury tool, it is possible to handle two types of nodes: basic events and gates (logic ports). Leaf nodes represent basic events. Mercury supports three types of gates: *and*, *or*, and *KooN*. The events leading to the top-event *failure* must be directly linked to a *gate*, making it possible to evaluate the probability of an event happening based on the probability obtained by joining basic events and child gates. Converting FT to RBD models is supported by the tool.

The EFM editor and evaluator enable the evaluation of availability, sustainability, and cost of cooling infrastructures and datacenter/clouds power, considering the power constraints of each component. Mercury supports five types of evaluations for EFM models. *Cost Evaluation* evaluates operational, acquisition, and total costs. *Exergy Evaluation* computes the sustainability through the exergy metric. Exergy estimates energetic efficiency. *Energy Flow Evaluation* evaluates the energy that flows through each device considering the power constraints of each one. *Combined Evaluation* provides an integrated evaluation of dependability, sustainability impact, and cost of cooling infrastructures and data center/cloud power [6]. *Combined Evaluation* provides an integrated evaluation of dependability, sustainability impact, and cost of cooling infrastructures and data center/cloud power [6]. *Flow Optimization (PLDA, PLDAD, and GRASP)* evaluates SPN, CTMC, RBD, and EFM models. Three optimization techniques were implemented for supporting this evaluation: power load distribution algorithm (PLDA) [17], power load distribution algorithm depth (PLDAD), and GRASP-based algorithm.

The Mercury scripting language was designed to facilitate the creation and evaluation of complex models. It allows greater flexibility in model evaluations using the Mercury engine. The language supports SPN, CTMC, DTMC, and RBD/DRBD formalisms.

Scripts can be executed by a command-line interface (CLI) or via an editor available inside the Mercury tool. The tool has a feature that automatically generates the script representing the selected model in the GUI. The advantage of using this language in conjunction with the CLI tool, using shell scripts, for example, is the possibility to automate an evaluation workflow. In addition, there are other advantages offered by the language which are not supported by modeling through the graphical interface:

- **Support for hierarchical modeling.** The resulting metric of a model can be used as input parameters for any other model, independent of the modeling formalism being adopted;
- **Support for hierarchical transitions on GSPNs.** This type of transition can be used as a way to reduce the complexity of models or to express a recurring structure in a model. It is important to highlight that for some tools [12] the support for hierarchical SPN models is only for coloured Petri nets;
- **Support for symbolic evaluations and experiments.** Parameters of a model can be defined as variables left open. Thus, these variables can be changed at the time of evaluation in order to measure the impact of these new parameter values on certain metrics;
- **Support for loop and conditional structures.** It allows the creation of nets with variable structures. The variables for controlling those structures can be treated as parameters of the model;
- **Support for phase-type distributions [18].** The family of phase-type distributions can be used to approximate any distribution that does not fit an exponential distribution. A number of approximate analysis techniques are based on matching the moments of continuous-time phase-type distributions. Models having non-Markovian properties may only be evaluated numerically through the adoption of phase-type approximation technique [19].

A script contains models, each one with its metrics and parameters, and a main section where values for input parameters are defined. Listing 1 presents a CTMC extracted from [20].

```
markov RedundantGC{
    state fu up; state fw; state ff; state uf up; state uw up;
    transition fw -> fu(rate = sa_s2);
    transition fu -> ff(rate = lambda_s2);
    transition ff -> uf(rate = mu_s1);
    transition uf -> uw(rate = mu_s2);
    transition uw -> fw(rate = lambda_s1);
    transition fw -> uw(rate = mu_s1);
    transition uw -> uf(rate = lambdai_s2);
    transition uf -> ff(rate = lambda_s1);
    transition fw -> ff(rate = lambdai_s2);
    transition fu -> uw(rate = mu_s1);
    metric aval = availability;
} main {
    lambda_s1 = 1/180.72; mu_s1 = 1/0.966902; mu_s2 = 1/0.966902;
    lambdai_s2 = 1/216.865; lambda_s2 = 1/180.721; sa_s2 = 1/0.005555555;
    print("Availability: " .. solve(model=RedundantGC, metric=aval));
}
```

Listing 1. Mercury script for a CTMC model.

4. New Functionalities and Updates

Several new features, updates, and bug fixes have been included in Mercury. These features include support for two new formalisms, namely DTMC and FT. In addition, the stationary and transient simulators were reimplemented. Improvements were also made to the usability of the tool, such as the possibility to assign a description to each component of a model. A description represents additional information about the component for the comprehension of the model under construction. Another improvement increases the readability of models. Once a component has been inserted, it is possible to read its properties in the drawing area by positioning the mouse cursor on it. A tooltip appears showing all properties of the component. Mercury provides this feature for all components of all supported formalism. As follows, new functionalities and improvements are presented.

The DTMC editor is one of the main features that have been added to Mercury. The analysis of DTMC models comprehends the computation of holding time and recurrence time [2], besides the usual state probabilities that are already obtained for CTMC models. The fault tree editor is another main feature that was included in the latest Mercury release. FT is a top-down logical diagram and it makes it possible to create a visual representation of a system showing the logical relationships between associated events and causes lead that may lead the evaluated system to a failure state. In the current version, Mercury supports *and*, *or* and *koon* logic gates. In the future, we intend to add suport to other logical operations such as *xor* and *priority and*.

The two main updates on the RBD editor were the implementation of sensitivity analysis and the change in the way the nodes are represented. Mercury computes partial derivative sensitivity indices for RBDs, which indicate the impact that every input parameter has on availability. Sensitivity analysis can only be performed when the model under evaluation has only exponential blocks.

Several improvements were performed on the SPN editor and evaluator. Regarding the drawing area, Mercury now supports two types of arc styles: rectangular and curved. An expression editor was implemented to make it easy to create large and complex guard expressions and metrics. The editor highlights parentheses, brackets, and braces as well as some keywords. Also, a reference updater was implemented to update all properties marking references to a definition when it is updated or removed. A definition is a variable that stores a numeric value. It may be attached to some properties of others SPN components. More than one property or expression may refer to the same definition. Definitions are useful for supporting experiments. In this case, by changing the value of a definition, it is possible to evaluate the impact of that change on an evaluated metric. Regarding the SPN evaluator, similar to immediate transitions, Mercury now supports prioritized timed transitions. It is important to highlight that in that case immediate transitions always take precedence to fire over timed transitions. Both SPN simulators were reimplemented. Besides several improvements, such as on the GUI and the possibility to export detailed information of the simulation, the stationary and transient simulators can detect the occurrence of rare events, which occur when the difference between the delays assigned to the transitions is very large.

- **Stationary Simulator.** Stationary simulation can be used when evaluating steady-state metrics of non-Markovian SPN models. Mercury implements the method of batch means [21]. This method comprises three steps: running a long simulation run, discarding the initial transient phase, and dividing the remaining events run into batches. In this new simulator, it is possible to follow the simulation step by step on the GUI. A large number of statistics are computed and displayed at the end of the simulation;

- **Transient Simulator.** Transient simulation may be adopted when evaluating metrics of non-Markovian SPN models considering a specific point in time. A transient simulation is composed of a set of replications where each replication is composed by a set of runs [21]. Each run executes from time 0 until the evaluated time t' is reached. A set of sampling points may be evaluated considering this time interval. When the current set of runs is finished, the value of each sampling point of the current replication is computed. A replication represents the mean values of the points in its set of runs. Mercury supports two methods for computing the value of each point:

  * **DES[10] + Linear Regression 1** computes the value of each sampling point at the end of each run using linear interpolation between two known points. When the required number of runs is executed, the obtained values for each point are stored. Also, the mean of the obtained values is assigned to the corresponding point in the current replication;
  * **DES + Linear Regression 2** calculates the value of each sampling point of the current replication when its set of runs has been executed. Unlike the first method, this technique involves the computation of each point of the current replication considering its entire set of runs. Linear regression is applied between multiple known points.

MTTA Simulator. The Mercury tool also provides a transient simulator, which evaluates the behavior of non-Markovian absorbing models. This simulator generates a large number of related statistics.

4.1. Supplementary Tools

An RVG is available in the Mercury tool, which is capable of generating random numbers from a range of probability distributions. The RVG module provides descriptive statistics from the generated data, and the sample data can be exported in order to be used in other applications. *Moment Matching* is another module from which it is possible to estimate what exponential-based probability distribution best fits the mean (first moment) and standard deviation (second moment) for a data sample.

5. Case Study

This section presents a case study to demonstrate the feasibility of adopting the Mercury tool for supporting the deployment of a cloud system. We investigated the gain in the availability by implementing a redundancy mechanism using a warm-standby strategy in the main component of a cloud architecture. Detailed information can be found in [20]. This case study considers an architecture with three nodes, where at least one node must be available for the cloud to work properly. An FT and a CTMC are adopted to represent

---

[10]Discret Event Simulation.

the hierarchical heterogeneous models. The FT describes the high-level components and is used to compute the total availability of the system, whereas the CTMC represents the components involved in the redundancy mechanism. Figure 3 shows the FT model with one non-redundant General Controller (GC) and its three nodes. It is important to stress that a warm-standby replication strategy cannot be properly represented by FT models, due to dependency between component states. In this case, a CTMC needs to be adopted to represent the redundant mechanism, with one active GC and one replicated GC host configured in warm-standby (see Figure 4). Through the computation of the availability of the redundant mechanism using the CTMC model, the availability of the cloud system can be known. Table 2 presents the measures obtained with Mercury. As we can see, there is a difference of 24 hours less in downtime when using redundancy.

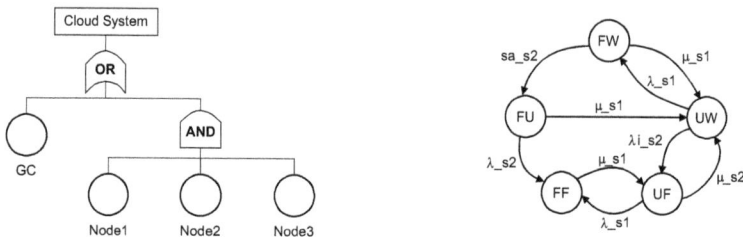

Figure 3. FT for the non-redundant cloud system.   **Figure 4.** CTMC to redundant system (two hosts [20]).

Table 2. Availability Measures of the System with and without Redundancy.

Metric	GC without redundancy	GC with redundancy
Steady-state availability	0.997192102190714	0.9999731974218314
Number of 9's	2.5516187019805443	4.571823428711702
Annual downtime	24.61 h	0.23 h

6. Conclusions

In this paper, we introduce the recent enhancements to Mercury. We evolved the Mercury tool aiming to support discrete-time Markov chain (DTMC) and fault tree (FT) formalisms. In addition, among other updates, both SPN simulators were reimplemented, and now stationary and transient simulations can detect the occurrence of rare events. One case study demonstrates the feasibility of applying Mercury for supporting infrastructure planning. Many research projects have been supported by Mercury and have been published in peer-reviewed journals and conferences. Mercury has proven to be a useful tool for what it has been designed. As future works, we intend to implement support for project creation where multiple models of the same formalism can be created in a single project file.

References

[1] Molloy MK. Performance analysis using stochastic Petri nets. IEEE Computer Architecture Letters. 1982;31(09):913–917.

[2] Trivedi K. Probability and Statistics with Reliability, Queueing, and Computer Science Applications, ed.: JHON WILEY &SONS. INC; 2002.

[3] Maciel PR, Trivedi KS, Matias R, Kim DS. Dependability modeling. In: Performance and Dependability in Service Computing: Concepts, Techniques and Research Directions. IGI Global; 2012. p. 53–97.

[4] Distefano S, Xing L. A new approach to modeling the system reliability: dynamic reliability block diagrams. In: RAMS '06. Annual Reliability and Maintainability Symposium, 2006.; 2006. p. 189–195.

[5] Vesely WE, Goldberg FF, Roberts NH, Haasl DF. Fault tree handbook. Nuclear Regulatory Commission Washington DC; 1981.

[6] Callou G, Maciel P, Tutsch D, Ferreira J, Araújo J, Souza R. Estimating sustainability impact of high dependable data centers: a comparative study between brazilian and us energy mixes. Computing. 2013;95(12):1137–1170.

[7] Johnson MA, Taaffe MR. Matching moments to phase distributions: Mixtures of Erlang distributions of common order. Stochastic Models. 1989;5(4):711–743.

[8] Silva B, Matos R, Callou G, Figueiredo J, Oliveira D, Ferreira J, et al. Mercury: An integrated environment for performance and dependability evaluation of general systems. In: Proceedings of Industrial Track at 45th Dependable Systems and Networks Conference, DSN; 2015. .

[9] Maciel P, Matos R, Silva B, Figueiredo J, Oliveira D, Fé I, et al. Mercury: Performance and Dependability Evaluation of Systems with Exponential, Expolynomial, and General Distributions. In: 2017 IEEE 22nd Pacific Rim International Symposium on Dependable Computing (PRDC); 2017. p. 50–57.

[10] Oliveira D, Matos R, Dantas J, Ferreira Ja, Silva B, Callou G, et al. Advanced Stochastic Petri Net Modeling with the Mercury Scripting Language. In: Proceedings of the 11th EAI International Conference on Performance Evaluation Methodologies and Tools. VALUETOOLS 2017. New York, NY, USA; 2017. p. 192–197. Available from: https://doi.org/10.1145/3150928.3150959.

[11] Sahner RA, Trivedi KS. Reliability modeling using SHARPE. IEEE Transactions on Reliability. 1987;36(2):186–193.

[12] German R, Kelling C, Zimmermann A, Hommel G. TimeNET: a toolkit for evaluating non-Markovian stochastic Petri nets. Performance Evaluation. 1995;24(1-2):69–87.

[13] Heiner M, Herajy M, Liu F, Rohr C, Schwarick M. Snoopy–a unifying Petri net tool. In: International Conference on Application and Theory of Petri Nets and Concurrency. Springer; 2012. p. 398–407.

[14] Grassmann WK, Taksar MI, Heyman DP. Regenerative Analysis and Steady State Distributions for Markov Chains. Oper Res. 1985 Oct;33(5):1107–1116. Available from: https://doi.org/10.1287/opre.33.5.1107.

[15] Marsan MA, Balbo G, Conte G, Donatelli S, Franceschinis G. Modelling with Generalized Stochastic Petri Nets. SIGMETRICS Perform Eval Rev. 1998 Aug;26(2):2. Available from: https://doi.org/10.1145/288197.581193.

[16] Kuo W, Zuo MJ. Optimal reliability modeling: principles and applications. John Wiley & Sons; 2003.

[17] Ferreira J, Callou G, Maciel P. A power load distribution algorithm to optimize data center electrical flow. Energies. 2013;6(7):3422–3443.

[18] Breuer L, Baum D. An introduction to queueing theory: and matrix-analytic methods. Springer Science & Business Media; 2005.

[19] Desrochers AA, Al-Jaar RY, Society ICS. Applications of petri nets in manufacturing systems: modeling, control, and performance analysis. IEEE Press; 1995. Available from: https://books.google.it/books?id=mL1TAAAAMAAJ.

[20] Dantas J, Matos R, Araujo J, Maciel P. Eucalyptus-based private clouds: availability modeling and comparison to the cost of a public cloud. Computing. 2015:1–20.

[21] Jain R. The art of computer systems performance analysis : techniques for experimental design, measurement, simulation, and modeling. New York: Wiley; 1991.

Intelligent Environments 2021
E. Bashir and M. Luštrek (Eds.)
© 2021 The authors and IOS Press.

doi:10.3233/AISE210076

A Mandami Fuzzy Controller for Handling a OpenHAB Smart Home

Pablo PICO-VALENCIA [a] and Juan A. HOLGADO-TERRIZA [b,1]

[a] *Pontificia Universidad Católica del Ecuador, Esmeraldas, Ecuador*
[b] *Universidad de Granada, Granada, Spain*

Abstract. Fuzzy control systems are widely used to carry out control actions where variables with a certain degree of ambiguity are involved. Such control actions are modelled based on fuzzy rules that are expressed in a language analogous to that used by humans to conceive their reality. In this sense, fuzzy controllers can be applied in emerging technologies such as the Internet of Things (IoT) aimed at controlling the real world through a set of objects interconnected to the Internet. This article presents a fuzzy controller of Mamdani-type developed in Matlab. This controller is integrated into a smart home– created by the OpenHAB tool– using RESTful services. The results of the evaluation reveal that the integration of the fuzzy controller with smart home can provide an accurate control of comfort specifically in this case. In addition, a better organization of the control rules was also possible using the proposed fuzzy controller especially when large systems are developed.

Keywords. fuzzy controller, Mamdani, internet of things, smart home, OpenHAB

1. Introduction

Fuzzy logic is an alternative to the classic logic that proposes the modelling of systems where variables are handled with a certain level of imprecision (i.e., temperature, air pollution and tank filling level). These kind of variables can take ambiguous values such as: very-low, low, medium, high, very-high [1]; empty, low, medium, full, critical [2], among others. Handling this type of variables could not be supported by a precise mathematical model to establish control actions [3]. However, the application of fuzzy logic based on fuzzy sets and rules for the creation of systems enables the execution of control actions using ambiguous variables more accurately.

A fuzzy logic based control system consists of three elements: a fuzzifier, an inference engine and a defuzzifier [4]. The fuzzifier allows the variables to be modelled by means of fuzzy sets. The elements of each involved fuzzy set have an associated membership degree and they use a certain membership function according to the nature of the problem (e.g., triangular, trapezoidal, Gaussian). After the fuzzification of input and output variables of the control system, the control rules are determined. These rules of type IF-THEN are used by the inference engine to perform fuzzy inference obtaining the value of the output variable(s). Finally, the output variable(s) are defuzzified to become

---

[1]Corresponding Author: Software Engineering Department, University of Granada, E.T.S. de Ingenierías Informática y de Telecomunicación, 18071 Granada, Spain; E-mail: jholgado@ugr.es.

real values from which control actions are performed on the system. To carry out this process, defuzzification methods such as centroid, bisector, min and max are used [5].

Recently, fuzzy control systems have been integrated into emerging paradigms such as the Internet of Things (IoT). In IoT ecosystems, the real world is controlled and tracked by monitoring the data provided by objects or things that are interconnected to the Internet [6]. However, it is also required to carry out control actions on the environment where humans daily live in order to satisfy their needs. In this sense, fuzzy controllers have been introduced into the IoT to control scenarios where imprecise data have to be sensed. Some of these systems are oriented for controlling automatic irrigation systems [7], lighting in smart homes [8], optimization the conditions of cultivation in greenhouses [9], controlling of pollution indoor [10], among others. These systems comply with the specifications of the most widespread fuzzy models, Mamdani or Takagi-Sugeno [11].

Since fuzzy controllers have been integrated into the IoT to control objects and communications, it is necessary to work on fuzzy controllers compatible with connected digital homes (smart homes) —scenarios where flexible and accurate control is required to satisfy the user needs. This has been one of the main motivations for this study whose objective is to integrate a Matlab fuzzy controller into a smart home built with a standard home gateway such as OpenHab [12]. Based on this integration via Restful web services, an evaluation of the controller was proposed, determining the accuracy to control smart home objects as well as the flexibility to create the control strategy; that is, defining how the set of rules was organized from which the IoT ecosystem is driven.

This paper is organized as follows. Section 2 describes how fuzzy controllers have been applied on connected digital homes based on the IoT. Section 3 describes the architecture of system as well as the process followed to integrate the fuzzy controller with the IoT. The results of the evaluation of the fuzzy controller are discussed in the section 4. Finally, conclusions and future work are outlined in section 5.

2. Literature Review

The main proposed fuzzy controllers for controlling household based on IoT scenarios showed that fuzzy controllers were developed in order to manage certain aspects related to comfort indoor, that is, lighting, thermal, and security. To do that, heterogeneous sensors of different providers (i.e., temperature, humidity, pH sensors) were installed at both indoor and outdoor of smart homes. Fuzzy controllers managed the data acquired by these sensors as main data input to perform the fuzzyfication, fuzzy inference and defuzzyfication tasks. Next, a description on the six recovered proposals after performing the scientific search are described in deep.

In most cases, the Mamdani-type fuzzy controller was used as the primary fuzzy control model [11]. In the management of thermal comfort, a first study, proposed in [13], presented a solution capable of optimizing the indoor thermal comfort condition according to seasonality (summer condition). The system developed by the researchers included IoT objects such as temperature, wind speed, fan, humidity sensors. From these devices it was possible to capture the contextual information on a smart home. Such information allowed the authors, using a linear regression model, to predict the Predictive Mean Vote (PMV) Index – a model that stands among the most disseminated models for evaluating indoor thermal comfort. Fuzzy controller used this index to compute the control signal by applying the predefined control rules. The model was developed in Java with the jFuzzyLogic library. The system was tested with an HVAC in a real building located in Jeju, Korea in 2017.

In the same line of thermal comfort, additional studies were proposed. In [14], a fuzzy system for controlling indoor temperature via a Mamdani model using the type-2 fuzzy logic was proposed. The developed system, programmed in IT2F environment, was oriented to manage the most appropriate indoor temperature with respect to outdoor ambient temperature, maintaining at the same time a comfortable climate for users. Among the modelled fuzzy variables, there were used the following: outdoor temperature and humidity. Managing the temperature of a specific room considering the outdoor conditions, a model was enabled to control the indoor temperature not only on humidity and temperature but also on the number of people in the room. This model is ideal for smart cities because is callable of saving energy and money.

In order to achieve the goals of saving energy and money, in [15] was proposed a similar controller capable to save around 40% of energy. Although authors in this study managed the same variables for the controller, they worked with different hardware and software technologies. For instances, researchers used the ThinkSpeak as an open IoT platform for interconnecting objects, Arduino UNO and Raspberry Pi as microcontrollers for connecting sensors, and jFuzzyLogic for modelling the fuzzy system.

Likewise, Mamdani type fuzzy controllers have also been applied to light comfort control in addition to temperature management for smart homes. Accordingly, a first study, proposed in [16], developed a fuzzy luminance control system loaded in an ESP 8266 microcontroller programmed in C language. They defined two input variables (error and error change between luminance set value and measured value) and one output variable (PWM duty cycle increment value). From these variables the system could effectively maintain a stable luminance level in an office environment as a part of a smart home. Thus, the system contributes positively to power saving efficiency.

Similarly to [16] but oriented to external systems such as intelligent cities, in [17], a Mamdani fuzzy system was proposed for controlling the intensity of street lights. This system developed in Matlab could adjust the streetlights intensity level based on the temporal and environmental conditions, that is, clear sky, cloudy, foggy and traffic density. Although the proposed system was adapted to the conditions of Islamabad in Pakistan, it could be adapted for operating according the same conditions to other cities of the world.

Finally, it was also highlighted two proposals focused on aspects related to security management applicable to IoT. A first study, proposed in [18], presented a framework called Fuzzy approach to the Trust Based Access Control (FTBAC) that can be used to calculate fuzzy trust values for any number of devices which makes it more suitable for scalable IoT. To do it, the fuzzy controller employed three variables that describe devices, that is, values of experience, knowledge and recommendation. This is a general framework that could be applied in devices of smart homes, but it has not been experimented.

Complementary, in [19], a trust aware access control system using fuzzy logic is described to ensure that only trusted users/devices access the data of sensor devices such as successful forward ratio, energy consumption rate and data integrity. From these data, the fuzzy controller was able to assist in the decision making of access control as well as calculate the context dependent trustworthiness of each device based on their previous behaviour. Therefore, this is a good alternative in smart homes because security and privacy are yet concerns in IoT.

3. Proposal

3.1. Scenario of IoT

The objective of the IoT scenario was controlling the room lighting level of a smart home. In this sense, the illumination indoor of the three rooms (bathroom, master room, living room) must be automatically controlled according the activity done in the room and the contextual conditions where the smart home is located. In order to develop this IoT ecosystem (Figure 1) it was used the OpenHAB platform [12] as smart home gateway. This platform provides a vendor and technology agnostic open-source automation software for integrating heterogeneous IoT devices available in a smart home. Thus, OpenHAB was selected because it serves as an integration way for all home automation needs and lets systems talk to each other across any vendor or protocol.

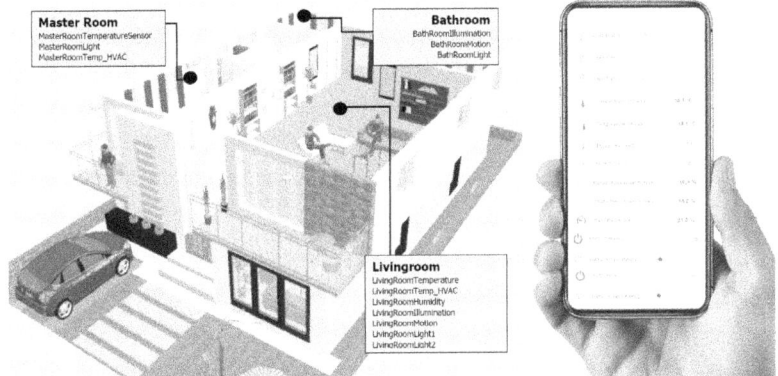

Figure 1. Scenario of IoT developed using OpenHAB.

The main IoT objects that were part of the described scenario, shown in Figure 1, included the following IoT objects: four light bulbs, two motion sensors and two illumination sensors. There were other objects as part of the scenarios but they were not used for lighting comfort. Files that implements the sitemap, items and rules of the studied scenario in OpenHAB are accessible in the project repository.

Each IoT object had associated a state and some data obtained from the linked sensor or actuator. For accessing to the status of the sensors, a Matlab Restful web service linked to the IoT sensor or actuator was invoked using the request-response model of the Hypertext Transfer Protocol Secure (HTTPS). A GET method was applied to obtain the status and data from sensors as data input of fuzzy controller. On the other hand, a PUT method was called to update the status of actuators, once the values are computed through the fuzzy controller. Both methods were accessed using uniform resource identifier (URIs) and a HTTPS client.

3.2. Fuzzy Controller Architecture

For the development of the proposed fuzzy controller in this study, the specifications of the general fuzzy controller model described in [4,9] were followed. This means that the

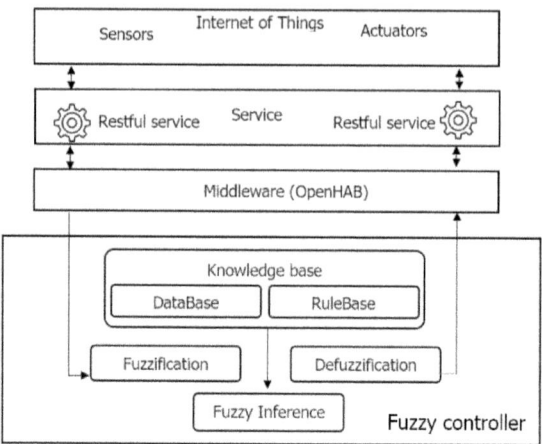

Figure 2. General architecture of the proposed fuzzy controller.

designed controller integrated the four basic components such as: a fuzzifier, a knowledge base, an inference engine and a defuzzifier. These ones are shown in Figure 2.

The schema of a fuzzy controller shown in Figure 2 defines a generic fuzzy controller that gets a set of inputs represented by fuzzy sets through the fuzzifier from the sensors of IoT ecosystem. From these variables, the rules were defined and based on them, the value of an output variable was inferred. This output variable was converted to a real value through the defuzzifier to determine the control actions to be applied to actuators. After defining the scheme of the fuzzy controllers to be developed, we determined the fuzzy logic library available for programming the fuzzy system in Matlab R2018b, that is, selecting the methods of the toolbox for fuzzy inference system (FIS).

3.3. Design of the Fuzzy Controller

To meet the control goals of the system (providing lighting comfort) one fuzzy controller was designed. This controller called *OpenHABFuzzyLigthingComfort* was oriented to control the lighting in the smart home illustrated in Figure 2, that is, in the bathroom and living room in our case. The motion and illumination sensors were used as input data to model the control actions. The fuzzy sets of the variables *illumination* and *motion*, fuzzified from both triangular and trapezoidal functions, are illustrated in Table 3.

After the fuzzyfication of the input and output variables of the proposed Mamdani controller, the fuzzy rules of the fuzzy controller were then formulated. For the controller *OpenHABFuzzyLightingComfort* 12 rules were defined. The antecedents (IF) and the consequent of the rules (THEN) are detailed in Table 1. It is important to emphasize that in both cases the AND operator was used to join more than two antecedents of the rules.

Finally, the centroid method was selected for the defuzzyfication of the output variable, because it is one of the most used to perform fuzzy control providing a more accurate value in relation to the other existing ones.

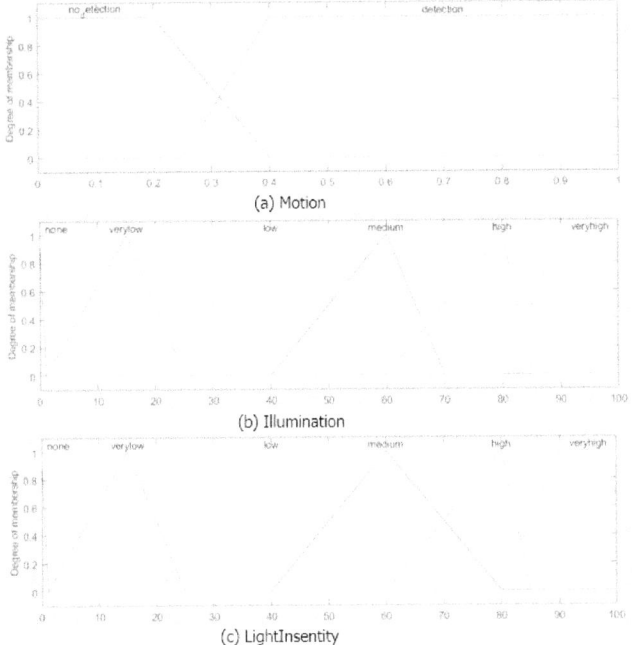

Figure 3. Fuzzy input and output variables for lighting control.

Table 1. Fuzzy rules for lighting control in the studied scenario.

Rule	Illumination (i_1)	Motion (i_2)	Lights (o_1)
1	none	no_detection	none
2	very_low	no_detection	none
3	low	no_detection	none
4	medium	no_detection	none
5	high	no_detection	none
6	very_high	no_detection	none
7	none	detection	very_high
8	very_low	detection	high
9	low	detection	medium
10	medium	detection	low
11	high	detection	none
12	very_high	detection	none

3.4. Development of the Fuzzy Controller

The controllers designed were developed using the library for fuzzy systems provided by Matlab R2018b. In general, the process for creating the fuzzy controller for handling lighting comfort (*OpenHABLightingComfort*) consisted in the creation of a fuzzy control, fuzzyfication of the two input variables such as motion and illumination, fuzzyfication of the output variable and finally, the definition of the rules that drive the controller as well as the setting of these rules to the previously fuzzy control created.

Once the fuzzy controller was created, it was executed continuously applying a control cycle on the studied modelled smart home. Listing 1 describes the implementation of

```
1   % READING DATA FROM THE ECOSYSTEM OF IOT
2   ep = "https://home.myopenhab.org/rest/items"
3   cycle=0;
4   while (MatFuzzyIoTGetData(ep,"LivingRoom_ONOFF","state")=="ON")
5     cycle=cycle+1;
6     disp('control_cycle' + cycle)
7     % CONTROL OF ILLUMITATION ON THE LIVINGROOM
8     LR_Illumination=MatFuzzyIoTGetData(ep,"LR_Illumination","state")
9     LR_Motion=MatFuzzyIoTGetData(ep,"Bathroom_Motion","state")
10    if (LR_Motion=="ON")
11      LR_Motion_=0.9;
12    else
13      LR_Motion_=0.28;
14    end
15    LR_Illumination_=str2double(LR_Illumination)
16    % COMPUTATION OF THE ANWER BASED ON THE FUZZY CONTROLLER
17    a=evalfis(fis,[LR_Motion_ LR_Illumination_]);
18    % APPLYING CONTROL ACTIONS OVER THE IOT ECOSYSTEM
19    MatFuzzyChangeIoTData(ep,"LR_Light1","state",a)
20    MatFuzzyChangeIoTData(ep,"LR_Light2","state",a)
21    pause(100)
22  end
```

Listing 1. Continuous cycle for controlling thermal comfort over the studied scenario of IoT.

illumination control for the living room called *OpenHABLightingComfortRun*. In line 4, the endpoint of deployed RESTful service is defined for reading input data from the IoT ecosystem such as illumination and motion (lines 10-17). Then, those data were used to compute the output variable using the fuzzy controller *OpenHABLightingComfort*, that is lightbulb intensity (line 19). Finally, the obtained value was used to change the state of the lightbulb of the IoT ecosystem (lines 21-23). In a similar way, the lighting was controlled on the bathroom. This script was executed while the operation switch installed on the smart home is in state "ON".

4. Results and Discussion

4.1. Evaluation of the Accuracy of Output

An evaluation of the system was carried out to validate its effectiveness in the control actions applied to the ecosystem of IoT under study. The results for all cases were coherent with the designed fuzzy controller. This implies the level of illumination in the studied scenario was as expected. Some particular examples changing data in the GUI of the IoT ecosystem, such as motion (M) and illumination (I) for controlling the lightbulb intensity (L) are described as follows:

The discourse universe of illumination and lightbulb intensity had a range from 0 to 100. For lightbulb intensity, we connected 4000 kelvin Hue light bulbs with 806 lumens while for illumination a Xiaomi smart light sensor was selected with a variability

Table 2. Results of the evaluation of the proposed fuzzy controller on the IoT ecosystem.

Illumination (i_1)	Motion (i_2)	Lights (o_1)
0	0.28	0
0	0.8	92.5
12	0.8	74.8
38	0.8	60
55	0.8	13.6
71	0.8	0
94	0.8	0

from 0 to 83000lux. The motion was detected by using a Hue motion sensor. Based on these ranges handled by the fuzzy controller, the values calculated by the defuzzyfication method allowed the application of automatic control processes adapted to reality and to the needs required in the rooms under study. However, the applied control is static because only a specific user pattern was considered. Therefore, it is recommended to include within the controller other variables related to user needs, that is, modes of use, user preferences, among others. This will enable the system adaptation to many more cases and to different user needs.

It is known that regarding fuzzy controllers, many works published in the literature have used Matlab's fuzzy logic library to develop controllers similar to the one proposed in this work. Its wide use has been one of the motivations for using this tool. The results obtained in this study allowed to determine the effectiveness of the controller in the IoT developed using OpenHAB. The level of intensity of the lightbulb managed from OpenHAB was carried out consistently and accurately according to the rules defined.

4.2. Evaluation of Response Time

An evaluation of the response time of the system was also carried out to execute the control cycle to manage light and thermal comfort in the studied IoT ecosystem. This evaluation was made on a Lenovo branded personal computer with a 2.50 GHz i7 processor and 16 GB of RAM. The computer is installed with the 64-bit Windows 8.1 operating system and the Java 1.8 virtual machine. Furthermore, the ecosystem of IoT, implemented with OpenHAB 2.0 ran on an HP personal computer with 1.6 GHz i5 processor and 8 GB of RAM. This ecosystem can be accessed via Internet because the cloud connector of OpenHAB was used.

In Figure 4, it can be observed that the response time required to run the light control system, which controlled the lightbulbs located in the bathroom and the living room, was 14.188 s. This result was obtained after calculating the mean after running the control loop *OpenHABLigthingComfortRun* for 10 iterations. The dispersion of the response times has a standard deviation of 0.902 s. All values were accessed over the cloud of OpenHAB using the HTTPS protocol over the Internet. The connection speed was 50 Mbps.

In summary, the developed system was executed efficiently. For a scenario such as an smart home that does not need hard real time, the times obtained are acceptable. However, such time may change when the system is running in an environment where IoT devices must be accessed on an intranet instead of the Internet. It is recommended that only those specific IoT objects that should be accessed from anywhere outside the au-

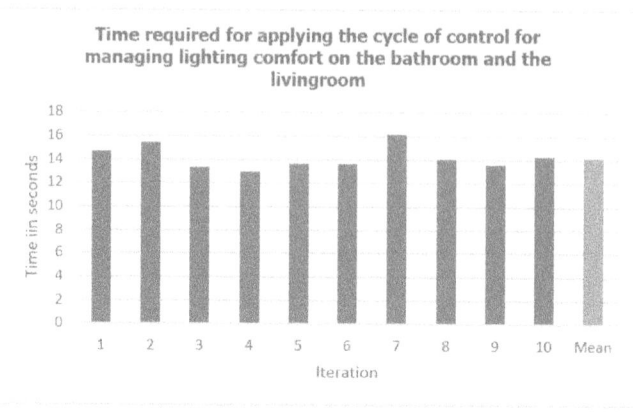

Figure 4. Time required for executing cycle of continuous control of lighting comfort.

tomated scenario may use the OpenHAB cloud connector in order to gain access to it. The remaining objects can be consumed by fuzzy controller via the intranet. These recommendations will considerably increase the overall efficiency of the developed system and can be applied to time-constrained tasks if needed.

5. Conclusions and Future Work

IoT ecosystems require intelligent mechanisms to carry out communications and control actions of the interconnected actuators and sensors as accurately as possible. To develop these mechanisms, Artificial Intelligence plays an important role. Techniques such as fuzzy control systems have helped designing and creating intelligent systems. In this sense, fuzzy logic is relevant because it allows dealing with variables that have some ambiguity. In the case of smart homes, the use of environmental conditions (i.e., temperature, humidity, lighting) supports some needs regarding the management of thermal, lighting and safety comfort to make its users feel more comfortable.

The use of IoT platforms, with services support, for the management of resources and communications between devices, contributes significantly to the integration of fuzzy controllers with IoT networks. Its use not only helps to capture data from the environment required by the controllers to carry out changes on actuators, but also allows the use of cloud computing to run the inference implemented by fuzzy controllers. This allows fuzzy controllers to be used in resource-constrained environments. As the results show, the application of fuzzy control does not require sophisticated resources. Furthermore, the possibility of applying control in undefined environments is not synonymous of decreasing in the effectiveness of control, which is generally done with simple rules.

As future work, we propose the development of an add-on oriented to the development of fuzzy controllers compatible with IoT in a way that abstracts the complexity of smart home systems when they are integrated in IoT networks as well as it facilitates the creation of additional smart scenarios such as hospitals, industry, city, smart university, among others. It also proposes the use of fuzzy variables to model not only conditions directly related to intelligent systems (i.e., temperature), but also user preferences (modes of use, preferences). This would facilitate the management of preferences in areas where there are multiple users, an aspect that is generally complex to model in order to satisfy more than single users.

References

[1] Ameen NM, Al-Ameri JAM. IoT-Based Shutter Movement Simulation for Smart Greenhouse Using Fuzzy-Logic Control. In: 2019 12th International Conference on Developments in eSystems Engineering (DeSE). IEEE; 2019. p. 635–39.

[2] Anduray RGJ, Irigoyen SMZ. Development of a fuzzy controller for liquid level by using raspberry pi and internet of things. In: 2017 IEEE Central America and Panama Student Conference (CONESCAPAN). IEEE; 2017. p. 1–5.

[3] Alcala-Fdez J, Alonso J. A Survey of Fuzzy Systems Software: Taxonomy. Current Research Trends and Prospects. 2015;24(1):40–56.

[4] Petritoli E, Leccese F, Cagnetti M. Takagi-Sugeno Discrete Fuzzy Modeling: an IoT Controlled ABS for UAV. In: 2019 II Workshop on Metrology for Industry 4.0 and IoT (MetroInd4. 0&IoT). IEEE; 2019. p. 191–95.

[5] Talon A, Curt C. Selection of appropriate defuzzification methods: Application to the assessment of dam performance. Expert Systems with Applications. 2017;70:160–74.

[6] Ng ICL, Wakenshaw SYL. The Internet-of-Things: Review and research directions. International Journal of Research in Marketing. 2017;34(1):3–21.

[7] Alomar B, Alazzam A. A Smart Irrigation System Using IoT and Fuzzy Logic Controller. In: 2018 Fifth HCT Information Technology Trends (ITT). IEEE; 2018. p. 175–79.

[8] Lu Y, Wang J, Bai X, Wang H. Design and implementation of LED lighting intelligent control system for expressway tunnel entrance based on Internet of things and fuzzy control. International Journal of Distributed Sensor Networks. 2020;16(5).

[9] Carrasquilla-Batista A, Chacón-Rodríguez A. Standalone Fuzzy Logic Controller Applied to Greenhouse Horticulture Using Internet of Things. In: 2019 7th International Engineering, Sciences and Technology Conference (IESTEC). IEEE; 2019. p. 574–79.

[10] Pradityo F, Surantha N. Indoor Air Quality Monitoring and Controlling System based on IoT and Fuzzy Logic. In: 2019 7th International Conference on Information and Communication Technology (ICoICT). IEEE; 2019. p. 1–6.

[11] Nguyen AT, Taniguchi T, Eciolaza L, Campos V, Palhares R, Sugeno M. Fuzzy control systems: Past, present and future. IEEE Computational Intelligence Magazine. 2019;14(1):56–68.

[12] Heimgaertner F, Hettich S, Kohlbacher O, Menth M. Scaling home automation to public buildings: A distributed multiuser setup for OpenHAB 2. In: 2017 Global Internet of Things Summit (GIoTS). IEEE; 2017. p. 1–6.

[13] Hang L, Kim DH. Enhanced Model-Based Predictive Control System Based on Fuzzy Logic for Maintaining Thermal Comfort in IoT Smart Space. Applied Sciences. 2018;8(7):1031.

[14] Jana DK, Basu S. Novel Internet of Things (IoT) for controlling indoor temperature via Gaussian type-2 fuzzy logic. International Journal of Modelling and Simulation. 2019;41(2):1–9.

[15] Meana-Llorián ea. IoFClime: The fuzzy logic and the Internet of Things to control indoor temperature regarding the outdoor ambient conditions. Future Generation Computer Systems. 2017;76:275–84.

[16] Xiaole Y, Chunlai Y, Zhe Y, Hongbo Y, Gang P. Power Saving LED Luminance Fuzzy Control with IOT Network. In: IOP Conference Series: Materials Science and Engineering. vol. 435. IOP Publishing; 2018. p. 012059.

[17] Sultan K. Fuzzy Rule Based System (FRBS) assisted Energy Efficient Controller for Smart Streetlights: An approach towards Internet-of-Things (IoT). Journal of Communications. 2018;13(9):518–23.

[18] Mahalle PN, Thakre PA, Prasad NR, Prasad R. A fuzzy approach to trust based access control in internet of things. In: Wireless VITAE 2013. IEEE; 2013. p. 1–5.

[19] Thirukkumaran R, Muthukannan P. TAACS-FL: trust aware access control system using fuzzy logic for internet of things. International Journal of Internet Technology and Secured Transactions. 2019;9(1-2):201–20.

36

Intelligent Environments 2021
E. Bashir and M. Luštrek (Eds.)
© 2021 The authors and IOS Press.
This article is published online with Open Access by IOS Press and distributed under the terms
of the Creative Commons Attribution Non-Commercial License 4.0 (CC BY-NC 4.0).
doi:10.3233/AISE210077

A Modular Architecture for Multi-Purpose Conversational System Development

Adrián ARTOLA [a,1], Zoraida CALLEJAS [a] and David GRIOL [a,2]

[a] *Dept. of Software Engineering, University of Granada, Granada, Spain*

Abstract. As the complexity of intelligent environments grows, there is a need for more sophisticated and flexible interfaces. Conversational systems constitute a very interesting alternative to ease the users' workload when interacting with such environments, as they can operate them in natural language. A number of commercial toolkits for their implementation have appeared recently. However, these are usually tailored to specific implementations of the processes involved for processing the user's utterance and generate the system response. In this paper, we present a modular architecture to develop conversational systems by means of a plug-and-play paradigm that allows the integration of developers' specific implementations and commercial utilities under different configurations that can be adapted to the specific requirements for each system.

Keywords. conversational systems, chatbots, modular architectures, natural language understanding, dialog management, conversational framework, human-machine interaction

1. Introduction

Intelligent Environments (IE) comprise a set of interconnected devices and sensors surrounding users to provide access to a plethora of information and services, which may create a great cognitive load in the users, specially in industrial settings (see e.g. [1]). Consequently, user empowerment can only be sustained in enhanced and more intuitive human-machine interactions.

Conversational systems have become very important to achieve this objective involving speech interaction and being able to process semantic and pragmatic knowledge [2, 3]. These interfaces have experienced a vast development in the recent years propelled by the widespread adoption of voice assistants and smart devices, the advances in Artificial Intelligence techniques and the increasing amount of data currently available to learn statistical models. These advances have created a whole new market for conversational systems, and in particular for IE assistants. The most renowned technological companies offer their language processing services both as assistants ready to be used

---

[1]A. Artola was with University of Granada during the realization of this work as part of his Master Thesis.

[2]Corresponding Author: David Griol, Periodista Daniel Saucedo Aranda SN, 18071 Granada, Spain; E-mail: dgriol@ugr.es.

(e.g. Amazon Alexa, Microsoft Cortana, Google Voice Assistant...) or as services that developers can employ to develop their own conversational systems or provide new skills to the already existing ones.

Such commercial tools offer different services, including natural language understanding and interaction management, which may vary in complexity. Task-oriented systems can be implemented easily with commercial toolkits such as DialogFlow or Amazon Lex. However, developers may find it difficult to combine the services from different vendors, specially natural language processing and dialogue management, as they are usually highly coupled in commercial systems.

Our aim is to offer a framework to develop multi-purpose conversational systems, which allows developers to use REST services to integrate their own implementations of specific modules of the system and also to combine them with the solutions provided by commercial toolkits.

The rest of the paper is organised as follows. Section 2 presents the state of the art about the different existing frameworks and architectures to develop conversational systems. Section 3 introduces our architecture as well as the terminology employed. Section 4 presents a practical implementation of our proposal, while Section 5 presents several configurations developed that show the appropriateness of the proposal to develop conversational systems in the same domain using different components from different vendors. Finally, Section 6 draws the conclusions and presents lines for future work.

2. Background

The typical pipeline to develop conversational interfaces consists of five components: automatic speech recognition, natural language understanding, dialogue management, natural language generation and text to speech synthesis [4].

Each component has specific purposes:

- The automatic speech recogniser receives the audio signal corresponding to the user's input and outputs a textual transcription. Typically, this module also provides confidence scores representing how confident the system is about the correctness of the returned text.
- The Natural Language Understanding module receives the text input and returns its semantic representation, as the perceived required task and the values for the necessary pieces of information required to perform it.
- The Dialogue Manager decides the next system action considering the semantic representation of the user's utterance, the previous dialogue history, the result of accessing the data repositories of the system, the specific regulations of the task, among others.
- The Natural Language Generator translates the action selected by the dialogue manager into one or more sentences in natural language (system prompt).
- Finally, the Text-to-Speech synthesizer translates the system prompt into an acoustic signal.

As mentioned before, there exist different frameworks for conversational systems development that try to accommodate some or all the previously described modules. McTear [3] presents a very complete and updated review of tools for developing dialogue systems, which are divided into tools for visual design, scripting tools, advanced toolkits and frameworks and research-based toolkits.

There is a huge variety of commercial toolkits offered by the largest IT companies. These toolkits offer all the necessary services to create a conversational system with high-level interaction and with improved possibilities to connect the developed system with different platforms, e.g., existing voice assistants, Telegram or Twitter. The most popular alternatives require the developer to define *intents* and *entities* for the understanding process, and dialogue management is determined according to the most relevant *intent* and the use of *slots* or active *contexts* that can be complemented with web services through *webhooks*.

Despite their flexibility, many times a considerable effort is required to handcraft a dialogue tree that is then coded into the system following the intents and entities format. Also these tools hinder the complexity and details of language and dialogue processing to developers, which is interesting to democratise conversational system development, but it may not be adequate for contexts in which developers want to have a broader control, including for example explainability of the decisions and security requirements, which are commonplace with IE interactions.

User confidence is key to IE [5]. As illustrated in [6], the capacity for self explainability is crucial for users to find IE trustworthy and reliable, specially when populated by smart assistants.

In academic settings, more sophisticated approaches are used for natural language understanding and dialogue management and can be used in controlled environments. This is the case of toolkits and implementation resources such as OpenDial [7], PyOpenDial [8], which also encompass a modular architecture, or the recent ConveRSE [9] and HRIChat [10], but are sometimes not straightforward to use in conjunction with commercial solutions.

Some of the previously mentioned alternatives do not offer the possibility of plugging different services for each of the modules that conform the conversational system. In fact, it is common to merge NLU and DM in commercial chatbot toolkits, as it is a trend to combine both options, specially when using machine learning to decide system responses.

However, in our approach, these two modules can be considered in isolation. This makes it possible to develop end-to-end systems in which both processes are performed at once, or to divide them into two independent services when it is necessary to have control over them. For example in hybrid systems where rules have to be applied into dialogue management. This is particularly interesting for IE as usually safety rules must be applied when operating the environments (e.g. always confirm when turning the oven on).

More versatile platforms like RASA [11] and DeepPavlov [12] provide more flexibility in the implementation. This paper presents exploratory work for a simple lightweight architecture that can be used to easily create conversational systems.

3. Proposed Architecture

We have chosen a service-oriented modular architecture based but not limited to the traditional pipeline. To foster interoperability, each part of the system is independent of the rest and the information shared is orchestrated by an *Information homogenisation module*. This new module receives the output of the Natural Language Understanding module and produces a technology-agnostic parsing into a *registry*, that can be subsequently translated into the format required by the rest of the services employed.

The dialogue history is stored as a board of *registries* that are categorised into *intents* (representing the intentions or actions required by the user) and *entities* (relevant pieces of information required to fulfil the task). The actions taken by the system are also incorporated into the board.

All the modules except the *Information Homogenisation (IH) Module* are divided into *infrastructure* and *superstructure*. On the one hand, the infrastructure is dedicated to connect all services and to interact with the container application. On the other hand, the superstructure is a specific Language Processing service that can be plugged and unplugged into the infrastructure.

Every service has a different format as input and output but all of them offer an interface using REST services. The advantage of doing this type of connection is that the developer is free to choose the best platform or programming language to develop each service.

Our infrastructure connects to the services using HTTP with REST calls and the message is sent in JSON format, so the modules have simple responsibilities:

- Generate a JSON with the input in the specific format required by the plugged service.
- Connect to the REST interface of the service with the required parameters such as API keys or any other authentication data.
- Receive the data from the REST service and parse it to the proper format required by the infrastructure.

The proposed framework can orchestrate new services either from commercial vendors or generated by the developer, just complying with the simple requisites described below. All services must be connected by REST calls so that the different modules are not coupled to the infrastructure. REST calls take the form of JSON messages, a de-facto standard in commercial systems.

The interaction with the container application can be done by directly calling the functions that the infrastructure offers or creating a new web service dedicated exclusively to the interaction with the infrastructure. In particular, the connection of the main modules of a conversational system works as follows.

To connect a Speech-to-Text service, the infrastructure will send to the new connector the path of the audio file. Depending on the service the connector will have to send the file in a different way. For example, Google Speech-to-Text requires a JSON message with the content of the audio file encoded in the message and IBM Watson Speech-to-Text just needs to upload the file on the request. Independently, the architecture will only require the connector to work sending the given audio file and returning the transcript text as shown in Fig. 1.

The Natural Language Understanding (NLU) service connector requires the output of the Speech-to-Text module, which is the transcribed user input. Generally, at least in the tested services (Google, Amazon and IBM), the request requires a JSON message with the user's query and the returned answer is also in JSON format. Thus, the Connector will have to send the transcribed text to the Natural Language Understanding service and this service will return a JSON with the *intents* and *entities*.

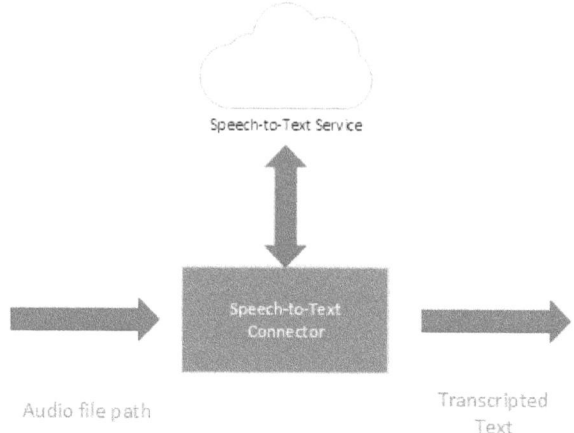

Figure 1. Scheme of the Speech-to-Text connector.

The developer will only need to implement the parser into the IH Module (see Figure 2) to generate a *registry* from the JSON answer of the NLU service.

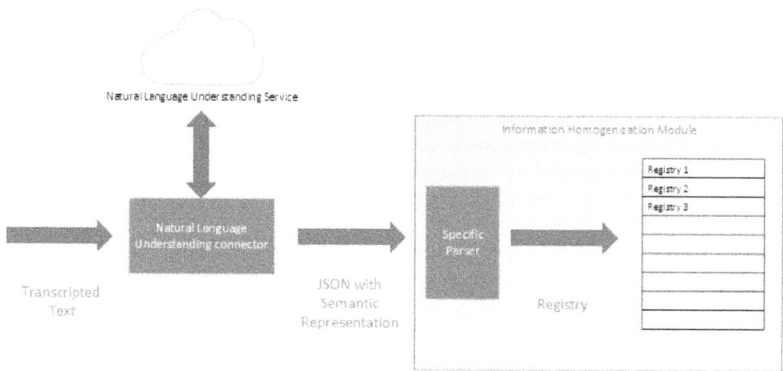

Figure 2. Scheme of the Natural Language Understanding connection and the Information Homogenization Module.

After being parsed, the generated *registry* is stored in the board of the IH Module. This way, the Dialogue Manager (DM) service can retrieve from the infrastructure the *registries* that the IH Module could store in the board as well as the previous system action, to use them as a basis for decision making, as shown in Figure 3. Then the next system action and the necessary data will be returned. The next action is stored in the IH Module for the next turn of the conversation.

In some cases, the next system action may also contain the generated natural language text for the user so the usage of the Natural Language Generation (NLG) module is optional and in the case it is used it will receive the DM output and will return the natural language answer. If the developer chooses to incorporate a NLG service, the output would be the set of sentences in natural language generated as system response.

In the case of the Text-to-Speech (TTS) module, the connector receives the text to be synthesised from the infrastructure and the TTS service returns the synthesised voice

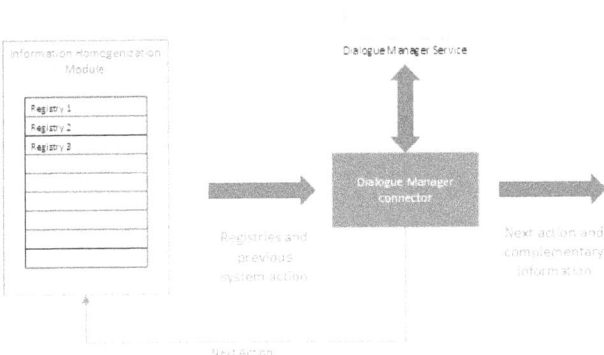

Figure 3. Scheme of the Dialogue Manager connector.

as an audio file to be played by the container application, this can be done including the path in the JSON response (as it is done in Google Cloud) or initiating the download of the audio file as a result of the REST request (as it is the case of IBM Watson Text-to-Speech).

4. Implementation of the Architecture

The infrastructure has been implemented in Java and already incorporates connections to the online services of the most important companies. We have tested and created connectors for the following services, but the proposal is not limited to them:

- Speech-to-Text module: Google Speech-to-Text and IBM Watson Speech-to-Text.
- Natural Language Understanding: Google DialogFlow and Microsoft LUIS. We also implemented our own Natural Language Understanding service that can be plugged to the infrastructure (see Section 4.1).
- Dialog Manager: we have implemented our own dialogue manager (see Section 4.2. It is worth noticing that commercial systems often have NLU and DM linked, so it is difficult to separate understanding capabilities from the dialogue management policy.
- Natural Language Generator: we have used a template-based generation.
- Text-to-Speech: Google Text-to-Speech and IBM Watson Text-to-Speech.

To show that it is not only possible to plug-in third party services, but also to easily implement and deploy the developers' own implementations, we have implemented and offer the connector for our own NLU and DM modules, which are described below.

4.1. Sample Implementation of a Natural Language Understanding Module

Our own solution for Natural Language Understanding has been developed in Python using Okapi BM25 [13] for *intent* detection and *Term Frequency Inverse Document Frequency* (TF-IDF) for *entity* detection. BM25 is often used by search engines to detect relevant results based on the user's input and for document-scoring [14]. We have used this probabilistic function to detect the *intents* from the input, including some improvements to make the result more accurate according to observed training phrases. BM25 can be trained in a similar fashion as commercial chatbot development systems, providing a corpus with example phrases for every *intent*. When the function is applied, it computes a score for every *intent*, so the one with highest score can be selected as the most probable.

For the purpose of finding relevant keywords and *entities* in the user's input phrase we use TF-IDF. For this task, we use the same corpus employed to train BM25 grouping phrases incorporating information about *entities*. TF-IDF computes the frequency of every word in the phrase. *Entity* detection requires a higher level of post-processing for which we have applied three filters: i) list of *stopwords* that are ignored; ii) a *dictionary* of words grouped by entity types (this filter can help identifying keywords but it is not so helpful when the same keywords belong to several entity types); and iii) *aliases* and synonym detection.

4.2. Sample Implementation of a Dialogue Manager

In current toolkits dialogue management is not usually a "pluggable" service. For example, Google DialogFlow is putting the Natural Language Understanding service, the Dialogue Manager service and the Natural Language Generator together and despite the fact that it is possible to use the Natural Language Understanding part as a service, the Dialogue Manager cannot be so easily decoupled.

We developed our dialogue manager using Python following the service format described for our proposal. The main task for the manager was to direct the conversation depending on the current *intent* and the *registry* history. The decision about how to continue the conversation is taken using Sklearn library's decision trees trained using a corpus that considers not only the previous system actions, but also the confidence of the natural language understanding module.

5. Implementation of Demo Systems

As a proof-of-concept we have developed two simple pizza ordering dialogue systems for home delivery, as a sample IE service. The systems use two different configurations of the architecture, involving different services for the same tasks. The infrastructure worked perfectly and the objective to combine different services and make them work together was performed successfully.

The first Demo uses the Google Speech-to-Text service, the Microsoft LUIS Language Understanding module, the Dialogue Manager we implemented with a simple Natural Language Generator and IBM Watson Text-to-Speech. The second Demo uses IBM Watson Speech-to-Text, the Natural Language Understanding service we implemented, the same Dialogue Manager we developed with the first demo system, and Google Text-to-Speech.

A single container application that worked for both demo systems was implemented in Java. The application had a button that the user had to press to speak, then the path of the recording was given to the infrastructure and it returned the answer audio file that the Java application played. With our framework it was possible to plug different solutions for the several modules implied and the result was transparent and worked seamlessly independently of the technology used.

6. Conclusions and Future Work

This paper has introduced an architecture for the creation of modular multi-purpose conversational systems that allows developers to combine already existing commercial services with their own solutions. This way, developers can focus on the implementation of
specific modules and the selection of the best-suited third-party alternatives. To generate
a functional system, the developer would only need to parametrise the infrastructure part
to connect with the proper REST services, and define the system *intents* and the *entities*.
To show the appropriateness of the framework as a practical development solution, we
have already included a repertoire of already existing solutions from Google, Amazon
and IBM and also generated our own Natural Language Understanding and Dialogue
Management modules successfully.

For future work we plan to retrieve the opinion of our prospective users, developers
with different expertise in the development of conversational systems, to validate our
proposal.

Acknowledgements

The research leading to these results has received funding from the European Union's
Horizon 2020 research and innovation programme under grant agreement No 823907
(MENHIR project: `https://menhir-project.eu`).

References

[1] Longo F, Padovano A. Voice-enabled Assistants of the Operator 4.0 in the Social Smart Factory:
Prospective role and challenges for an advanced humanmachine interaction. Manufacturing Letters.
2020;26:12–16.

[2] Adamopoulou E, Moussiades L. Chatbots: History, technology, and applications. Machine Learning
with Applications. 2020 Dec;2:100006.

[3] McTear M. Conversational AI: Dialogue Systems, Conversational Agents, and Chatbots. Morgan &
Claypool; 2020.

[4] McTear M, Callejas Z, Griol D. The Conversational Interface: Talking to Smart Devices. 1st ed. New
York, NY: Springer; 2016.

[5] Hornos MJ, Rodrguez-Domnguez C. Increasing user confidence in intelligent environments. Journal of
Reliable Intelligent Environments. 2018 Jul;4(2):71–73.

[6] Autexier S, Drechsler R. Towards Self-explaining Intelligent Environments. In: 2018 7th International Conference on Reliability, Infocom Technologies and Optimization (Trends and Future Directions) (ICRITO); 2018. p. 1–6.

[7] Lison P, Kennington C. OpenDial: A Toolkit for Developing Spoken Dialogue Systems with Probabilistic Rules. In: Proceedings of ACL-2016 System Demonstrations. Berlin, Germany: Association for
Computational Linguistics; 2016. p. 67–72.

[8] Jang Y, Lee J, Park J, Lee KH, Lison P, Kim KE. PyOpenDial: A Python-based Domain-Independent
Toolkit for Developing Spoken Dialogue Systems with Probabilistic Rules. In: Proceedings of the 2019
Conference on Empirical Methods in Natural Language Processing and the 9th International Joint Conference on Natural Language Processing (EMNLP-IJCNLP). Hong Kong, China: Association for Computational Linguistics; 2019. p. 187–192.

[9] Iovine A, Narducci F, Semeraro G. Conversational Recommender Systems and natural language:: A
study through the ConveRSE framework. Decision Support Systems. 2020 Apr;131:113250.

[10] Nakano M, Komatani K. A framework for building closed-domain chat dialogue systems. Knowledge-
Based Systems. 2020 Sep;204:106212.

[11] Bocklisch T, Faulkner J, Pawlowski N, Nichol A. Rasa: Open Source Language Understanding and Dialogue Management. arXiv e-prints. 2017 Dec;1712:arXiv:1712.05181.

[12] Kuratov Y, Yusupov I, Baymurzina D, Kuznetsov D, Cherniavskii D, Dmitrievskiy A, et al. DREAM technical report for the Alexa Prize 2019. In: 3rd Proceedings of Alexa Prize; 2019. Available from: `https://m.media-amazon.com/images/G/01/mobile-apps/dex/alexa/alexaprize/assets/challenge3/proceedings/Moscow-DREAM.pdf`.

[13] Amati G. LIU L, OZSU MT, editors. BM25. Boston, MA: Springer US; 2009. Available from: `https://doi.org/10.1007/978-0-387-39940-9\_921`.

[14] Robertson S, Zaragoza H. The Probabilistic Relevance Framework: BM25 and Beyond. Foundations and Trends in Information Retrieval. 2009 Apr;3(4):333–389.

Intelligent Environments 2021
E. Bashir and M. Luštrek (Eds.)
© *2021 The authors and IOS Press.*
This article is published online with Open Access by IOS Press and distributed under the terms
of the Creative Commons Attribution Non-Commercial License 4.0 (CC BY-NC 4.0).
doi:10.3233/AISE210078

The Effect of Uncertainty in Whole Building Simulation Models for Purposes of Generating Control Strategies

Amar Kumar SEEAM [a,1], David LAURENSON [b] and Asif USMANI [c]

[a] *Middlesex University Mauritius, Mauritius*
[b] *The University of Edinburgh, Scotland, United Kingdom*
[c] *Hong Kong Polytechnic University, Hong Kong*

Abstract. Buildings consume a significant amount of energy worldwide in maintaining comfort for occupants. Building energy management systems (BEMS) are employed to ensure that the energy consumed is used efficiently. However these systems often do not adequately perform in minimising energy use. This is due to a number of reasons, including poor configuration or a lack of information such as being able to anticipate changes in weather conditions. We are now at the stage that building behaviour can be simulated, whereby simulation tools can be used to predict building conditions, and therefore enable buildings to use energy more efficiently, when integrated with BEMS. What is required though, is an accurate model of the building which can effectively represent the building processes, for building simulation. Building information modelling (BIM) is a relatively new method of representing building models, however there still remains the issue of data translation between a BIM and simulation model, which requires calibration with a measured set of data. If there a lack of information or a poor translation, a level of uncertainly is introduced which can affect the simulation's ability to accurate predict control strategies for BEMS. This paper explores effects of uncertainty, by making assumptions on a building model due to a lack of information. It will be shown that building model calibration as a method of addressing uncertainty is no substitute for a well defined model.

Keywords. building model, calibration, uncertainty, ESP-r

1. Introduction

Intelligent Buildings have recently become a focus of attention for sustainable living, with their promises of optimal efficiency and comfort, using zero energy to solve these issues. The literature describes these buildings as being self-sufficient, even self-organising and *sentient* and in some cases can be true prosumers, a promising theory supported by the development of an emerging smart grid infrastructure, where buildings produce more energy than they consume. Variables which can be adjusted in a building to maintain adequate comfort in an energy efficient way, often relate to the thermal, lighting and air

---

[1] Corresponding Author: Middlesex University Mauritius, Coastal Road, 90203, Uniciti, Mauritius; E-mail: a.seeam@mdx.ac.mu.

quality characteristics of the building. These variables are required to be constantly monitored so that they can be manipulated to make the building comfortable to work or live in, and there are a wide range of computer controlled systems that are used to achieve this. We are now at the point where building models can be coupled with BEMS to help aid in prediction of control outcomes. However, before a building model can be used to generate predictive control strategies or perform energy auditing when coupled to a building management system, it needs to be accurately modeled and calibrated with the data gathered from previous Buiding Information Model (BIM) data, and monitored data from sensors instrumented in the building that is to be modeled. Notably current building energy management systems are not as dynamic as they could be, and in even in real world use, are often neither correctly used, nor optimised for energy efficiency. Changes in the environment (both internal and external) can affect the operation, and over time settings drift to inefficient boundaries, leading to situations that make occupants uncomfortable and a waste of energy. Simulation assisted control (SAC) utilises pre-existing (i.e. white-box) building models, which fully represent the building in terms of their geometry, operations and constructions and can have their energy performance and thermodynamic and airflow behaviour predicted using building energy performance simulation (BEPS) tools, such as ESP-r [1], used in this study. These predictions can then be used to formulate energy efficient control strategies, such as *optimum heat startup*, taking full consideration of all potential physical processes in a building, and are not constrained by the range of experience learned by Model Predictive Control (MPC) techniques [2,3,4] from training data, which would otherwise only consider a subset of the building's true energy performance. Building information modelling (BIM) is an emerging discipline which can potentially aid in providing the required information needed to create BEPS models. Essentially BIM is an extension of 3D CAD, with supplementary building specific information, though they require further translation in order to represent the additional nuances required in BEPS models, such as the processing of architectural geometry into thermal boundaries and zones. A lack of information or approximation of the geometry or constructions may require calibration to a set of data to address uncertainty (therefore equivalent to a grey box model, which assumes some level of knowledge). Converting BIMs to be used in BEPS tools currently requires a degree of human intervention during the translation (i.e. semi-automated), due to the intricacy required in assigning BEPS specific details. If the process is automated without human intervention, an approximate model may be produced, which may require further calibration with measured data to tune the model and *'fill the gaps'* with BEPS specific details not contained in the original BIM. These approximations may create various levels of uncertainty that lead to a model that is further divergent from reality, but may appear plausible in some cases, if not fully investigated. Furthermore, there is the issue of quality of data in terms of information provided to create the building model. That is, calibration can be used as a means to reduce uncertainty, when faced with a lack of information, which can rectify the model according to the data provided, but may be limited in scope and application since there may be a dependency on the measured data. In other words, calibration may not be the best approach when creating models, particularly if detailed data can be attained. The next section provides some background to the problem; Section 3 will contextualise the methodology used, Section 4 describe the results of introducing uncertainty and the final section will conclude the paper. The main contribution of this paper is to demonstrate that calibration is no substitute for a highly detailed model, with accurate construction data. That is to say, it will be demonstrated that calibration as a method of addressing uncertainty is no substitute for a well defined model, and calibration eventually will lead to skewed results, rendering a BEMS ineffective for the purposes of accurate prediction.

2. Background

Before a building model can be used to generate predictive control strategies or perform energy auditing when coupled to a building management system, it needs to be accurately modeled and calibrated with the data gathered from previous BIM data, and monitored data from BEMS sensors instrumented in the building that is to be modeled. The procedures to calibrate a building model using measured BEMS data will be discussed. Building simulation calibration is important to yield an accurate usable model. The accuracy of the model can be determined by comparing simulated and measured data using several metrics. Calibration is not a trivial problem, as energy models are complex with many interactions [5]. In the area of building simulation research, this is a deep and challenging problem, largely dependent on the quality of measured data available [6]. Most calibration methods reported in the literature pertain to commercial buildings which typically have higher-stake retrofit measure considerations than residential buildings [7]. That is the potential that could yield significant improvements to retrofit more efficient HVAC is greater for commercial applications.

3. Calibration Methodology

Calibration involves modifying model input parameters, in a sytematic way, until the model has passed a threshold to be deemed calibrated. It has been said that calibration, relies on user knowledge, past experience, statistical expertise, engineering judgement, and an abundance of trial and error [8]. There are several approaches and methodologies which can be followed. The ASHRAE 1051-RP project [9] was an attempt to define a method to improve the process of calibrating whole building energy simulation models using monthly utility data. 1051-RP used the ASHRAE Guideline 14 [10] which states goodness of fit criteria (CV(RMSE)) to assist calibration, specified under section 5.2.11.3 (Modeling Uncertainty), and is typically the main methodology applied in similar studies. CV (RMSE) (Coefficient of Variation of the Root Mean Squared Error) measures the differences between simulated (s) and measured (m) values, at each timestep i, for a total number of timesteps, n. A lower value indicates less variance and hence higher quality model. CV(RMSE) aggregates time specific errors into a single dimensionless number.

Coefficient of variation of the root mean square error (CV(RMSE))

$$RMSE = \sqrt{\frac{\sum_{i=1}^{n} (m_i - s_i)^2}{n}}$$

$$CV(RMSE) = \frac{RMSE}{\bar{m}}.100$$

CV(RMSE) of 15% is acceptable for calibration models and 30% for hourly models. Hourly data gives the most accurate results, though is the most difficult to capture;

monthly data can also be acceptable depending on the application, but can mask inaccuracies that can appear at hourly or daily resolutions [11]. The coefficient of determination, R^2 has also seen use in work by Tahmasebi for optimisation, [12], to assess models' goodness-of-fit for simulation assisted control, based on temperature calibration. R^2 provides an indication of how well observed outcomes are replicated by the simulation model. The coefficient of determination ranges from 0 to 1, with a value of 1 indicating that the model is a perfect fit. ASHRAE guideline 14 is widely used to help calibration and clearly defines statistic indexes as thresholds for calibrated model. The majority of literature in the area of building model calibration have thus adopted these conditions and statistical measures.

Coefficient of determination R^2

$$R^2 = \frac{\Sigma m_i s_i - \Sigma m_i \Sigma s_i}{\sqrt{(n\Sigma(m_i^2 - (\Sigma m_i)^2)(n\Sigma s_i^2 - (\Sigma s_i)^2)}}$$

4. Results

In this section, uncertainty will be introduced into an ESP-r building model (based on [13]) and an attempt at calibration will made using measured data. In this model, we have a whole domestic house, with scheduled heating systems in an Office space and Living Room. This will represent an analogy to a grey box model, which assumes some level of building knowledge. The effects of uncertainty will be analysed, by making assumptions on the building model due to a lack of information. It will be shown that uncertainty in a building model needs to be minimised by having as much information as possible, and that building model calibration as a method of addressing uncertainty is no substitute for a well defined model.

4.1. Reasons for Uncertainty

There could be number of reasons for uncertainty in a model.

1. A basic translation from a BIM. Geometry translation from BIM to BEPS models can lead to a loss of information.
2. Lack of source data about the building construction.

In either case, calibration is required to adjust the model accordingly, by comparing the output of the model with measured data. Calibration techniques and approaches will also be covered and discussed. There can be several sources of uncertainty in a model. The following sections will describe the types of uncertainty introduced in the model according to those identified by [14] in an *"Analysis of uncertainty in building design evaluations and its implications"*.

4.2. Specification Uncertainty

Relates to a lack of information on the exact properties of the building, such as the building geometry. In the uncertain model the roof has been removed (Figure 1), approximating the geometry of the model.

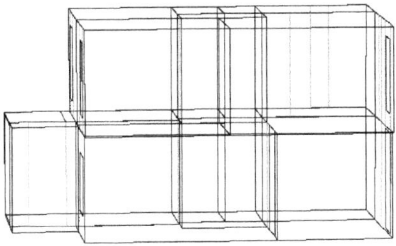

Figure 1. Geometric approximation - no roof.

4.3. Parameter Uncertainty

There can be degree of a uncertainty for each input parameter, for example, material properties.

4.4. Modelling Uncertainty

Arises from simplifications and assumptions that have been been introduced in the development of the model. Thus in the uncertain model there is no fluid flow model for air applied, though scheduled airflow is explored, further simplifying the building dynamics and physical processes.

4.5. Model Calibration

To reduce uncertainty, calibration techniques need to be applied. Calibration involves modifying model input parameters, in a systematic way, until the model has passed a threshold to be deemed *"calibrated"*, according to criteria set out by the American Society of Heating, Refrigerating, and Air-Conditioning Engineers (ASHRAE) in ASHRAE Guideline 14 [10] which uses CV(RMSE) to assist calibration, specified under section 5.2.11.3 (Modelling Uncertainty). The ASHRAE guidelines are often used to benchmark building models in the majority of calibration and validation studies for building simulation. According to the criteria, a CV(RMSE) of 15% is acceptable for calibration models using monthly data and 30% for hourly models. Hourly data gives the most accurate results, though is the most difficult to capture; monthly data can also be acceptable depending on the application, but can mask inaccuracies that can appear at hourly or daily resolutions [11].

4.6. Calibration Outcome

This section shall present the goodness of fit results, graphically and statistically for the calibration (before and after). The base case simulation is based on the initial values for the wall constructions (20mm wool external, 12mm wool internal) as the starting point for the calibration.

Figure 2, shows the simulated and measured data with a CV(RMSE) of 14.8%, which is acceptable according to ASHRAE guidelines, which requires models to be under 30%. Days one - four (0 to 100 hours), represent the weekdays, showing the schedul-

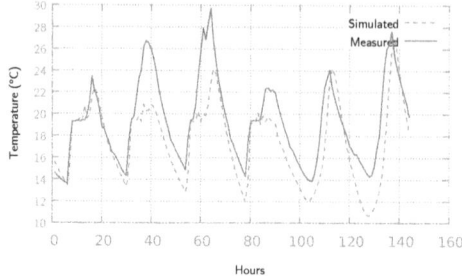

Figure 2. Temperature : Base Case : March 20th-25th 2012 : Office, Setpoint 19.3°C [Original Database values, 20mm wool external,12mm wool internal].

ing of the heater from 6am to 5pm to maintain a heating setpoint of 19.3°C. The last two days represent the weekend, when the heater was off and temperature variations are due to the external climate only. Temperatures which rise above the setpoint of 19.3°C can be attributed to solar gains and an increase in external ambient temperature In this case, the simulator does not adequately represent this phenomenon for days two and three, though there is good agreement for day one. For this particular day, the rise in temperature is closely matched with an identical gradient for both measured and simulated data, as the heating system is actuated to reach setpoint of 19.3°C. The setpoint is maintained until midday, when the temperature rises steeply due to external gains. The simulator repre-sents this phenomenon matching the measured well, along with the drop in temperature. However this is not the case for subsequent days where the measured data shows temper-ature peaks reaching nearly 30°C on the third day, and gradient drops in temperature that are slightly steeper. The last two weekend days do follow the trend of the measured data, reasonably well, but the simulator again drops to a lower temperature by as much as 3°C by the sixth day. Figure 3 shows the measured and simulated data with a CV(RMSE) that is very high at 439%, which is significantly outwith ASHRAE guidelines. The simulated values demonstrate that the model heater is having to work harder to maintain a setpoint, with repeated hourly actuations.

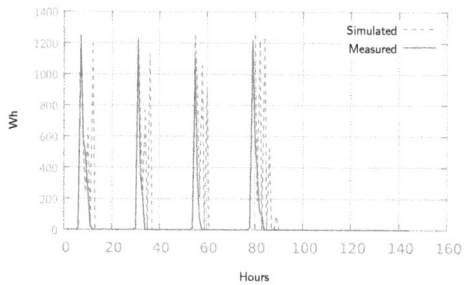

Figure 3. Heat Energy : Base Case : March 20th-25th 2012 : Office, Setpoint 19.3°C [Original Database values, 20mm wool external,12mm wool internal].

4.7. Calibrating for the Lowest CV(RMSE) for Energy Consumption

Since the CV(RMSE) for the base case temperature response is within ASHRAE guidelines, calibration on energy consumption will be explored, in an attempt to lower it towards an acceptable level.

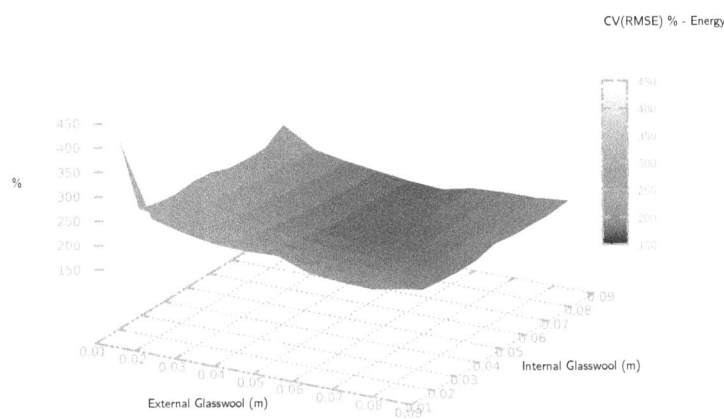

Figure 4. 3D Surface plot : Calibration Period 1 : March 20th-25th 2012: Office, Setpoint 19.3°C.

The surface plot shown in Figure 4 shows the results of the calibration run for the automated calculations of CV(RMSE) for energy consumption of the Office heater and depicts the relationship between the CV(RMSE) and glasswool internal and external thickness. The highest error occurs with the lowest amounts of glasswool internal and external thickness. The plot is largely flat indicating the tuning of these parameters is overall ineffective. The results of the calibration give a CV(RMSE) for energy and temperature at 135.86% and 12.1% respectively. This revises the external insulation glasswool thickness = 59mm and internal insulation thickness = 59mm. Figure 5 shows the measured and simulated temperature data when the model is set to these parameters. Simulated values now follow the trend of the measured data much more closely, with the 2.7% improvement in CV(RMSE) for temperature compared to the pre-calibration case. In particular, the solar gains affecting the model is more evident with the model demonstrating overheating curves that closely match the measured trends. However looking at the first day, the simulator is demonstrating higher sensitivity to overheating compared to the previous pre-calibration case. In subsequent days though the simulator represents the overheating phenomenon more closely; in particular day three reaches peak temperature to within 1°C (though the rise in temperature is delayed by several hours). Day four's simulated profile is almost a perfect match to the measured data, with the simulated overheating occurring at the same time, and a gradient drop in temperature that is near identical. The following two weekend days are also closely matched, though the simulator's rise in temperature is slightly delayed in comparison.

Though 135% is a high CV(RMSE) for energy response, compared to the ASHRAE guidelines, the load profile of the simulated values is consistent with the measured data,

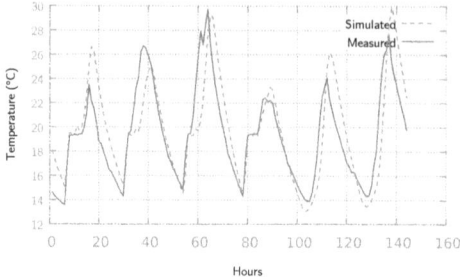

Figure 5. Temperature : Calibration Period 1 : March 20th-25th 2012 : Office, Setpoint 19.3°C [Calibrated Database values, 59mm wool external, 59mm wool internal].

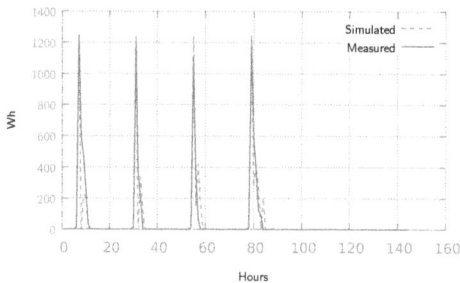

Figure 6. Heat Energy : Calibration Period 1 : March 20th-25th 2012 : Office, Setpoint 19.3°C. [Calibrated Database values, 59mm wool external, 59mm wool internal]

as shown in Figure 6. Here the limitation of performing calibration based on CV(RMSE) as per guidelines, at the hourly level for electrical heating loads is evident. However the CV(RMSE) for temperature response yields a low 12.1% for CV(RMSE), which is well within the guidelines. Following a potentially well matched initial load profile as seen in Figure 6 and Figure 5, the heating works the hardest first thing each morning to reach setpoint, but later actuations can significantly vary between time periods. Clearly this due to the fact CV(RMSE) compares predicted with measured data point to point, which may be appropriate for hourly temperatures but not highly variable energy delivery. [15] also recognised that evaluating calibration accuracy at small time scales (or scales where conditions are very variable) using CV(RMSE) is not appropriate. Graphically and statistically, the simulated temperature profile of this room could suggest this model is calibrated. In previous calibration studies (such as those carried out by Tahmasebi *et al.*) who only considered a subset of the building and an averaged zone temperature profile for a single floor of a building, a match can indeed be attained to the measured data, however may not be the case when taking a wider view across the whole building - the effects of ceiling and floor dynamics must also be considered. Furthermore, graphical analysis is equally important, particularly for temperature response, since a small change in CV(RMSE) can actually lead to a significantly better fit to the measured data when shown graphically against the simulator.

The importance of considering whole house dynamics is highlighted by looking at the adjacent Living Room. Figure 7 shows the hourly measured v simulated temperature data with the chosen calibrated values, and demonstrates how the uncertain model is failing at predicting the temperature response for this zone.

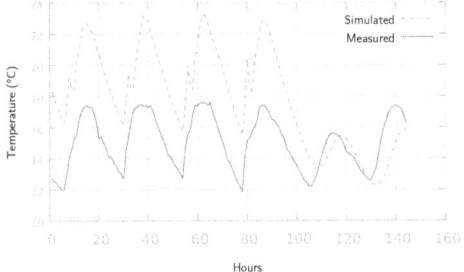

Figure 7. Temperature : March 20th-25th 2012 : Living Room, Setpoint 17.3°C [Calibrated Database values, 59mm wool external, 59mm wool internal].

Figure 7 shows a large disparity between the temperatures of the measured and simulated values by as much as 6°C. The simulator is consistently overheating the zone, suggesting that there may issues with heat transfer, as it appears the heat is not escaping to allow the temperature to equalise to setpoint.

Furthermore, in terms of energy response shown in Figure 8, the measured values indicate that the heating remains on for the duration of the day, thus showing the heater having to work harder to maintain setpoint. In contrast though, the simulator heat load is minor in comparison, though the simulator temperature response indicates significant overheating, further demonstrating how ineffective the uncertain model is, since it is using a fraction of energy compared to what was measured. As for modelling uncertainty, airflow has not been considered. This could lead a modeller to apply ESP-r's standard scheduled airflow technique in an attempt to 'force' heat transfer. An example of this is shown in Figure 9 with the application of a scheduled airflow rate of 2.5 Air Change Rate (ACH). This results in the simulator temperature response being more erratic, though the differences between peak temperatures between the simulated and measured data has reduced. There could be a temptation to further manipulate the ACH rate in the absence of an air flow model, but this would certainly lead to a model far removed from reality.

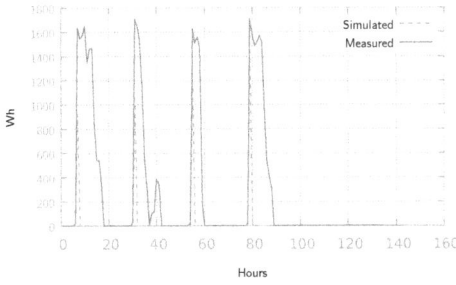

Figure 8. Heat Energy : March 20th-25th 2012. Living Room, Setpoint 17.3°C [Calibrated Database values, 59mm wool external, 59mm wool internal].

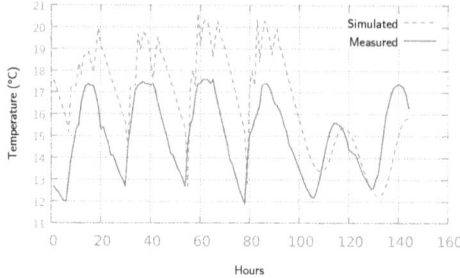

Figure 9. Temperature : March 20th-25th 2012 : Living Room, Setpoint 17.3°C [Calibrated Database values, 59mm wool external, 59mm wool internal] 2.5 airchanges/hour.

5. Conclusion

Due to the complexity of interactions in a building model, calibration should only be used to determine a few uncertain parameters. In particular when applying ASHRAE guidelines for temperature fit, a model may appear calibrated when looking at the temperature response of an individual zone, but may not be the case upon deeper investigation of other zones. The uncertain model makes some assumptions on the structure of the house. The most prevalent assumption, is that there is no roof zone, which has been approximated as an external boundary. There is uncertainty in the choice of unknown parameters (insulation thickness), which can lead to a fit to the data, even if they are out of the actual range (maximum end of range chosen to be 90mm, whereas the actual was 100mm). Finally in terms of modelling uncertainty, a simplified approach to introducing airflow has been applied to investigate if the characteristics can be improved, which can be seen as attempt at 'fudging', and not conducive to produce a reasonable solution. The problem with this model is most apparent when observing the large disparity in temperature and energy response contained in the the Living Room results, and demonstrates some of pitfalls when relying on calibration to try and achieve goodness of fit to tune the model. A satisfying solution may be achieved for one zone in the model, as can be seen with the high goodness of fit with the temperature response in the Office zone, but on closer inspection, may not be the case in other zones, as shown in the Living Room zone results. Furthermore, the difficulty with calibrating on energy use when using electrical heating power has been demonstrated, and that CV(RMSE) may not be the most ideal metric to ascertain 'goodness of fit', when doing hourly comparisons of heat delivery. Though it may be possible to further adjust parameters on the model to achieve a better fit, and may not necessarily represent reality, and therefore may not be able to predict adequately when presented with other data sets and use cases.

References

[1] Strachan PA, Kokogiannakis G, Macdonald IA. History and development of validation with the ESP-r simulation program. Building and Environment. 2008 Apr;43(4):601–609. Available from: http://www.sciencedirect.com/science/article/pii/S0360132306003313.
[2] Bengea SC, Kelman AD, Borrelli F, Taylor R. Model Predictive Control for Mid-Size Commercial Building HVAC : Implementation , Results and Energy Savings. HVAC&R Research. 2014;20(1).

[3]   Lehmann B, Gyalistras D, Gwerder M, Wirth K, Carl S. Intermediate complexity model for Model Predictive Control of Integrated Room Automation. Energy and Buildings. 2013 Mar;58:250–262. Available from: http://www.sciencedirect.com/science/article/pii/S0378778812006561.

[4]   Široký J, Oldewurtel F, Cigler J, Prívara S. Experimental analysis of model predictive control for an energy efficient building heating system. Applied Energy. 2011 Sep;88(9):3079–3087. Available from: http://www.sciencedirect.com/science/article/pii/S0306261911001668.

[5]   Clarke J.   Energy Simulation in Building Design.   Routledge; 2001.   Available from: http://www.elsevier.com/books/energy-simulation-in-building-design/clarke/978-0-7506-5082-3.

[6]   Li S, Song Z, Zhou M, Lu Y. Sensor data quality assessment for building simulation model calibration based on automatic differentiation. In: 2013 IEEE International Conference on Automation Science and Engineering (CASE). IEEE; 2013. p. 752–757. Available from: http://ieeexplore.ieee.org/articleDetails.jsp?arnumber=6654061.

[7]   Robertson J, Polly B, Collis J. Evaluation of Automated Model Calibration Techniques for Residential Building Energy Simulation Evaluation of Automated Model Calibration Techniques for Residential Building Energy Simulation. NREL; 2013. September.

[8]   Reddy TA, Maor I, Panjapornpon C. Calibrating Detailed Building Energy Simulation Programs with Measured DataPart I: General Methodology (RP-1051). HVAC&R Research. 2007 Mar;13(2):221–241. Available from: http://www.tandfonline.com/doi/abs/10.1080/10789669.2007.10390952.

[9]   Reddy A, Maor I. ASHRAE Research Project 1051- RP Procedures for Reconciling Computer-Calculated Results With Measured Energy Data; 2006. iii.

[10] ASHRAE. Measurement of Energy and Demand Savings. ASHRAE; 2002.

[11] Raftery P, Keane M, Costa A. Calibrating whole building energy models: Detailed case study using hourly measured data. Energy and Buildings. 2011 Dec;43(12):3666–3679. Available from: http://www.sciencedirect.com/science/article/pii/S0378778811004415.

[12] Tahmasebi F, Zach R, SchußM, Mahdavi A. Simulation Model Calibration : An Optimization-based Approach. In: International Building Performance Simulation Association; 2012. p. 386–391.

[13] Seeam A, Laurenson D, Usmani A. Evaluating the potential of simulation assisted energy management systems: A case for electrical heating optimisation. Energy and Buildings. 2018;174:579–586.

[14] de Wit S, Augenbroe G. Analysis of uncertainty in building design evaluations and its implications. Energy and Buildings. 2002 Oct;34(9):951–958. Available from: http://www.sciencedirect.com/science/article/pii/S0378778802000701.

[15] Ruiz Flores R, Lemort V. Calibration of Building Simulation Models: Assessment of Current Acceptance Criteria. 2014 Sep;Available from: http://orbi.ulg.ac.be/handle/2268/168878.

3rd International Workshop on Intelligent Environments and Buildings (IEB'21)

Intelligent Environments 2021
E. Bashir and M. Luštrek (Eds.)
© *2021 The authors and IOS Press.*
This article is published online with Open Access by IOS Press and distributed under the terms
of the Creative Commons Attribution Non-Commercial License 4.0 (CC BY-NC 4.0).
doi:10.3233/AISE210080

Preface to the Proceedings of the 3rd International Workshop on Intelligent Environments and Buildings (IEB'21)

Mohammed BAKKALI[a,b,1]
[a] The Bartlett School, University College London, UK
[b] The International University of Rabat, Morocco

The 3rd International Workshop on Intelligent Environments and Buildings (IEB'21) will be held within the 17th International Conference on Intelligent Environments (IE'21) in Dubai (UAE) on 21st-22nd June 2021. Smart infrastructure is a key element for developing and managing smart cities, buildings, materials, industrial processes and so on. Smart approaches help to seal economical competitiveness and environmental sustainability. Intelligent environments require cross disciplinary approaches that encompass a multitude of disciplines including architecture, design, engineering, urban planning and other different aspects standing from decision making to local economical fabric. Continuous processes and complete cycles are required from core sciences to innovation, products' business plans and their management. Integrated design for cities, buildings and technologies with multifaceted and tailored solutions and options related to the specifications of territories. Issues such as enhancing building energy efficiency, aiming for lower $CO2$ emissions, higher air quality, higher indoor and outdoor life quality and comfort, health, efficient emergency management are very relevant challenges nowadays. In this year's edition, we will cover topics such as smart mobility, smart city, smart campus, lighting technologies, new computational approaches for urban design, net zero energy buildings, co-integration of oil and commodity places and building and urban energy modelling. The event will provide an overview of current research, developments and ongoing work on smart cities, building science and technology, and will offer a platform for discussions about future expectations. During the event, renowned researchers and innovators on smart approaches will give oral presentations and discuss their new achievements.

---

[1] Corresponding Author: The Bartlett, UCL Institute for Environmental Design and Engineering, Central House, 14 Upper Woburn Place, WC1H 0NN London, UK; E-mail: mohammed.bakkali.10@ucl.ac.uk.

Intelligent Environments 2021
E. Bashir and M. Luštrek (Eds.)
© 2021 The authors and IOS Press.

doi:10.3233/AISE210081

From Smart Buildings and Cities to Smart Living

Felipe Samarán Saló[a,1]
[a] Universidad Francisco de Vitoria, Madrid, Spain

Abstract. We are drowning in information while starving for wisdom said the two-time winner of the Pulitzer Prize Edward Osborne Wilson. We are attached 24 hours to our "smart" phone that gets us closer to those who are far away and keeps us apart from those who are near, and we want to live in "smart" buildings and "smart" cities where systems are used and data can be gathered to save energy, create comfortable ambiances, regulate traffic, or deal with our waste products, but we surely need to reconsider what "Smart" living is all about. From the roman empire to the actual high-rise buildings, through the modern movement leaded by Le Corbusier, each time has used the technology available, but neither of the great master pieces of architecture and urbanism is remembered by the technology that made it possible but by Its ability to seduce the minds, hearts and souls of its habitants and the generations that came after them. It is essential to know that we truly need, to clarify WHY and WHAT FOR we do things before we solve or engage ourselves in HOW we make it happen.

Keywords. Information, wisdom, smart living, architecture, urbanism

---

[1] Corresponding Author: Universidad Francisco de Vitoria, Madrid, Spain; E-mail: f.samaran@ufv.es

Intelligent Environments 2021
E. Bashir and M. Luštrek (Eds.)
© *2021 The authors and IOS Press.*

doi:10.3233/AISE210082

Personalized Circadian Light: A Digital Siblings Approach

C. Papatsimpa [a,1] and Jean-Paul Linnartz [b]

[a] *Eindhoven University of Technology, the Netherlands*
[b] *Signify, the Netherlands*

Abstract. In this paper, we introduce a human-centric lighting control system optimized to support sleep and circadian coordination. We present an approach to optimize lighting that combines a "digital siblings" approach, i.e., a stochastic extension of a digital twin. It estimates and optimizes parameters in experimentally validated models of circadian and sleep regulation with a novel optimization algorithm for optimal timing of light exposure. We acknowledge that people have varying preferences for the lighting levels throughout the day based on their personal preferences and schedules and explicitly include this as a key parameter in our optimization strategy. Our results show that with the suggested lighting schedules, alignment of circadian rhythmicity to the desired sleep-wake schedule can be achieved with minimal disruption to people's daily lives.

Keywords. Circadian light, sleep, digital twin, particle filter, personalized model

1. Introduction

Human centric lighting that enhances health, performance and well-being has attracted the attention of the lighting industry in the recent years. Although the pathways through which light affects humans have been studied extensively [1], [2] to date little of these insights have been translated into practical lighting control systems. Nonetheless, there is strong evidence that better lighting systems can significantly improve wellbeing and health. Light is the main time cue for the human biological clock. This internal pacemaker regulates a number of physiological systems including hormone production, heart-rate and body temperature, thus influences our daily sleep-wake pattern, alertness and cognitive performance. Evolution has shaped our physiological processes to follow the natural cycle of light and darkness. Yet, indoor life and electric light have muddled our natural light exposure patterns. Nowadays, people are typically exposed to low daytime lighting and receive excessive exposure to light during the night. This un-natural light exposure acutely suppresses melatonin and sleepiness and delays the circadian clock inducing adverse effects for physiology and cognitive performance [3]. Moreover, social demands often oblige us to set an alarm that is out of phase of our propensity rhythm. In fact, a large part of the population is estimated to have social jet-lag [4], that is, circadian rhythms that are out of phase with people's daily schedules. Circadian rhythm sleeping disorders such as insomnia, inefficient sleep and mismatch between

---
[1] Corresponding Author, Charikleia Papatsimpa, Electrical Engineering Department, Signal Processing Systems Group, Eindhoven University of Technology, Groene Loper 3, 5612 AE Eindhoven, The Netherlands; E-mail: c.papatsimpa@tue.nl.

sleep and circadian rhythmicity are associated with adverse mental and physical outcomes [5]. These trends in general population health combined with the indoor lifestyle is one of the triggers for healthy building design to consider human centric lighting. Well-timed artificial light has potential as an effective time cue for the human biological clock [6], [7]. Yet, despite being technologically feasible, wide-scale adaptation of human centric lighting control as a tool that promotes health, well-being and sleep is still lacking. The main barriers that need to be overcome are **1) Precise monitoring of an individual's actual circadian state**; the individual differences in how humans perceive lighting, attributed partly to genetic variations in clock genes [8] and environmental influences [9], create a challenge for the translation of circadian research into lighting control. **2) Assessing circadian time using non-clinical and non-invasive sensor data**; the current techniques for circadian monitoring are considerably invasive **3)** The creation of automated lighting control systems demands **optimization algorithms based on quantified models on how humans process the light input** that can be executed by automated control systems.

Our recent work in [10] effectively addresses the first two challenges showing the feasibility of circadian phase estimation based on non-invasive light and actigraphy observations. We developed a digital siblings approach based on the statistical framework of a particle filter that not only estimates the circadian state but also adaptively calibrates model parameters to account for inter-individual differences in circadian response to light. We now address the third challenge by proposing an optimization algorithm for optimally adjusting the timing and levels of light exposure. We introduce an iterative optimization algorithm that derives personalized lighting schedules based on quantified models of the circadian and sleep mechanism. Ideally, people are exposed to the natural light-dark cycle, receiving bright daytime light and avoiding light exposure after sunset. Yet, we acknowledge that such lighting schemes are impractical to co-exist with modern lifestyle. After a day at work or school, people typically spend time with family or performing other leisure activities. We thus propose a novel optimization algorithm that takes into consideration personal lighting preferences. We exploit the fact that light is more biologically effective at certain times of the day than others, i.e., depending on when in the circadian trajectory it is administered, light is able to either phase advance or delay the human circadian clock. These phase shifts can be larger or smaller depending on the timing and magnitude of light exposure. The algorithm automatically finds when it is more effective to tune the light levels in order to exert the maximal circadian effect with the least disruption to a person's private life.

2. Circadian phase estimation

The problem of setting the light to optimize sleep and circadian functioning prerequisites the accurate estimation of the underlying circadian state and understanding of the mechanisms that regulate sleep in humans. In this section, we briefly review the underlying mathematical models of the circadian pacemaker and sleep regulation and describe the statistical framework we developed to estimate the circadian state using non-invasive light and actigraphy data.

2.1. Models of sleep and circadian regulation

In order to model the sleep mechanism, we adopt the Phillips–Robinson model [11]. According to the model, sleep occurs because of a flip-flop switch between wake-promoting (MA neurons) and sleep-promoting (VLPO neurons) that inhibit each other. Spontaneous wake-up occurs at the time (t_{w_sp}) that the firing rate of wake-promoting neurons, Q_{MA}, surpasses a threshold value Q_{th}, while sleep onset occurs at the time (t_{s_sp}) that Q_{MA} drops below the threshold value. The neuronal population is described by neuron mean cell body potential dynamics

$$\tau_{VLPO}\dot{V}_{VLPO} + V_{VLPO} = -\nu_{VLPO-MA}\,Q_{MA} + D_{VLPO} \tag{1}$$

$$\tau_{MA}\dot{V}_{MA} + V_{MA} = -\nu_{MA-VLPO}\,Q_{VLPO} + D_{MA} \tag{2}$$

The mean cell body potential is related to the firing rates of neurons by the firing function

$$Q_j = \frac{Q_{max}}{1 + \exp\left(\frac{\vartheta - V_j}{\sigma}\right)} \tag{3}$$

where $j = VLPO, MA$. Switching between sleep and wake occurs because of a drive to the VLPO that has both homeostatic and circadian components, the homeostatic drive H with sensitivity ν_{vh} and circadian drive C with sensitivity ν_{vc} respectively,

$$D_{VLPO} = \nu_{vc}C + \nu_{vh}H + D_0 \tag{4}$$

The homeostatic component of the drive is the homeostatic sleep pressure H which represents the sleep dept which increases with Q_{MA}

$$\chi\dot{H} + H = \mu_H Q_{MA} \tag{5}$$

while the circadian component C is approximated by

$$C = 0.5(1 + 0.8x - 0.55y) \tag{6}$$

Light input affects sleep by directly steering the circadian component C. In this paper, the dynamics of the circadian system are represented using the Jewett-Forger-Kronauer (JFK) model [12]. The circadian oscillation is described as a modified van der Pol oscillator with state variables x and y that oscillate according to

$$\dot{x} = \frac{\pi}{12}\left[y + \mu\left(\frac{1}{3}x + \frac{4}{3}x^3 - \frac{256}{105}x^7\right) + B\right] \tag{7}$$

$$\dot{y} = \frac{\pi}{12}\left\{\frac{1}{3}By - \left[\left(\frac{24}{\tau\,0.99729}\right)^2 + kB\right]x\right\} \tag{8}$$

Light enters the model as illuminance I. After an initial filtering operation, light input is transformed into the light drive signal B. The resulting photic drive depends on the ratio of activated photoreceptors n and the state variables x and y, suggesting that the circadian system has varying sensitivity to light throughout the day.

$$\alpha = a_0 \sqrt{\frac{I}{9500} \frac{I}{I + 100}}. \tag{9}$$

where α describes the rate of photoreceptor activation following light exposure I

$$\dot{n} = 60[\alpha(1 - n) - \beta n] \tag{10}$$

$$B = G\alpha(1 - n)(1 - bx)(1 - by) \tag{11}$$

All model parameters values are listed in Table 1.

Table 1. Model parameter values

Circadian process	Value	Homeostatic process	Value
μ	0.13	μ_H	4 nMs
k	0.55	v_{vc}	2.9 mV
a_0	0.1	χ	4.5 h
β	0.007	v_{vh}	1 mVnM^{-1}
G	37	D_0	10.2 mV
τ	24.2h	$\tau_{v,m}$	10 S
b	0.4	$v_{MA-VLPO}$	2.1 mVs
		$v_{MA-VLPO}$	1.8 mVs
		D_{MA}	1.3 mV

The models are based on general population data and model parameter values have been determined by fitting average responses to physiological processes. Skeldon et al. [13] showed that changes in circadian and homeostatic parameters can explain the observed inter-individual differences in the timing and duration of sleep. Since we want to predict individual responses to the light input, we make a plea to consider parameter value characterization in our system design. Given the limited size of our training dataset, and in order to avoid over-fitting, we initially chose the circadian parameter τ (representing the intrinsic circadian period in humans) to be calibrated based on real user data as there is strong evidence that the impact of light exposure on sleep timing strongly depends on the intrinsic circadian period in humans [14].

2.2. Digital siblings framework

To verify and compare the impact of multiple alternative light recipes, developing a digital twin of the human circadian pacemaker mechanism would be attractive. However, a deterministic approach of a digital twin is out of reach for our purpose: as the observations are inherently noisy and incomplete, we need a stochastic approach. We interpret the problem of generating a digital twin as one of characterizing the parameters and estimating the state of the human pacemaker. To this end, the statistical approach of a particle filter is an appropriate framework. Instead of a deterministic update of the state of the physical model based on input from the monitoring system, we generate a large set of samples (particles), each representing a possible state, thus consisting of possible values resulting from noisy input and uncertain model parameters. We interpret every particle as a digital sibling of the real human. Siblings slightly differ from each other and from the real human not only in key model parameters, but also in their experiences, such as differences in their exposure to light. These differences represent uncertainties and noise in our ability to monitor the human. That is, we use the particle filter in a way that it accounts for both "nature and nurture" of the siblings, by estimating model parameters and state, respectively. The siblings evolve and re-incarnate: particles are filtered and resampled according to their likelihood given the observed data. During cycles of executing the particle filter algorithm, siblings that have parameters that do not fit the model are discarded and replaced by siblings that inherit parameters (and states) that fit the observations better and (hopefully) converge towards the real human. The particle filter propagation and measurement update operations are schematically presented in Figure 1.

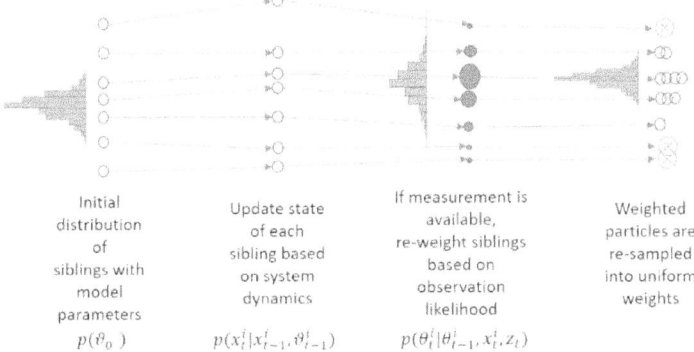

| Initial distribution of siblings with model parameters | Update state of each sibling based on system dynamics | If measurement is available, re-weight siblings based on observation likelihood | Weighted particles are re-sampled into uniform weights |

$$p(\vartheta_0)$$ $$p(x_t^i|x_{t-1}^i, \vartheta_{t-1}^i)$$ $$p(\theta_t^i|\theta_{t-1}^i, x_t^i, z_t)$$

Figure 1. Schematic depiction of particle filter operations. State variables are combined in state vector x and individual-specific parameters are captured in ϑ with initial probability distribution $p(\theta_0)$. $p(x_t^i|x_{t-1}^i, \vartheta_{t-1}^i)$ represents the prediction update step of the particle filter, realized by evaluating the system dynamics in equations (1)-(11). The term $p(\theta_t^i|\theta_{t-1}^i, x_t^i, z_t)$ represents how the new parameter distribution depends on the (previous) parameter, state, and observation z_t at instant t.

We begin with an initial particle distribution. 1) Each particle (sibling) is an instantiation of a random variation of initial model parameters drawn from realistic distribution of such parameter values in humans and is exposed to an instantiation of random variation of the measured light exposure drawn from a distribution of the error mechanisms of the light sensor. 2) The prediction step propagates each sibling based on the system's dynamics as described by the set of equations (1)-(11). 3) If a measurement is available, each sibling is evaluated against its fit to observations of the human response (e.g. wake-up time). 4) A resampling operation then discards siblings in the areas of low probability and reincarnates the siblings with parameters and states that have a better fit. This repeating process results in a set of values for the model parameters that update with every new observation. A more elaborate description of the mathematical framework can be found in [10]. This digital twin framework ensures a personalized representation of individual people that is connected and continuously updated by data from the real user.

3. Light optimization

The problem of setting the light to optimize circadian aspects does not fit a traditional control systems approach. The impact of light exposure on the circadian mechanism is not instantaneous but requires evaluation of the process for several hours after the actual timing of light exposure, the history of light exposure of several days also has an influence. Thirdly, user preferences might drive the shape of the optimization, as light that is biologically optimal is not always preferred by the user. Here, we present our solution to this complex optimization problem.

3.1. User satisfaction

The circadian impact of light needs to be weighed against immediate acute needs to perform tasks or to feel comfortable. These acute needs depend on the individual light preferences, on the activities performed and possibly on many other aspects. We follow the common observation that each user has its own preferred illuminance level $\xi = [\xi_1, ..., \xi_N]^2$, where we take the preferred illuminance level ξ_i [in lx] to be dependent on the i-th hour. Here, N denotes the total duration of the observation. Prior to the tests, users had the possibility to self-select different lighting levels depending on what lighting they found comfortable based on their schedules. We are aware of limitations as human tend not to intervene with light settings unless they are dissatisfied. Nonetheless, lacking better data, we used these observations as being representative for the most preferred illuminance levels throughout the day. The human eye senses brightness approximately logarithmically over a moderate range, thus, following [15], we model user satisfaction for lighting level I_i as a log-normal shaped function according to

$$U(I_i) = \exp\left(-\frac{(\ln I_i - \ln \xi_i)^2}{2\sigma^2}\right) \tag{12}$$

Here, parameter sigma σ describes the tolerance for illuminance of each individual; some people are only satisfied with a small range of illuminance values while others are more tolerant.

3.2. Mathematical formulation of the lighting control system

We consider a dimmable lighting system denoted as $I = [I_1, ..., I_N]$, where I_i is the illuminance level [lx] of the i-th hour, $0 \leq I_i \leq I_{max}$ and I_{max} corresponds to the illuminance level when the lighting system is fully turned on. The goal is to optimize vector I, i.e., the light levels, such that we simultaneously minimize the social jet-lag and lighting discomfort. We quantify social jet-lag as the timing difference between the unconstrained wake-up times (t_{w_sp}) and the enforced (by social constraints e.g. alarm) wake-up timing (t_{enf}), by controlling the light levels in I. Given the circadian system dynamics (1)–(11) and the user preference model (12), the lighting control problem may be formulated as a constraint optimization problem as:

---

[2] In this paper, the vectors are represented in bold.

$$\boldsymbol{w}^* = argmin \quad f(\boldsymbol{I}) = \|t_{w\_sp} - t_{enf}\| \tag{13}$$

That is subject to:

$$0 \leq I \leq I_{max},$$

$$U(I_i) > \alpha,$$

Here, $f(\boldsymbol{I})$ is the objective function. The enforced wake-up time t_{enf} is the time that the user sets his/her alarm, while the spontaneous wake-up time t_{w_sp} is derived by the mathematical model described by the set of Equations (1)–(11), using $\boldsymbol{I}$ as the input in Equation (9). The first constraint is a physical requirement that the dimming level should be between 0 and I_{max}. The second constraint ensures users' comfort by offering satisfying lighting conditions to their preference, i.e., user satisfaction function is not allowed to drop below a certain threshold (e.g., $\alpha = 0.6$ ensures that a user is 60% satisfied with the lighting level).

3.3. Solution to the optimization problem

The relationship describing the system dynamics (1)–(11) is highly non-linear, and the impact of light $\boldsymbol{I}$ on the wake-up time t_{w_sp} is not direct, but requires the evaluation of the process for several hours afterwards. As a result, a closed-form expression for $\boldsymbol{I}^*$ cannot be found analytically. To solve this, we follow a numerical approach. The iterative optimization algorithm exploits the fact that light is more biologically effective at certain times of the day than others by tuning the light levels at the time slots that are able to introduce larger shifts in the spontaneous wake-up times but exert the least disruption to the user. We use an iterative scheme that operates by introducing a small change to the light level step by step in an appropriately chosen time interval. The underlying principle is to take the interval that additional light has the highest biological impact and least user disruption. Formally, in the k-th iteration step, we change the lighting level to the time slot i_k that then minimizes the weighted sum of the impact on wake-up time and user disruption, i.e., we chose

$$i_k = \underset{i}{\operatorname{argmin}} (1 - \gamma)f(I_i) + \gamma(1 - U(I_i)) \tag{14}$$

Here, $U(I_i)$ represents the satisfaction of the user for a lighting level I_i, consequently, minimizing the objective function $1 - U(I_i)$ minimizes the disruption of the user. Parameter γ weights the input of each objective function. The searching is stopped when the improvement drops below a certain threshold.

4. Results

The results presented here are based on data from a field study with 15 participants (average age 70.9±4 years). The participants were asked to join the study for 7 days following their normal daily routines. Each participant wore a Philips Actiwatch Spectrum Pro measuring actigraphy (sleep onset and offset times) and light intensity.

First, we applied the digital sibling framework to the field data in order to estimate the intrinsic circadian period of each participant (model parameter τ). Our results confirm the large inter-individual variability in circadian period. In fact, in our limited sample size, the estimated intrinsic circadian period was distributed with 24.12h $\pm$ 0.21, matching clinically measured distribution of intrinsic circadian period in humans [16]. People do not only show variability in genetic predistortions, e.g., variability in τ, but also in the light levels they are exposed throughout the day. Data from the field study, presented in Figure 2, reveal the large inter- and intra-individual differences in the daily patterns of light exposure, i.e., the timing and magnitude of light exposure varied both between participants, but also from day to day within the same participant. The digital siblings approach results in a personalized to every participant circadian model that enables more accurate predictions of how users respond to light exposure.

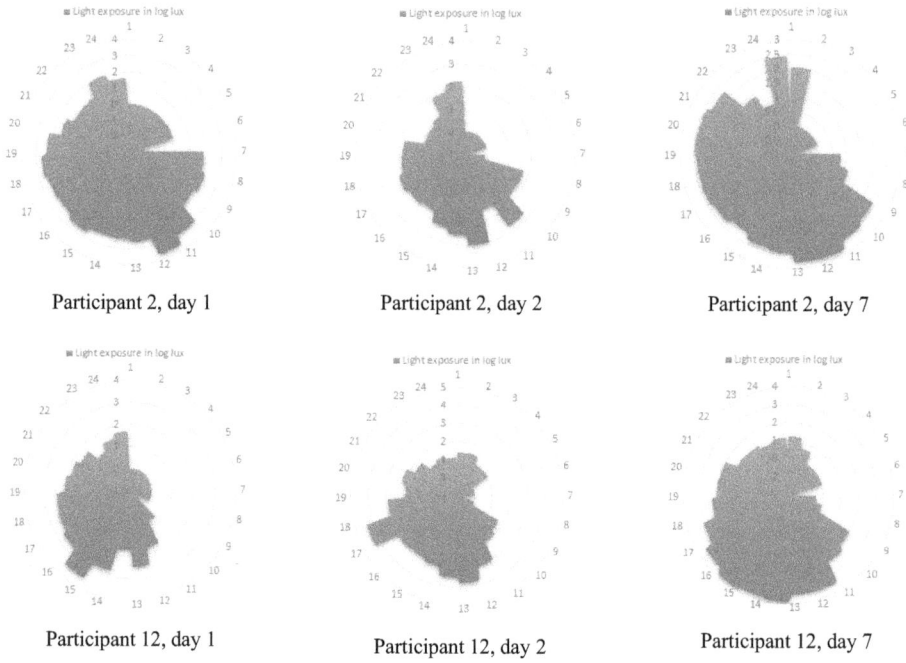

Figure 2. Inter- and intra-subject variability in daily light exposure [in log lux] throughout the 24 hour day. Illustrative examples through polar plots in 2 study participants.

The efficacy of lighting control could potentially be optimized by individual tailoring, based on personal light preferences and considering the internal rhythms of users. As an illustrative example, we present optimization results for a single participant. This participant is characterized by an average intrinsic circadian period (τ = 24.21), however, he typically receives bright light during the late evening and night (see Figure 2) which considerably delays his sleep cycle. In fact, the participant typically wakes-up at 8.5h $\pm$ 0.69 and sleeps at 0.6h $\pm$ 0.23. If this participant sets an alarm at 07:30 (t_{enf} = 7.5), this would mean that his biological rhythm is ~1 hour misaligned with his daily schedule. We thus wish to optimize the lighting levels to correct this circadian misalignment. Since most people spend a large part of their days in indoor public settings, they usually have little ability to control their lighting conditions. We thus

$$0 \le I \le I_{max},$$

present optimization results for a domestic setting (early morning and late evening and night) as a light intervention is easier and more practical to realize in such settings. For the simulations, tolerance σ is set to 0.6 and the preferred illuminance level ξ is set to the hourly daily average illuminance prior to the lighting intervention. The resulting lighting schedule, presented in Figure 3a, is characterized by increased light levels in the early morning, as soon as the participant wakes up, and dim light levels during the late evening and night, i.e., the time of the day that light introduces the largest delays in the biological clock. As seen in Figure 3b, the new lighting setting is estimated to gradually shift the sleep pattern of the participant, resulting in earlier wake-up and sleep times compared to the sleep schedule without any lighting intervention. In fact, in the last day of the intervention (day 7), the sleep cycle is estimated to be advanced by approximately 48 minutes resulting in the desired wake-up time (alarm at 07:30).

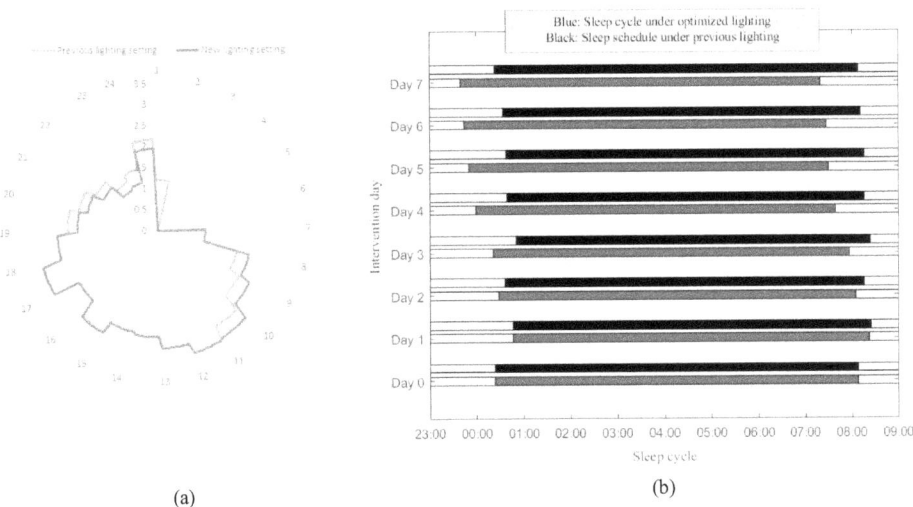

(a)   (b)

Figure 3. a) Algorithm suggested lighting schedule superimposed with pro-intervention daily average light exposure. b) Resulting daily shift in sleep cycle compared with sleep pattern prior to the intervention. Results are presented for an example participant with intrinsic circadian period $\tau = 24.21$, and tolerance $\sigma = 0.6$.

5. Conclusions

A well-functioning internal clock is of utmost importance for good sleep and optimal functioning during wake. Disturbances in sleep-wake patterns due to the absence of a strong time cue or an ill-timed signal for the circadian system can result in a lack of energy, cognitive deficits and compromised sleep quality. Despite the insights of how light affects human physiology, the benefits of these insights are not (yet) captured by practical lighting control systems. The natural intra-individual and inter-individual variability of internal time advocates against "one-size-fits-all" lighting interventions and highlights the need for a personalized approach to human-centric lighting. Here, we present a lighting control system optimized to provide personalized lighting that supports sleep and aligns the circadian mechanism to people's daily schedules. Our optimization strategy combines a "digital siblings" framework of the circadian mechanism, i.e., a stochastic extension of a digital twin, with a novel optimization algorithm that takes into consideration both the biological effects of light and personal lighting preferences. We arrive at well-timed and feasible lighting schedules that are able to gradually shift the sleep pattern of participants and align their sleep cycles to their daily schedules. Schedules are feasible in the sense that these fit within the personal tolerances to light level deviations.

References

[1]   I. Provencio, I. R. Rodriguez, G. Jiang, W. P. Hayes, E. F. Moreira, and M. D. Rollag, "A novel human opsin in the inner retina," J. Neurosci., vol. 20, no. 2, pp. 600–605, Jan. 2000, doi: 10.1523/jneurosci.20-02-00600.2000.

[2]   T. Roenneberg, E. J. Chua, R. Bernardo, and E. Mendoza, "Review Modelling Biological Rhythms," Curr. Biol., vol. 18, pp. 826–835, 9AD, doi: 10.1016/j.cub.2008.07.017.

[3]   C. Cajochen et al., "Evening exposure to a light-emitting diodes (LED)-backlit computer screen affects circadian physiology and cognitive performance," J. Appl. Physiol., vol. 110, no. 5, pp. 1432–1438, May 2011, doi: 10.1152/japplphysiol.00165.2011.

[4]   M. Wittmann, J. Dinich, M. Merrow, and T. Roenneberg, "Social jetlag: Misalignment of biological and social time," in Chronobiology International, 2006, vol. 23, no. 1–2, pp. 497–509, doi: 10.1080/07420520500545979.

[5]   K. G. Baron and K. J. Reid, "Circadian misalignment and health," Int. Rev. Psychiatry, vol. 26, no. 2, pp. 139–154, 2014, doi: 10.3109/09540261.2014.911149.

[6]   R. N. Golden et al., "The efficacy of light therapy in the treatment of mood disorders: A review and meta-analysis of the evidence," American Journal of Psychiatry, vol. 162, no. 4. pp. 656–662, Apr-2005, doi: 10.1176/appi.ajp.162.4.656.

[7]   M. Münch et al., "Blue-Enriched Morning Light as a Countermeasure to Light at the Wrong Time: Effects on Cognition, Sleepiness, Sleep, and Circadian Phase," Neuropsychobiology, vol. 74, no. 4, pp. 207–218, Jul. 2017, doi: 10.1159/000477093.

[8]   D. A. Kalmbach et al., "Genetic basis of chronotype in humans: Insights from three landmark gwas," Sleep, vol. 40, no. 2. Associated Professional Sleep Societies,LLC, 01-Feb-2017, doi: 10.1093/sleep/zsw048.

[9]   M. Hébert, S. K. Martin, C. Lee, and C. I. Eastman, "The effects of prior light history on the suppression of melatonin by light in humans," J. Pineal Res., vol. 33, no. 4, pp. 198–203, Nov. 2002, doi: 10.1034/j.1600-079X.2002.01885.x.

[10]   J. Bonarius, C. Papatsimpa, and J.-P. Linnartz, "Parameter Estimation in a Model of the Human Circadian Pacemaker Using a Particle Filter," IEEE Trans. Biomed. Eng., pp. 1–1, Sep. 2020, doi: 10.1109/tbme.2020.3026538.

[11]   A. J. K. Phillips, P. A. Robinson, and A. Phillips, "A Quantitative Model of Sleep-Wake Dynamics Based on the Physiology of the Brainstem Ascending Arousal System," J. Biol. Rhythms, vol. 22, no. 2, pp. 167–179, 2007, doi: 10.1177/0748730406297512.

[12]   M. E. Jewett, D. B. Forger, and R. E. Kronauer, "Revised limit cycle oscillator model of human circadian pacemaker.," J. Biol. Rhythms, vol. 14, no. 6, pp. 493–9, Dec. 1999, doi: 10.1177/074873049901400608.

[13]   A. C. Skeldon, G. Derks, and D. J. Dijk, "Modelling changes in sleep timing and duration across the lifespan: Changes in circadian rhythmicity or sleep homeostasis?," Sleep Medicine Reviews, vol. 28. W.B. Saunders Ltd, pp. 96–107, 01-Aug-2016, doi: 10.1016/j.smrv.2015.05.011.

[14]   A. S. Lazar et al., "Circadian period and the timing of melatonin onset in men and women: Predictors of sleep during the weekend and in the laboratory," J. Sleep Res., vol. 22, no. 2, pp. 155–159, 2013, doi: 10.1111/jsr.12001.

[15]   C. Papatsimpa and J.-P. Linnartz, "Personalized Office Lighting for Circadian Health and Improved Sleep," Sensors, vol. 20, no. 16, p. 4569, Aug. 2020, doi: 10.3390/s20164569.

[16]   J. F. Duffy et al., "Sex difference in the near-24-hour intrinsic period of the human circadian timing system," Proc. Natl. Acad. Sci. U. S. A., vol. 108, no. SUPPL. 3, pp. 15602–15608, Oct. 2011, doi: 10.1073/pnas.1010666108.

Intelligent Environments 2021
E. Bashir and M. Luštrek (Eds.)

doi:10.3233/AISE210083

A Smart Campus Template

Juan Carlos AUGUSTO[a,1]

[a] *Research Group on Development of Intelligent Environments*
Department of Computer Science, Middlesex University London, UK

Abstract. We highlight a lack of models and theories associated with the Smart Campus concept and also an absence of processes to support its design and development. This paper provides a first approach to a theory and a set of design principles to guide their development. The theory and principles are flexible enough to be easily adapted and adopted by any organization interested in developing a Smart Campus.

Keywords. Smart Campus, Intelligent Environments, Human-centric Computing

1. Introduction

Ask yourself the question: which has been the latest worldwide adopted revolution in technology-mediated education? Mostly likely we will agree on: PowerPoint-style of digital slides. Later on "smart" whiteboards became a product, some educational organizations acquired them and in some countries they are fairly common, however they did not offer much in the way of smartness in the clever sense rather facilitated digitalization of content. This for "smart education" and "smart classrooms" [1, 2, 3]. When we escalate the look at "Smart Campus" level, the concept is mentioned on the web, however, it has been mostly hijacked by certain organizations which have done some new equipment acquisition, again, abusing a technical term which is meant to signify something more substantial. Uptake from Universities have been slow. Universities are busy exporting innovation outside their walls and often forget to incorporate technical innovation themselves.

Despite the occasional bad use of the term it is a worthy concept to adopt, we can apply it properly, and reap the rewards of innovating in this direction. It can use state of the art technology to introduce efficiencies for students, staff, and administration. Smart Campus can help connecting people and resources. It can be designed to address in a balanced and ethical way the preferences and needs from all main University stakeholders. It could be a positive stimulus both internally and externally for more innovation.

The following articles provide a picture of the state of the art at an international level. Janelle et al., [4] investigated the spatially enabled campus through a multi-disciplinary team including technological, institutional, and social perspectives. For the specific organization conducting that introspective assessment it triggered the collection of ideas and evidence to support future development in that direction and emphasized their internal priorities on: sustainability, knowledge sharing, cost effectiveness, student involvement and learning, safety, and other perspectives. Kwok [5] Highlights the role of data, procedural knowledge, and system integrations as resources as well as privacy concerns and the time it takes to develop the concept, as challenges. It emphasizes the importance of networks, computers, systems, processes, people and knowledge

---

[1] Corresponding Author: Juan Carlos Augusto; E-mail: j.augusto@mdx.ac.uk.

extraction processes as part of the essential infrastructure required to materialize the concept. Sari et al. [6] provides an example of a smart campus design from Indonesia focused on the use of IoT for the development of smart campus services limited to education, parking and rooms automation. Bi et al. [7] provides an example from China focusing on the use of Building Information Modeling (BIM) and of 3D Geographic Information System (3DGIS). Chan and Chan [8] looks at the latest development of Smart Campus and, especially, Smart Libraries, focusing on how innovation in those areas create positive impact on those who use their services. Muhamad et al. [9] investigates the kind of technology used, the models used, the services created, and the perceived benefits. It highlights the growth in activity within this area of innovation and the diversification of services beyond teaching and learning, for example sustainability and management support. This article highlights the need for services to be reactive and dynamic, and highlights the challenges in the areas of interaction and interoperability. Min-Allah and Alrashed [10] distills a Smart Campus approach from Smart City ones: People into Community, Planet into Campus Infrastructure, Prosperity into Sustainability/Employability, Governance into Administration/Management, and Propagation into Replicability/Innovation. They notice the absence of a single vendor/provider of services, the absence of global standards and the possibility to attract investments through small scale projects with clear business returns to make the project self-sustainable.

Despite these isolated proposals and initial explorations there is still little consensus on the fundamental principles and underpinning theories, no methods and no tools, which are immediately helpful to develop this concept. This paper aims at providing a first unified conceptual view that build up on previous experiences and captures the essence of the main concepts at stake and hopefully act as a community discussion starter.

2. Smart Campus as an Intelligent Environment

Here we take the view that a Smart Campus is an instance of an Intelligent Environment [11], in this case one created to support those who make use and interact with the campus.

Definition: *"The 'Smart' technology, associated software, and processes, which facilitate the main objectives of the campus."*

Here 'smart' is understood as sensing/actuation technology which supports context-aware decision-making. This is related to the following areas in Computer Science: Ubiquitous Computing, Pervasive Computing, Intelligent Environments, Internet of Things, Ambient Intelligence, and their unifying principle: context-awareness [12].

Important components are sensors, actuators, and interfaces linked by networks, and supplemented by middleware which facilitate the context-awareness. These can also be complemented by AI software, mobile computing, robotics. etc.

There is no shortage of technological resources, for example: (passive) screens (relying information), 'smart' whiteboards, advanced interfaces (e.g., voice processing, image processing, haptics, etc.), sensing technology (e.g., to measure use of a resource), smart phones and their apps, data (storing/analysing/visualization), mixed reality, virtual presence, robots, artificial intelligence (e.g., real-time context-awareness, machine learning, etc.).

On one hand tools shape what we can do and how, however technology in an intelligent environment is only relevant to the extent it contributes to provide to the stakeholders the services they expect [13]. Determining what the stakeholders expect is the starting point, however these systems are so complex they require some iterations through suggesting services, mapping to infrastructure and checking these with stakeholders until they converge to a desirable, and also feasible, project (hence the bidirectional arrows in figure 2.1).

Figure 2.1 Fundamental Smart Campus Triad

We expand on the triad stakeholders-services-technology of important inter-related concepts in the subsections below, giving stakeholders more importance as they will greatly affect what services are created and which infrastructure is considered.

2.1 Stakeholders

We advocate here for a stakeholders-led system creation, such as the User-centred Intelligent Environments Development Process [14]. In terms of identifying stakeholders, obviously this can be conducted at all sort of granularity levels depending on how specific the services are. At an initial stage of a Smart Campus an obvious higher level partition of stakeholders can make emphasis on the main groups of activities and responsibilities, for example (figure 2.2):

- Teaching and Learning related: these can comprise students of taught courses and those delivering the material to be learnt,
- Research and Innovation related: these could be trainees such as PhD students and also supervisors and lab assistants, and
- Decision-making and Support-related: these comprise different groups of staff with the task of supporting the operational activities of the organization at all levels, this includes, for example, senior managers and technicians.

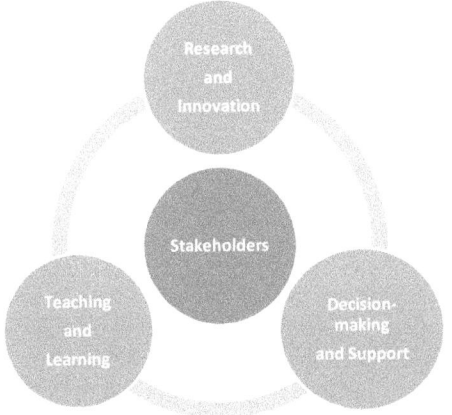

Figure 2.2. Core Smart Campus stakeholder groups

We emphasize these should not be taken as the definitive stakeholder groups, only as an illustration of the concept and there are interesting subtleties to consider. On one hand these categories can be looked at a finer grain, for example, distinguishing amongst undergraduate and postgraduate students. Additional subtleties include that some stakeholders can be in more than one category, for example a member of staff can be pursuing further studies and being a stakeholder both as staff and as student, or a member of staff can have decision-making roles such as Director of Program or Research Degrees Coordinator which are complementary to the usual staff responsibilities. On the other hand stakeholders categories seem static however they are dynamic when looked at a granularity of months/years and people can move through stakeholder categories. For example, from student to alumni, from student to staff, from teaching staff to decision-maker as Head of Department.

2.2 Services

Services are the specific benefits which stakeholders ultimately get from the system. Each University may be able to offer different services and also may have different internal priorities, so it is not possible to provide a prescription, however, some options associated with campus activity linked to the most generic stakeholder groups can be:

- To support learning activities: accessing information, facilitating learning and assessment, health and wellbeing of students, etc.
- To support innovation activities: creativity, training, coaching/mentoring, connecting and collaborating, etc.
- To support decision-making activities: connecting with students/staff, optimizing services, keeping services operational, support in identifying and reacting to critical events, etc.

We are using these three categories as a way to illustrate the overall concept without getting ourselves lost in too many details, however every University have to work out their stakeholders organization which best represent their situation and focus.

Echoing what we stated about stakeholders changing stakeholder group, services associated with those individuals may have to travel with the person to the next stakeholder category s/he is entering to. And of course services themselves may change with time as the organization changes priorities and resources availability.

2.3 Technological infrastructure

Ultimately the enabler of the Smart Campus concept will be the use that humans give to the selected technology. There are plenty of options now on technologies, examples of recent technologies which are tempting to consider in relation to various services are: (passive) Screens (relying information), "smart" whiteboards, advanced interfaces (e.g., voice/image processing, haptics), sensing technology (e.g., to measure use of a resource), smart phones and their 'apps', data (storing/analyzing/visualization), mixed reality, virtual presence, robots, artificial intelligence (e.g., real-time context-awareness, and machine learning).

In fact there are so many interesting gadgets and devices and people is heavily bombarded with news from companies about each of these that often projects end up being technology led. Many projects which are not carefully managed end up as a cocktail of gadgets and systems with low acceptance and short life. Following a more stakeholders centred approach and making technology the consequence of decisions instead of the cause should increase the chances to end up with a safer Smart Campus system.

3. Landing the concept

A Smart Campus, like any Smart Environment, is heavily based on its physical spaces. A campus typically has the following spaces: classrooms, lab rooms, offices, meeting rooms, corridors, food areas, socialization areas, toilets, walk ways, storage areas, sport areas, etc. Some of these areas are more often used by different groups of stakeholders, some areas are more specialized and attract users with more specific aims, others will provide services to all, for example a food area. Areas in the campus can be identified according to their priority intended use, services available and perhaps also which stakeholders can have access to them, which also has implications with security. For example, consider Figure 3.1 and assume the outer shape represents the campus, the inner six rectangles represent six different buildings and the smaller four rectangles inside represent rooms. Let us assume the services clusters are those related to: Teaching and Learning related activities (represented with "T"), Research and Innovation related activities (represented with "R") and Decision-making related activities (represented with "D"). This represents different rooms have services expected to cater for different focal areas of the organization. For simplicity we considered only three here, however, naturally there could be more areas with complementary functions associated, for example, areas for interaction and socialization such as 'common rooms', cafes, eating areas, faith related areas, and of course there are also areas which are meant to be used by everyone such as corridors and patios.

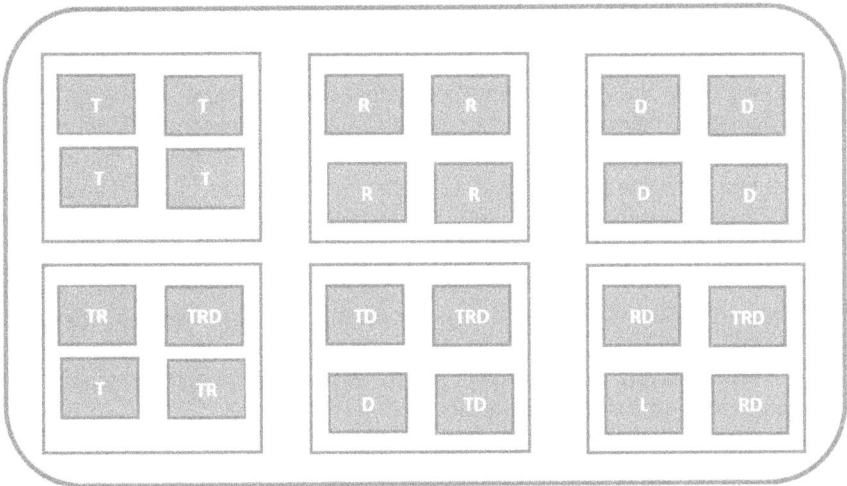

Figure 3.1. Combinations of "Smart Spaces"

Next we consider some scenarios which we use to illustrate how the concepts above can be interlinked. Again, they are not intended to be prescriptive, rather act as examples and motivators for future developers to adapt to their specific project. Figure 3.2 shows an extract of the previous figure where we highlight spaces dedicated to provide services of a specific type, as we clarified before in practice spaces can, and often will, combine services of different categories serving different groups of stakeholders.

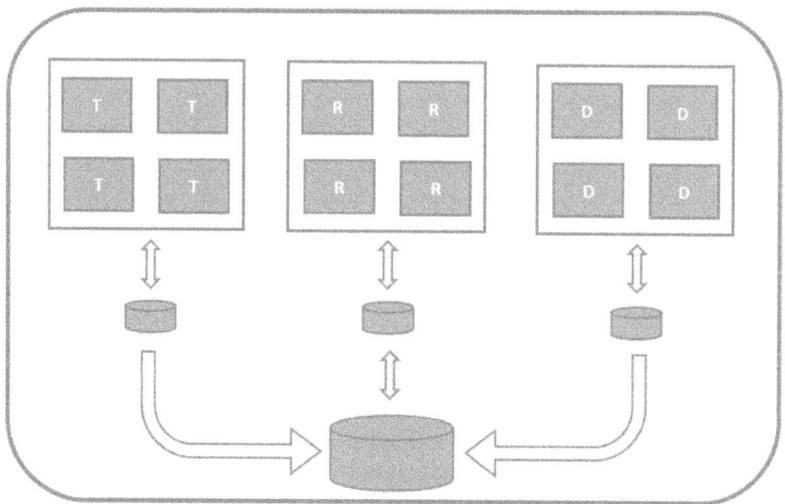

Figure 3.2. Services template for scenarios

3.1. Scenario 1

Main focus: to support stakeholders related to learning and teaching activities.
Services:
- Students are alerted through their an app the lecture starts in a certain room, they are warned today the lecture takes in a different place, the orientation app helps them find the place from where they are.
- Once in the classroom students attendance is confirmed.
- The lecturer starts the class the smart whiteboard help retrieving previous digital material, drawing and writing newly generated material, storing in digital form this new material. Content is multi-media (e.g., text, audio, video, drawings, documents, augmented reality files). Material can be now accessed in a multimedia channel.
- Side screens allows Distance Learning students to connect and participate.
- App to capture disruptive students (contextualized recording only operable by staff).
- When room idle is empty lights go off, when out of expected academic expected use, security can be alerted if noisy.

Possible Infrastructure to achieve them:
- Smart whiteboard;
- Extra screen to support Distance Learning;
- Recording and Storage of selected material; and
- Light management through movement sensors.

3.2. Scenario 2

Main focus: to support stakeholders related to innovation activities.
Services:
- Support spontaneous collaboration or funding call driven.
- Apps to connect colleagues by technical interest ('theme pals' nearby).
- Library services to bring relevant information more easily.
- Multi-touch screens on mobile frames:
  o To concentrate "creativity apps";
  o To support "out of the box" ideas;
    – For formation; and
    – Progression.
  o Simultaneous collaboration; and
  o Easy conversion of collaboration into easily transportable digital records of interaction.

Possible Infrastructure to achieve services:
- Flexible access to information and knowledge repositories; and
- Creativity stimulating environments.

3.3. Scenario 3

Main focus: to support stakeholders with higher decision-making responsibilities.
Services:
- It should provide:
  o centralized and distributed; and
  o synchronous and asynchronous communication.
- Issues tracking (challenges and progress highlighting).
- Meeting and interaction facilitation.
- Solutions support:
  o Relevant Data facilitation; and
  o Creative brainstorming facilitation.
- Planning support.
- "Firefighting" support.

Possible Infrastructure to achieve services:
- Data analytics;
- Data visualization; and
- Creativity support.

4. A Reusable Template

How this fit in the wider Computer Science landscape? We can see the concept of Smart Campus as an instantiation of the wider concepts of Ubiquitous Computing related systems, a family of areas which developed in the last three decades using sensing to produce systems which are useful to humans in practical daily life situations, especially here we explore more deeply the definition in section 2 at system level.

Most distinctive to these systems is their input data coming from sensors, which complements that more traditional from interfaces and databases, updating in real-time the contexts relevant to the services the system is supposed to deliver. These "Smart" systems also use actuators to produce an effect in the physical environments where they operate. By "sensors" here we mean artifacts which can translate parameters from the physical world into digital information and by "actuators" we mean artifacts which can translate digital information into manifestations in the physical world. Smart systems by their very nature are created to benefit humans in specific environments and contexts [12] and are meant to consider the system stakeholders preferences and needs and optimize certain stakeholders satisfaction represented by a function Ω (Fig. 4.1).

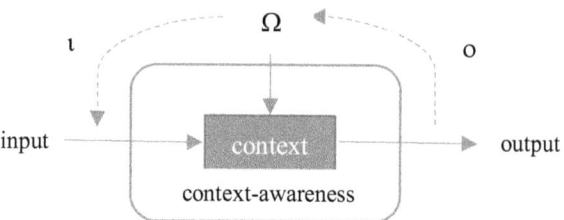

Fig 4.1 Context-aware system reactive to services optimization function Ω

These systems, in whatever way have been implemented and regardless of how consciously their developers were based in these premises above always try to implement a system which provides some value to some stakeholders. Ideal full global satisfaction may be impossible as sometimes the preferences of some stakeholders conflict with those of other stakeholders. Some systems have well defined hierarchies and that defines how certain services are given priority over others. There may be a number of optimal and suboptimal levels of system performance.

Like all control systems, these systems can have dynamic feedback loops ("o", omicron, in Figure 4.1) informing updates on the optimization function Ω and adjusting the allowed inputs ("ι", iota, in Figure 4.1). Often environments have multiple groups of stakeholders with overlapping and also different service expectations and for a system to satisfy all stakeholders a global optimization function needs to represent these and be managed by the system. That is Ω can be defined as a function $\Omega(\omega_1, ..., \omega_n, O_1, ..., O_m)$ defined on a number of stakeholders groups $\omega_1, ..., \omega_n$ and a number of Organizational Objectives $O_1, ..., O_m$. Each ω_i is the outcome of a function taking into account each stakeholder s_j of that stakeholder group, collectively or as an addition of individual's functions and each organizational objective. Each O_j is set by the organization (with feedback from other stakeholders), they can be revised periodically and provide the main comparison reference to understand how well the system is serving stakeholders. Examples of organizational objectives could be: to produce innovation, to train and graduate new professionals, and to have a positive impact in society, we leave them generically indicated in the examples that follow, we will come back to them at the end of this section.

In the reminder of this section we offer some indications on how these feedback loops work for the three scenarios outlined in the previous section. For example, in Scenario 1: "o" could channel the recent organization desire of augmenting health over other service clusters, now $\Omega(\omega_{learners}, \omega_{teachers}, \omega_{decision-makers}, O_1, ..., O_m)$ will take in the results of each of these stakeholder polls resulting on an adjusted "ι" from what it was (let us say, "safety > health > pedagogy > sustainability") into "health = safety > pedagogy > sustainability" which is used then to affect the decisions from the context-awareness modules leading to for example, more messages in screens about health related issues around areas designated as having a role in teaching and learning ("T" type in figure 3.1). Another example, for Scenario 2: "o" reflects now the aim of the organization to augment inter-faculty scientific cooperation amongst colleagues so $\Omega(\omega_{innovators}, \omega_{decision\text{-}makers}, O_1, \ldots, O_m)$ will tune in the system input accordingly through "ι" from what it was (e.g., "information finding > creativity > cooperation > sustainability") into "cooperation > creativity > information finding" and amongst the practical consequences of this adjustment the context- awareness modules will prioritize selecting funding calls which are labelled as multidisciplinary for circulation and encourage technical "meetups" amongst relevant colleagues. Again as a way of an example, for Scenario 3: "o" can reflect the adjusted priority from the higher strategic decision-makers of the University that the organization should have increased emphasis on knowledge transfer activities so $\Omega(\omega_{students}, \omega_{innovators}, \omega_{decision\text{-}makers}, O_1, \ldots, O_m)$ will tune in the system input through "ι" to encourage participation on such activities by placing aim related activities to be favoured "(Knowledge Transfer > ... other-organizational-aims ..." and amongst the practical consequences of this adjustment the context- awareness modules will prioritize the external dissemination of highlights on successful innovation and coaching events with academic and marketing staff.

Figure 4.2 shows a complementary view of the process, highlighting how different stakeholders provide feedback which is then used to measure how well the experience of the stakeholders is aligned with the overarching organizational objectives and this is used by the system to make decisions that can help improve the experience of stakeholders.

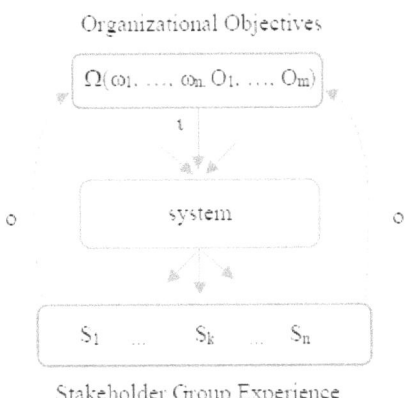

Fig 4.2 A wider view of the organizational feedback loop

One way to implement Ω is through a set of metrics M in the system to measure indicators in "o" of how close is the stakeholders' experience ω_i from what the organization would like them to have. Once a gap is identified a number of candidate actions A known to have a beneficial effect on ω_i can be selected. The outcome "ι" should have as a side effect in the decision-making model that actions A are favoured.

This system only tries to automate basic services however the proposed Smart Campus concept will still be driven by humans and will be for humans to benefit, they decide on the main goals, and the way to achieve them and when to change these.

5. Conclusions

We reviewed the technical literature on Smart Campus, noticing there is an over emphasis on technology and a lack of models and recipes which other colleagues can adapt and use. Here we proposed a model centred on stakeholders as the guiding principle to select which services make sense and which technology, acceptable by stakeholders, can facilitate those services.

Even the technologies and services deployed so far are mostly: underused, isolated, and uncoordinated. Although Smart Campus can grow from simple services upwards, it is desirable they do that as part of a more holistic strategy which needs to have been defined by stakeholders (this includes everyone in the University at all levels). After that process the Smart Campus concept can incorporate existing services, help to create additional ones, and blend them together.

We offer also a higher level view of the Smart Campus concept as an instance of so called Smart/Intelligent Environments. This model provides a theory behind the Smart Campus concept as a system which can aim towards a global satisfaction level to the campus community with regards to certain objectives considered as higher priority by the interaction with stakeholders.

References

[1] Augusto J.C. Ambient Intelligence: Opportunities and Consequences of its Use in Smart Classrooms. Innovation in Teaching and Learning in Information and Computer Sciences (Italics), 2009; 8(2):53-63, Taylor and Francis.

[2] Currie E., Harvey P. H., Daryanani P., Augusto J. C., Arif R., Ali A. An investigation into the eficacy of avatar-based systems for student advice. EAI Endorsed Transactions on e-Learning, 2016; 16(11):e5.

[3] Yang, J., Pan, H., Zhou, W. et al. Evaluation of smart classroom from the perspective of infusing technology into pedagogy. Smart Learn. Environ. 2018; 5, 20.

[4] Janelle, D. G., Kuhn, W., Gould, M., Lovegreen, M. Advancing the Spatially Enabled Smart Campus, Final Report. UC Santa Barbara: Center for Spatial Studies. 2014.

[5] Kwok, L. A vision for the development of i-campus. Smart Learn. Environ. 2015; (2) 2.

[6] Sari M., Ciptadi P., Hardyanto R. Study of Smart Campus Development Using Internet of Things Technology. IOP Conference Series: Materials Science and Engineering, 2016; (190), IAES International Conference on Electrical Engineering, Computer Science and Informatics, Semarang, Indonesia. IOP Publishing Ltd.

[7] Bi T., Yang X., Ren M. The Design and Implementation of Smart Campus System. Journal of Computers 2017; 12(6), 527—533.4

[8] Chan, H., Chan, L. (2018) Smart Library and Smart Campus. Journal of Service Science and Management, 11, 543-564.

[9] Muhamad, W., Kurniawan, N. B., Suhardi, S., Yazid, S. Smart campus features, technologies, and applications: A systematic literature review. In Proceedings of 2017 International Conference on Information Technology Systems and Innovation, (ICITSI). 2018; 384-391. Institute of EE Inc.

[10] Min-Allah N., Alrashed S. Smart campus—A sketch. Sustainable Cities and Society. 2020; (59). Elsevier.

[11] Augusto J.C., Callaghan V., Kameas A., Cook D.J., Satoh I. Intelligent Environments: a manifesto. Human-centric Computing and Information Sciences, 2013; 3:12. Springer.

[12] Augusto J. C., Quinde M. J., Oguego C. L., Gimenez-Manuel J. G. Context-aware Systems Architecture (CaSA). To appear in Cybernetics and Systems, Taylor and Francis. 2020.

[13] Augusto J. C., Muñoz Ortega A. User Preferences in Intelligent Environments. Applied Artificial Intelligence, 2019; 33(12):1069-1091, Taylor and Francis.

[14] Augusto J.C. A User-Centric Software Development Process. In Proceedings The 10th International Conference on Intelligent Environments (IE'14), 2014; 252-255. Shanghai. IEEE Press.

Intelligent Environments 2021
E. Bashir and M. Luštrek (Eds.)
© *2021 The authors and IOS Press.*
doi:10.3233/AISE210084

Innovative Evaporative Cooling System Toward Net Zero Energy Buildings

Andreu Moià-Pol [a,1] Victor Martínez-Moll[a], Susana Hormigos[a], Andrey Lyubchik[b]
a Department of Industrial Engineering and Construction, UIB, Spain
[b] REQUIMTE, Universidade Nova de Lisboa, 2829-516, Caparica, Portugal

Abstract. The SSHARE project will develop innovative self-sufficient envelope for buildings aimed at net zero energy, thereby contributing to the European technology. Envelope is a combination of two breaking through technologies: HUNTER-Humidity to Electricity Convertor and Advanced Radiant Panel for Buildings that will cool or heat the building, depending on the time of year, imitating perspiration of living beings and using only water as both thermal and electric energy supply. Successful realization of the project is assured by implementing a coordinated network of knowledge sharing in materials science, chemistry and mechanical engineering; by solidifying the state-of-the-art understanding in nanoelectronics and energy efficiency, and by applying bottom-up nanoengineering approaches via an international and inter-sector collaboration of highly qualified researchers from Portugal, Spain, Ukraine, Belarus, Tajikistan, Uzbekistan, Azerbaijan and the Joint Institute for Nuclear Research Russian Federation. Technological (panels fabrication) as well as fundamental (renewable energy) issues will be assessed by this multidisciplinary consortium. This paper explains the basis and principles for the development of a new generation of building materials and hence the creation of net zero building. Sharing the culture of research and innovation, the SSHARE project will allow applying recent advancements in nanotechnology science and mechanical engineering to address ""Plus Energy Houses"" EU 2050 concept.".

Keywords. Renewable Energy, Humidity to electricity, Evaporative cooling, Zero Energy Building

1. Introduction

There is a lot of research in using Thermal Energy Storage as a passive technology whose objective is to provide thermal comfort with minimal use of Heating Ventilation and Air Conditioning (HVAC) energy [1]. When high thermal-mass materials are used in buildings, passive sensible storage is the technology that allows the storage of a high quantity of energy, providing thermal stability inside the building.

Materials typically used in the construction are rammed earth, alveolar bricks, concrete, or stone. Standard walls, use sensible storage to achieve energy savings in buildings in winter, reducing thermal losses from the wall. For example, with Trombe walls the air between wall and glass could be heated and thus introduced into the room with a natural draught due to the chimney effect of the heated air an effect valid at summer and winter. Other systems can be used alongside to reduce the energy consumption, like geocooling

---

[1] Department of Industrial Engineering and Construction. University of Balearic Islands, Ctra. Valldemossa km. 7, 5, 07122 Balearic Islands (Spain), e-mail: andreu.moia@uib.es.

or geoheating. The University of Balearic Islands has developed a radiant system to recover the latent heat for indoor applications [2]. The same system now is tested for outdoor use, aimed to use evaporative cooling of a permeable surface, imitating the effect of the perspiration of living beings, using only water as thermal energy supply.

Sprinkling water or irrigating on permeable surfaces can effectively reduce the convective sensible heat flux discharge and increase the latent heat flux. [3] In the majority of the Mediterranean countries historically the buildings are designed with a water storage due to the low annual precipitation and periods of drought, and many houses by law have a large rain water storage tank. Combining this storage with other technologies (solar, heat pump,..) through a proper system, could allow all the energy from solar radiation to be used for heat and cool the building. The heat can be dumped into a tank and use the water storage like a big buffer tank in the cooler days: this being more efficient than the geothermal systems. [4] The water in the basement could be between 5-20°C according to the place and the period. These temperatures are adequate to be used for refrigeration or heating of the building internally and externally. The new system of energy recovery using humidity, presents the possibility of generating electric energy through conversion of moisture adsorption energy on the surface of ZrO_2 based on nanopowder systems into electrical charges. Investigated tablets were estimated to produce a maximum power density of about 10 nW/cm^2 [5], developed in the HUNTER project that now is used in the SSHARE project [8].This project is a collaboration of researchers from Portugal, Spain, Ukraine, Belarus, Tajikistan, Uzbekistan, Azerbaijan and the Joint Institute for Nuclear Research Russian Federation.

Figure 1. Average temperatures in the Cities of SSHARE project.

The water is used for storage of energy during the year due to the high specific heat. At summer, during the hottest days (28-45°C), the temperature of the storage water according to the location is between 4-18°C, similar to the annual average temperature, see at table 1. With these low temperatures it could be used to refrigerate the building. Calcium Silicate radiant panels have been tested in a seminar room in the Balearic Islands University with success during 4 years [2] using a closed circuit. The properties of the absorption materials could be used to design open systems for radiant surfaces. The only energy consumption present to this system is a periodical electric pumping of this water from the basement to the ceilings. At the new system of recovering energy from humidity, the maximum power density produced by investigated tablets was estimated around 10 nW/cm^2 [5] but by increasing the density of the ZrO_2 this could arrive up to 1 W/m^2.
In Europe, it has been estimated that around 1.4 million GWh/year can be saved and 400 million tons of CO_2 emissions avoided, in buildings and in industrial sectors by more extensive use of heat and cold storage [6].

Table 1. Climate conditions of the members of the SSHARE project

Country	City	Average Temperature (°C)	Precipitation (l/m² year)	Maximum Temperature (°C)
Spain	Palma	18.2	449	38.0
Portugal	Lisboa	17.5	774	44.0
Azerbayan	Baku	4	1252	28.0
Ukraine	Kiev	4.9	619	39.4
Tajikistan	Dusambe	8.2	568	45.0
Uzbekistan	Taskent	14.8	427	44.6
Belarus	Minsk	6.7	690	35.8
Russia	Moscow	5.8	707	38.2

2. Experimental results in a Façade.

Tests with calcium silicate panels in combination with a water system will be carried out during the next months at the exterior on the Façade of a building. Other materials will be studied according to the location of the project and their typical constructions, in order to evaluate the evaporation rate of each permeable material.

Table 2. List of selected solid–liquid materials for sensible heat storage

Material	Temperature Range (°C)	Density (kg/m³)	Specific Heat (J/(kg·K))	Latent Enthalpy (kJ/kg)
Ice	-10-0°C	920	2050	333
Water	0-100°C	1000	4190	2260
Sand	0-50°C	1555	800	
Rock	0-50°C	2560	879	
Brick	0-50°C	1600	840	
Concrete	0-50°C	2240	880	
Wood	0-50°C	740	1760	
Mineral wool	0-50°C	40	800	
Calcium Silicate	0-50°C	300	1003	
Silica bricks	0-50°C	1820	1000	

An experimental building (see figure 2) made of wood situated in the Balearic Islands has been adapted in order to test the advantages of the project, and indeed different construction materials would be tested at the exterior walls simultaneously. The dimensions of the building are 2.82x1.92x2.65. With a total surface of 36 m². The initial studied façade will be the south with a total surface of 7,5m². There is a Heat Pump combined with PV panels. In order to maintain the temperature and test all the conditions, different sensors have been installed inside and outside.

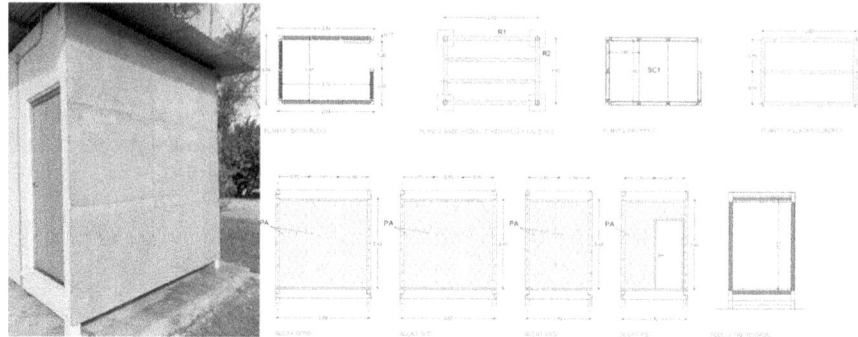

Figure 2. Experimental building.

$$\dot{Q}_{wall} = \left[(\tau\alpha)_0 F_R G_T - U_L (T_{wall} - T_{amb}) \right] S_a \qquad (1)$$

The walls are gaining energy due to the solar radiation according to the material, mass, absorptivity and weather conditions (temperature of the air, solar radiation wind,..). In Palma de Majorca a light façade (100 kg/m²) could exceed the 50°C, when the environmental temperature is almost 30°C.

Irrigating the wall during the hours of maximum solar radiation at summer can reduce the temperature of the wall by the consumption of water. This results into two effects: a sensible energy from the low enthalpy of the water to the permeable material of the external side of the wall and a second one with the latent heat of the evaporative effect. The flow of the water (kg/hour) and the evaporation (kg/hour) will be tested so as to gain a better understanding of the thermal behavior of permeable walls under dry and wet conditions. It has been proven by other studies that the evaporation will be higher according to the Humidity, wind and solar radiation. The relative humidity and wind velocity have a linear rank correlation with the evaporation [9].

$$Q_{transmision} = U_L \cdot S \cdot (T_{wall} - T_{in}) - Q_{water}$$

$$Q_{water} = Flow \bullet 4,19(T_{wall} - T_{water}) + Evaporation(2,29) \qquad (2)$$

At the moment the tests carried are the following; Test 1: The preliminary simulated results and some real tests demonstrate us that the heat transmission could be reduced to a 20%, using less than 0,8 kgH₂O/m²day. Test 2: With more water and a porous material the transmission could be reduced an 80%, using less than 2,3 kgH₂O/m²day, the external side of the wall can arrive to the wet-bulb temperature (WBT), according to the Humidity conditions and porosity of the material.

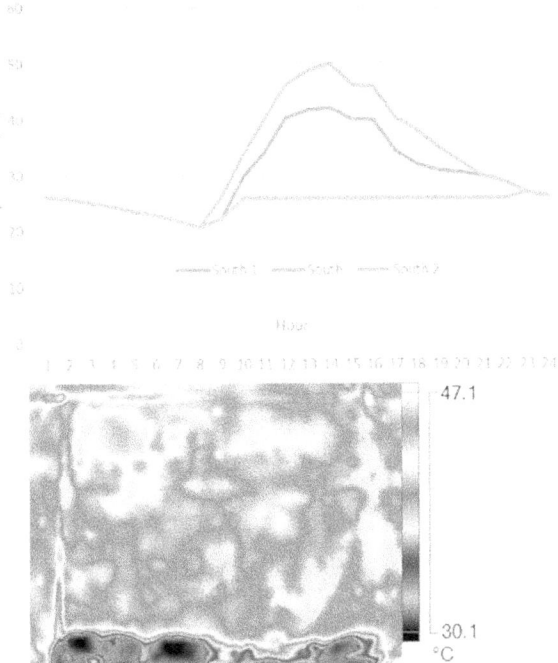

Figure 3. Temperatures of the south-east façade with different scenarios and first test.

With the preliminary results, the maximum amount of electricity for pumping with the new SSHARE System would be 1 W/m^2 *7,5 m^2 which means 7 W.

At normal summer weather days, this power can provide a water flow of 36 kg/h, in the initial testing, only was needed a maximum water flow of 0,7 kg/h m^2 to reduce the temperature of the external wall. At extreme summer days a maximum of 0,23 m^3 could be used. As the precipitation in Majorca is very low - near 449 l/m^2 per year - for the studied building the water storage necessary could be about 2,43 m^3 . This amount is enough to reduce the consumption with the proposed system, and destine the rest for other purposes.

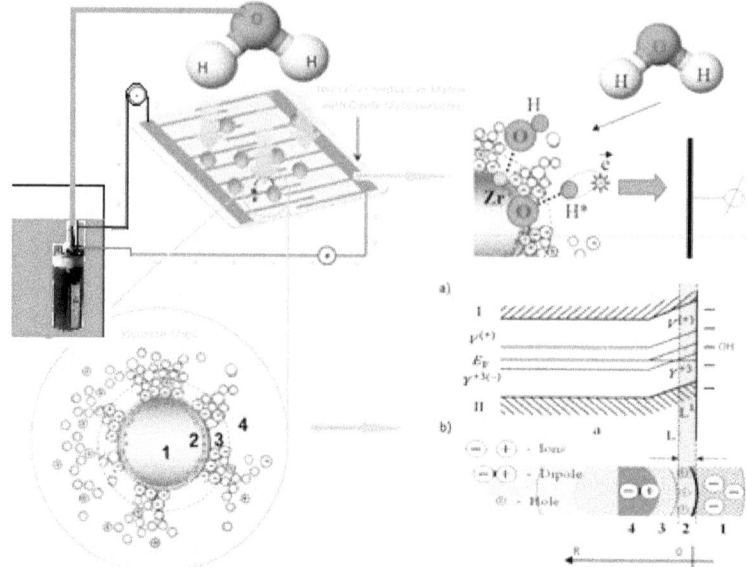

Figure 4. Schematic representation of ionic atmosphere surrounding a ZrO_2 nanoparticle [8]

Conclusions

The SSHARE project is expecting to arrive to zero energy buildings using water to reduce the energy consumption of the HVAC and recovering the produced energy.
The water has to be considered as a renewable energy source, with the preliminary results, the recollected water is more than used water which makes it a sustainable system.

Acknowledgments

This work was supported by the grant SSHARE Grant agreement ID: 871284 from MSCA-RISE-2019 - Research and Innovation Staff Exchange

References

[1] A Comprehensive Review of Thermal Energy Storage. Ioan Sarbu and Calin Sebarchievici. Sustainability. MDPI, Basel, Switzerland 14 January 2018.
[2] Solar Seminar Room in the University of Balearic Islands with a New Advanced Radiant System (DOI: 10.18086/eurosun2018.06.07). EUROSUN 2018.Rapperswil, Switzerland. September 2018
[3] Experimental investigation on the influence of evaporative cooling of permeable pavements on outdoor thermal environment. Junsong Wang, Qinglin Meng, Kanghao Tan, Lei Zhang, Yu Zhang.(https://doi.org/10.1016/j.buildenv.2018.05.033) Building and Environment. Elsevier May. 2018.
[4] Solar and heat pump systems. An analysis of several combinations in Mediterranean areas. Andreu Moià Pol, Víctor Martínez Moll, Miquel Alomar Barceló, Ramon Pujol Nadal. January 2012 Strojarstvo. 449-454, Rijeka, Croatia.
[5] Experimental evidence for chemo-electronic conversion of water adsorption on the surface of nanosized yttria-stabilized Zirconia. A. Lyubchyk, H. Águas, E. Fortunato, R. Martins, O. Lygina, S. Lyubchyk, N.

Mohammadi, E. Lähderanta, A. S. Doroshkevich, T. Konstantinova, I. Danilenko, O. Gorban, A. Shylo, V. K. Ksenevich, and N. A. Poklonski. H2020-MSCA-RISE-2015 (http://www.worldscientific.com/doi/abs/10.1142/9789813224537_0059)

[6]   International Renewable Energy Agency (IRENA). The Energy Technology Systems Analysis Programmes(ETSAP): Technology Brief E17; International Energy Agency: Paris, France, 2013. Available online:http://www.irena.org/Publications (accessed on 8 July 2015).68.

[7]   Influence of a pulsed magnetic field on the electrical properties of nanopowder system based on zirconia. Artem Shylo1, Aleksandr Doroshkevich, Tetyana Konstantinova, Igor Danilenko. (DOI 10.12776/ams.v23i3.970) Acta Metallurgica Slovaca, Vol. 23, 2017, No. 3, p. 208-214

[8]   http://hunter-greenenergy.com/about-us-sshare/

[9]   Wind Evaporation on Wet Surfaces under uncertainty Conditions. J.M. Gozálvez-Zafrilla et.al. Universidad Politécnica de Valencia, Spain. Excerpt from the Proceedings of the COMSOL Conference 2010 Paris.

[10] Study on building integrated evaporative cooling of large glass-covered spaces. Dereck.Vissers. Department Building Physics and Systems at Eindhoven .University of Technology. Eindhoven. NL 2011

Intelligent Environments 2021
E. Bashir and M. Luštrek (Eds.)
© 2021 The authors and IOS Press.
This article is published online with Open Access by IOS Press and distributed under the terms
of the Creative Commons Attribution Non-Commercial License 4.0 (CC BY-NC 4.0).
doi:10.3233/AISE210085

Computational Approaches for Urban Design Within the MENA Region

Jaouad Akodad[a], Mohammed BAKKALI[b] Mounir GHOGHO[c]

[a,b,c] *UIR Technology of Information and Communication Laboratory, Rabat, Morocco*
[a] *National School of Architecture, Rabat, Morocco*

Abstract. This paper focuses on the use of computational tools to develop a data driven approach for an analytical study about different urban systems. This "framework" examines urban Big Data in the old medina of Sale in Morocco. The computational tools are more effective to provide insights within complexity, becoming a key to generate more efficient solutions throughout the design process. The findings of this study highlight the potential of a data driven approach to explore analytical aspects and move further to generative design using algorithms.

Keywords. Parametric, Computation, Big Data, Urban Analytics

1. Introduction

The technology of information and communication (TIC) had an important impact in the development of technology as well as the design practice. Technology is always redefining new boundaries by the introduction of new tools that simulates architects and designers' creativity, allowing them to explore new concepts and innovative methods.
The introduction of the TIC in the 60's for the design field had a significant impact by replacing the hand drawing tools with computational aided design (CAD). Afterwards, these tools had been improved as a base for architectural design, interior design, urban design and so on and so forth. They have a strong ability to produce efficient results and add a fresh perspective to these fields. The integration of this technologies on the design practice remains largely underexploited, presumably because technology is advancing faster than the construction industry. In order to benefit from the true potential of computation, this has to be implemented from the early stages of the design process, not only as a drawing tool, but introducing new possibilities in the use of computation in both the analytical phase and the design process through developing a very strong bond leading to a generative design where analytical data is exploited smartly. The possibilities derive largely from the increase of data control and creating new computer-oriented approaches that benefit from the computer processing power.

---

[1] Corresponding Author: National School of Architecture, Rabat, Morocco; E-mail: jaouad.akodad@e.enarabat.ac.ma.

2. Methodology

This paper provides an analysis study of the historical medina of Sale in Morocco. Through computation, urban data will be implemented in a networked system (see, figure 1), generating insights about how the city's related systems respond and perform vis-à-vis end-users' for today and tomorrow. The paper will start by setting the sources of data to be used (data mining), and how they are managed in a dynamic way using algorithms and visual programming (VP). The following chapter will focus on the construction of interlinked relationships between various form of data in order to produce meaningful visualisation of these systems. The goal is to provide a "framework" based on computer and human interactions and to seek to produce computational modelling so as to better examine urban big data in a frame of networked urbanism.

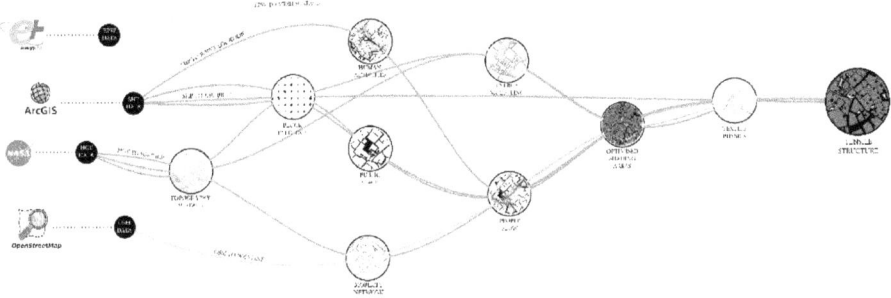

Figure 1. neural network methodology

3. Data mining and the city

Nowadays, cities become more complex and more dynamic to be apprehended making it unbearable for people to manually process data. At the same time, Big Data is increasing considerably, becoming more available and elaborated, and providing massive information about all types of behaviours and patterns within cities. Information extracted from Big Data and synthetized through Data Visualisation will enhance the level and accuracy of decision making and experts knowledge regarding different design choices. From the perspective of architects, urban designers and planners, and considering the tools with which they work in buildings and other design types, many aspects of data can be supported in the analytical phase such as geo-referenced data, which gives insight about spaces and their related information. It will support an evolved understanding of complex urban systems such as infrastructure networks, mobility, urban typo-morphologies, urban activities, energy modelling, topography and so on and so forth.

3.1. Image processing of historical maps

In the past centuries, historical maps were largely produced manually, giving valuable information about land use at that time. Such cartographies can still provide useful data for historians, geographers, and planners to comprehend different historical aspects.
Much research has been made concerning historical maps. Jessop and Rumsey and Williams [4]. investigated the use of Geographic Information Systems (GIS) to analyse historical maps and layering them with other spatial data. Balletti [5] proposes some efficient methodologies for quantitative analysis of the metrical content of historical maps. This chapter proposes an investigation of the historical maps in the specific site of the medina of Sale, mainly produced under the French protectorate. This work aims to study the urban evolution of the spot through Image processing. In the first stage, a data mining is proceeded. The goal is to collect a considerable number of cartographies, which will be scanned and stored in digital files. In the context of the Medina of Sale, the cartographies highlighted few concerns on maps from the 18th century until 2018, a period when the historical site went through major transformations. The cartographies are rescaled afterwards and overlapped in a matrix of pixels. Cartography corresponds to a large number of symbols, colours and observations displayed on a constrained screen with limited resolution. Based on this view, building symbols can be selected through their pixel's values. The next step consists in writing an algorithm that maps each square in the matrix with the number of overlapping layers. For example, a square was (clarify here please) 1 layer of building is overlapped, will correspond on value of 1, with 2 layers the value will be 2 and so on and so forth. Finally, a colour gradient is applied to visualise properly the results depending on the values of each pixel. The use of image processing in a complex site proved an effective result on representing urban evolution. The innovative use of image processing allows the computation of high number of maps and visualise meaningful information, and providing experts with useful knowledge on the city's growth. The results may be improved using additional cartographies. Some obstacles had to be overcome while extracting information from images of historical maps. The layering of image maps had to be slightly adjusted because of the accuracy of the hand-made cartographies provided. Also, the level of accuracy was highly related to the image resolution.

Figure 2. Mapping Urban Evolution (left) and Topography visualization (right)

3.2. Topography

Shuttle Radar Topography Mission (SRTM) data are used to reconstruct the topography of large regions, which consist of Digital Elevation Models generated by NASA and the National Geospatial-Intelligence Agency. SRTM topographic data consists in 90-meter (295-foot) pixels spacing for elevation's information of many part of the world. Figure 2 illustrates the generation of topographic levels of Sale city that correspond to (N34 W006). Within the visual programming platform of Rhinoceros, many scripts are provided to help convert Hgt (explain Hgt) files into geometrical surface, and then, applying horizontal contours in order the visual the topography curves.

3.3. Morphology

Two dimensional (2D) urban morphology was provided from CAD files and managed into a Shapefile (explain a shapefile in what it consists of) to contain the elevation data. The idea is to extrude each plot with the corresponding elevation value in order to construct a 2.5D model (see, figure 3).

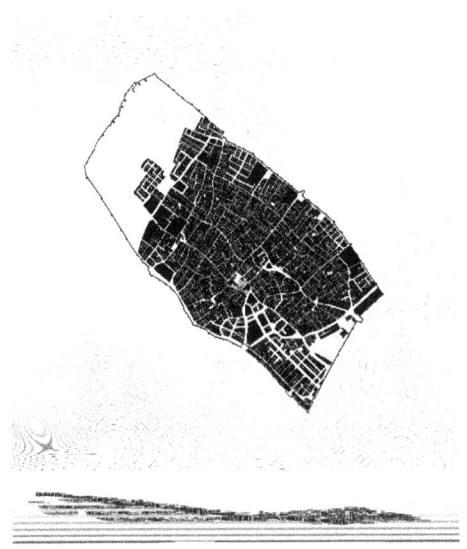

Figure 3. Morphology visualization

3.4. Infrastructure Network

Open Street Map in a non-profit organization providing geographic diverse user-based data of addresses, places' names, and route planning. Once the data has been collected, it is entered into an algorithm that manages the infrastructure network. Figure 4 visualizes how mobility is distributed from highways to pedestrian pathways.

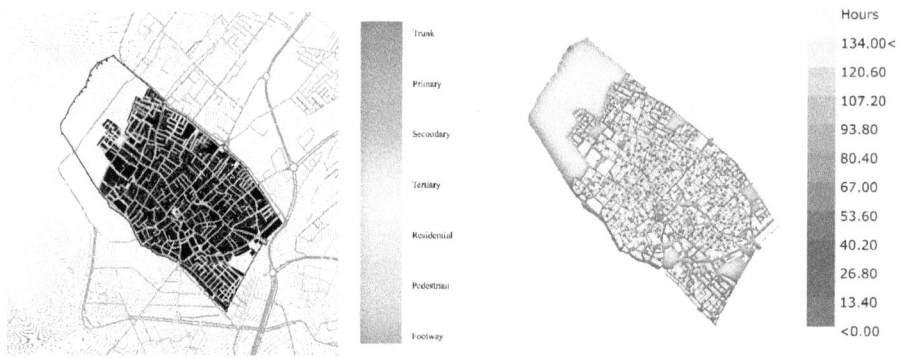

Figure 4. Mobility Network visualization (left) and Sun Hours analysis (All days of December - right)

3.5. Energy analysis:

Recently, many tools allowed new approaches of energy modelling and thermal simulation. Ladybug is an open-source plugin implemented in Rhinoceros/Grasshopper to help designers explore relationships between environmental data and buildings geometry models. The process of making energy models within the design platform is more effective to support complex geometries, i.e. the urban morphology of a historical medina. The using of energy modelling to endorse environmental analysis within architectural practices has been noticed to reveal more useful solutions when coupled with design processes. The generation of graphical data such as the sun path, sunlight hours, radiation studies, and more are proceeded by providing specific inputs. In one hand, Energy Plus with weather data (EPW) of the city of Sale. In the other hand, the geometry of the city, which is the 2.5D model of the medina. In order to achieve more effective results, a projection will be applied to the geometry geo-located at the surface of the topography levels (see, figure 4).

3.6. Local Activities

Applied to the context of the medina of Sale (see, figure 5), we provide various types of local activities, integrated in a database-oriented Geographic Information System (GIS). To increase control of this massive data, a classification is made to construct five major themes: Artisanal, retail, cultural, and touristic activities.

Figure 5. Local activities visualization

4. Generative process through a networked system

This chapter focus on identifying interlinked relationships between the various sources of data to explore how the city is performing, by integrating them into a networked system based on algorithms. According to this view, the city is seen as a set of complex interactions between its different urban components, which is mainly incorporated in the concept of the Smart City. The urban Big Data become a tool to better inform planners about how the city is responding by giving insights from very large and dynamic datasets. The proposed method proposes consecutive phases, gathering data coming from different sources, and implemented into an algorithm. They are summarized in three main phases:

- Definition of thematic activities.
- Investigation of public spaces.
- Determination of people flow.

Decomposing and analysing parts of the medina is significant to reveal patterns of its urban systems. In this process, the traditional techniques may not be capable to evaluate such complex activities neither to deal with different parts acting simultaneously. Meanwhile, using computational tools can process such complexity, and manage human activities into thematic classification to form six main themes: Retail, Artisanal, Tourism, and Culture. The thematic classification can serve to investigate the behaviour of public spaces that mainly serve to the growth of informal activities. The term ''Informal activity'' is this study is not considered as a spatial activity taking place randomly around the public space, but more related to a harmonious growth, primarily fed by the existing urban systems. Thus, the evolution of informal activities through time is related to city patterns, that are specific to a domain from where data is collected. By integrating the distribution of the thematic activities into the algorithm, a zone of people moving in and out can be predicted by connecting all the locations of each thematic classification. The people flow visualization is based also on the existing urban roadways that provide information about several hierarchical levels of the mobility network. Furthermore, the people flow will serve to determine the zone where informal activities self-evolve, while public places are considered as the centre of growth of each thematic activity. Whenever a block level is null and its size is larger than $200m^2$, the urban typology is referenced as public place. The investigation can also examine the potential of each public space to become the centre of a thematic activity. Through computational processing, the visual analytic explores how the city is performing by setting relationship between the public places, the people flow, and the location of classified activities. The next stage on the design process consists of the generative process based on the network system output and additional parameters, such as the energy modelling that will serve mainly to evaluate the thermal simulation on the path of people flow. The application of energy simulation in this process aim to study the solar shading in public spaces, in order to optimize long term outdoor thermal comfort. Therefore, the existing building and topography must be considered when creating shading outdoor areas. In addition, the EnergyPlus provides Data (EPW format) from exact locations to study the solar path of all periods of a day, month, or the whole year. The analytical study results on areas that necessities sufficient shading while avoiding excessive shading that are already created from the surrounding environment. Finally, the resulting areas that need to be shaded, are converted to mesh surfaces, and implemented to a plugin (Kangaroo) within grasshopper platform, to apply physics of textile. The surrounding buildings will serve as anchors, by connecting the mesh surface to the closest building corners. Additional parameters are also taken into account such as gravity force, and the spring strength. Tensile structures are the result of the generation process, which are applied to each thematic path, generating a multitude of shapes mainly related to environmental parameters. visual analytics served to specify areas that necessities a design intervention, rather than applying the design to the whole site (see, figures 6 - 8).

5. Results

This study represents the findings derived from the use of computational approach applied to the urban design framework. The method gathers visual analytics of urban Big Data which furthermore serves as inputs to a generative design. The historical context of the Sale medina demonstrates the relevance of this approach to reveal the complexity found within its urban systems. The early stage of the design consists in data visualisation of urban pattern such as the topography, mobility network, urban typologies, and human activities. It is essential to take into consideration those urban patterns in every stage of the urban design, and to inform the design generation using a data and computation. What is more is the capacity to increase control of the data collected, while integrating them into a networked system, within the Rhinoceros/Grasshopper platform. As demonstrated in this study, the environmental data serves as input to develop, generate, optimized membrane structures in the public space of the medina, based on a data driven process. The key to a more efficient designs combines the computer processing constructed to manage connections between the information collected from the analytical study and inform the design formation. The visual analytics served to specify areas that necessities a design intervention, rather than applying the design to the whole site.

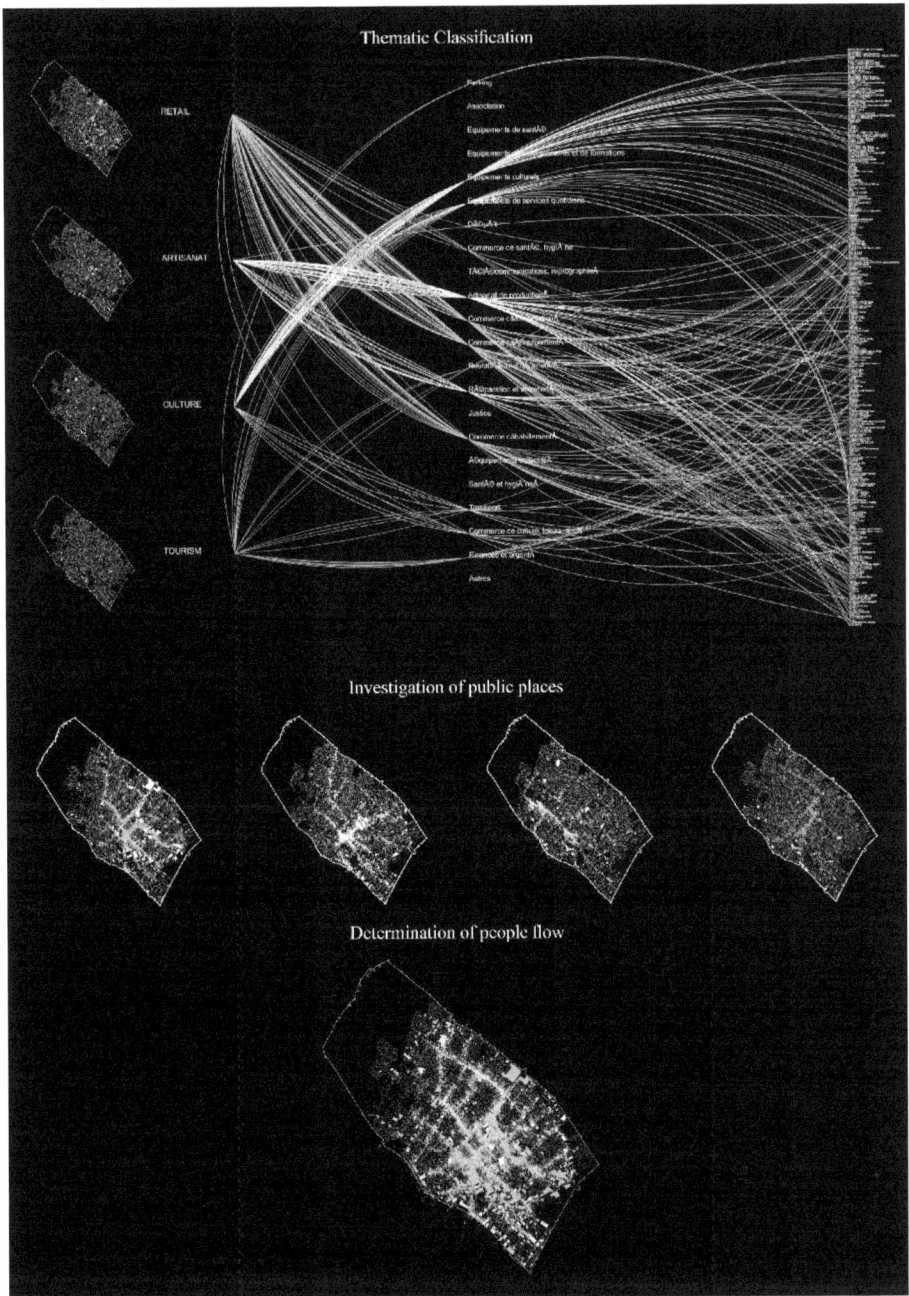

Figure 6. Visual Analytics

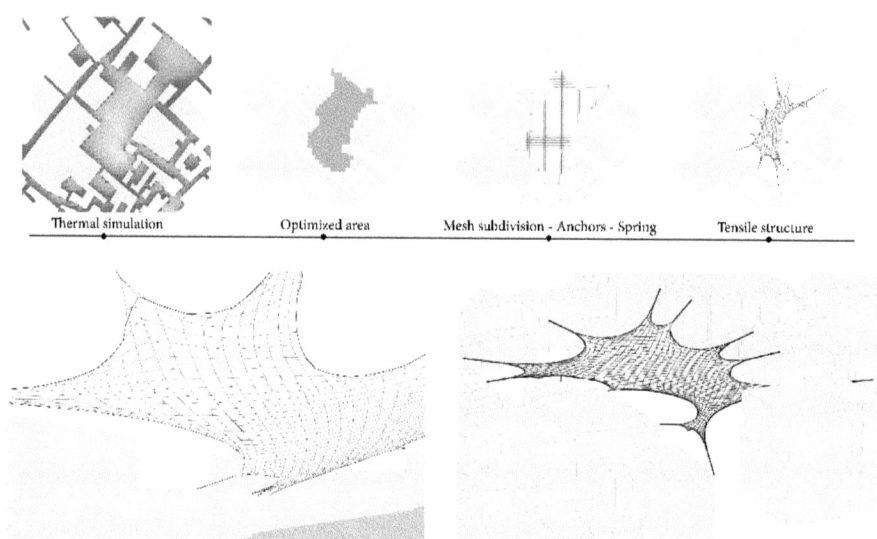

Figure 7. Generative design of tensile structure

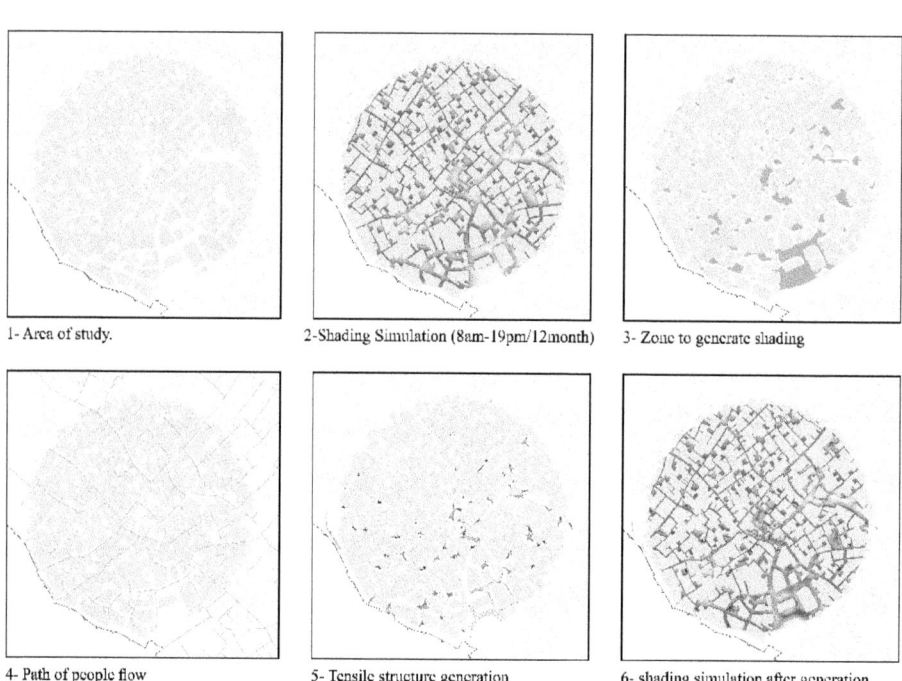

Figure 8. Step by step of the design generation process

Conclusion and future work

The framework based on algorithms and visual programming shows its ability to better control large amount of data city from various sources, within a unified interface. It is become essential to consider a data driven approach in the design methodology, especially with the increase of data production that are giving added value in term of informative insights and knowledge to guide decision making and inspire for further limitless development that can be found in its application. Urban Big Data should be increasingly available in the forthcoming years. The analysis phase could integrate image base data (like satellite images). It could be a powerful tool to manage the city patterns use of image recognition, and better inform urban planning and designers. Computing processing could also be extended in the generative process of urban planning and design. For instance, the use of shape grammar shows a potential to work with geometry and rules. These rules could be linked to urban parameters and informed from city data, providing the ability to generate shapes according to this specific area.

Acknowledgment

This study would not be possible without the support from the UIR Technology of Information and Communication Laboratory under the PV Build Project. I would like to thank my supervisors Prof. Mohammed BAKKALI and Dr. Mounir GHOGHO for their guidance and support throughout the study.

References

[1]   Ackoff, R. L. From data to wisdom. Journal of Applied Systems Analysis. 1989: 16, 3-9.
[2]   Carlo Ratti, Paul Richens. Urban Texture Analysis with Image Processing Techniques. The Martin Centre for Architectural and Urban Studies, University of Cambridge. UK; 1999;
[3]   Christina Martelli, Emanuelle Bellini. Using Value Network Analysis to support Data Driven Decision making in Urban Planning. Conference: Proceedings of the 2013 International Conference on Signal Image Technology and Internet-Based Systems. 2013 Dec:112-6
[4]   Jessop, M (2006). Promoting cartographic heritage via digital resources on the Web. ePerimetron vol.1/3: p. 246 – 252.
[5]   Balletti, C. Georeference in the analysis of the geometric content of early maps. ePerimetron vol.1/1: p. 32 – 42. (2006).
[6]   Mostapha Sadeghipour Roudsari. Ladybug: A parametric environmental plugin for grasshopper to help designers create environmentally conscious design. 2013 Jan: 3128-3135.
[7]   Pınar ÇALIŞIR ADEM, Gülen ÇAĞDAŞ. Interpretation of urban data in the complex pattern of traditional city: The case of Amasya. Vol 15 No 1, March 2018: 23-38.
[8]   Rob Kitchin. Data Driven, Networked Urbanism. SSRN Electronic journal. 2015.

Intelligent Environments 2021
E. Bashir and M. Luštrek (Eds.)
© *2021 The authors and IOS Press.*
This article is published online with Open Access by IOS Press and distributed under the terms
of the Creative Commons Attribution Non-Commercial License 4.0 (CC BY-NC 4.0).
doi:10.3233/AISE210086

A New Approach of Smart Mobility for Heavy Goods Vehicles in Casablanca

Mohammed Mouhcine MAAROUFI[a,1], Laila STOUR[a], Ali AGOUMI[b]

[a] *Process Engineering and Environmental Laboratory, Faculty of Science and Technology of Mohammedia, B.P. 146, Mohammedia, Hassan II University, Casablanca, Morocco.*
[b] *Civil Engineering, Hydraulics, Environment and Climate Laboratory, Hassania School of Public Works, Casablanca, Morocco.*

Abstract. Managing mobility, both of people and goods, in cities is a thorny issue. The travel needs of urban populations are increasing and put pressure on transport infrastructure. The Moroccan cities are no exception and will struggle, in the short term, to respond to the challenges of the acceleration of the phenomenon of urbanization and the increase in demand for mobility. This will inevitably prevent them from turning into smart cities. The term smart certainly alludes to better use of technologies, but smart mobility is also defined as "a set of coordinated actions intended to improve the efficiency, effectiveness and environmental sustainability of cities" [1]. The term mobility highlights the preponderance of humans over infrastructure and vehicles. Faced with traffic congestion, the solutions currently adopted which consist of fitting out and widening the infrastructures, only encourage more trips and report the problem with more critical consequences. It is true that beyond a certain density of traffic, even Intelligent Transport Systems (ITS) are not useful. The concept of dynamic lane management or Advanced Traffic Management (ATM) opens up new perspectives. Its objective is to manage and optimize road traffic in a variable manner, in space and in time. This article is a summary of the development of a road infrastructure dedicated to Heavy Goods Vehicles (HGV), the first of its kind in Morocco. It aims to avoid the discomfort caused by trucks in the urban road network of the city of Casablanca. This research work is an opportunity to reflect on the introduction of ITS and ATM to ensure optimal use of existing infrastructure before embarking on heavy and irreversible infrastructure projects.

Keywords: smart mobility, intelligent transport systems, advanced traffic management, heavy goods vehicles.

---

[1] Corresponding Author: Process Engineering and Environmental Laboratory, Faculty of Science and Technology of Mohammedia, B.P. 146, Mohammedia, Hassan II University, Casablanca, Morocco; E-mail: mmmaaroufi@gmail.com.

1. Introduction

In Casablanca, the traffic problem is one of the major challenges that it must win, whether decision makers choose the solution of the smart city or not. An efficiency gain in terms of circulation and mobility could lead to significant savings. But what seems obvious now is that conventional traffic management would not make it possible to absorb sustainably the consequences of congestion during peak hours. Traffic congestion will inevitably lead to a deterioration of the urban framework and road safety conditions and will hurt the city's competitiveness. The intensification of HGV flows from or towards the port on the Casablanca routes accentuates congestion, causes roadway degradation, worsens various nuisances, and pollution (noise, air, visual). Moreover, it contributes to the consumption of public spaces and assistance in the mortality of users of alternative modes. At the same time, the existence of industrial and commercial blocks in the city center, backed by storage functions, generates a high demand for transport and parking. Based on this fact, the deviation of HGV flows from the city center involves a major urban challenge. The optimization of logistics for the delivery of goods has vital importance for the competitiveness and attractiveness of the city, improvement of the quality of life, accessibility and road safety. The case of the separation of the urban traffic and the HGV traffic circulating between Casablanca port and Zenata dry port is a relevant example where smart mobility could provide efficient solutions allowing to avoid building costly tunnels.

2. Literature Review

In 2014, the National Sustainable Development Strategy 2030 (SNDD 2030) of Morocco identified the transport sector as the third energy consumer in Morocco. It accounts for 16% of total emissions and 28% of emissions from energy. Morocco was among the first countries to have embarked on the Mobilize Your City (MYC) initiative during the COP22 in Marrakech in 2016 [2]. Sustainable mobility is defined as "a transport policy which seeks to reconcile accessibility, economic progress, and the reduction of the environmental impacts of the selected transport systems" [3]. The transport of goods is also concerned with ambitions to optimise existing networks and improve nearby exchange platforms, allowing efficient transfers between different modes of transport. Morocco is the first country to have initiated an adaptation of the global macro-roadmap for the transformation of transport based on the Paris Process on Mobility and Climate (PPMC). The Moroccan Roadmap of 2018 recommends the creation of vertically and horizontally integrated, sustainable industrial zones close to consumption and connected to mass transportation modes. Defragmented and shortened supply chains reduce the need for transport and eliminate unnecessary trips. Over the past decade, Morocco has seen significant progress and reforms in the areas of the environment, sustainable development, and the fight against climate change. Several sectoral strategies, including transport and logistics, integrate these environmental dimensions. The economic stakes are high as the cost of air pollution in Morocco accounts for more than 10 Md DH (1% of GDP) [4].

Morocco is implementing an integrated national strategy for the development of the logistics sector by 2030 with clear and quantified macro-economic, urban, and environmental objectives. Since sustainable development is at the heart of this strategy, its objectives contribute to a reduction of around 35% in CO_2 emissions resulting from the transport of goods by road [5]. To achieve these objectives, pooling flows of goods has been considered as a primary solution. The creation of 3000 ha of logistics platforms by 2030 is among the main levers for reducing delivery costs and the carbon footprint of the import/export supply chain, thereby improving the quality of life, accessibility, and competitiveness of urban communities [6]. In 2019, Casablanca was selected by the Institute of Electrical and Electronics Engineers (IEEE), to be part of the IEEE Smart Cities Initiative. The city was recognised for innovative projects aimed at the transformation to a smart city and intentions to invest in the human and financial capital of the city. Aiming to tackle economic, urban, and environmental challenges the Casablanca's Urban Mobility Plan (UMP) considered the following trend scenario [7]: the energy consumption: MAD 4.2 billion (2004) compared to MAD 9 billion in 2019, the cost of congestion: MAD 114 million in 2004 against MAD 3.4 billion in 2019, the cost of pollution: MAD 319 million in 2004 against MAD 1 billion in 2019. To address these challenges, the priority actions recommended in the Casablanca's UMP include the creation of logistics lanes for HGV.

3. Research methods

The average annual daily traffic of the road circulating between the port of Casablanca Multi-Flow Logistics Zone (MFLZ) of Zenata, resulting from an automatic count established by a permanent post, is 21000 vehicles [8]. A metering campaign made it possible to quantitatively and qualitatively a load of directional traffic at crossroads and in the section during rush hour. The maximum Peak Hourly (HP) traffic in a section is around 3300 Vehicles in both directions, 10% of which are HGV of 13 m. The strongest hypothesis of the National Ports Agency (ANP) considers that 100% of container HGV (3200) and 100% of non-container HGV from port activity (5600) will pass through the northern service in both directions by day. The dimensioning HP traffic is 1100 HGV (36% of HGV of 13 m and 64% HGV of 17 m) [9]. To quantify the impact of the proposed lane dedicated to HGV in Crossroads, dynamic simulation is carried out by Aimsun software. This simulation makes it possible to visualize the circulation of vehicles and pedestrian crossings. The generation of vehicle traffic on the main road and on the secondary roads was carried out by taking the above traffic data. Several replications were launched to obtain an average of over one hour. Each replication generates traffic randomly over time while respecting the Origin/Destination matrix. Thus, each replication has variations in traffic making it possible to observe different traffic conditions (local congestion, repetitive calls on secondary axes, absence of pedestrian calls).

4. Research results

4.1. Analysis and proposal for the development of a lane reserved for HGV

Six variants can be considered. They are presented as follows:
- Variant 1: Road in 2x3 lanes: mixed traffic between HGV and urban traffic.
- Variant 2: Dedicated corridor for HGV: partial separation of traffic.
  - Sub Variant 2.1: Dedicated central corridor for HGV in 2x2 lanes.
  - Sub Variant 2.2: Dedicated central corridor for HGV in 2x1 lanes.
  - Sub Variant 2.3: Dedicated side corridor for HGV in 2x1 lanes).
  - Sub Variant 2.4: Two dedicated bilateral lanes for HGV.

- Variant 3: Dedicated corridor for HGV in 2x2 lanes and uneven junctions: total separation of traffic.

The chosen variant must meet the following requirements:

a. An optimized impact on the expropriation and the networks;
b. Design compatible with the adjacent cornice project from an urban point;
c. Fluidity and protection of pedestrians heading towards the beach;
d. Capacity on the current section of lanes reserved for HGV;
e. Maintained operation in the event of accidents in the lanes reserved for HGV;
f. Secure traffic at intersections;
g. Travel time promoting the competitiveness of the logistics area;
h. Fluidity of vehicles and improvement of capacity on highways;
i. Fluid and unrestricted management of traffic during the works;
j. Limited equipment maintenance;
k. Respect the cost allocated to the project.

The sub-variant 2.2 is the only one to meet the above requirements. Its feasibility was subsequently studied. The Cross-Type Profile (see Fig. 1) of this variant is as follows:

- 2x2 lanes of urban traffic at the lateral level + 3rd Turn Left lane;
- Central corridor dedicated to HGVs in 2x1 lanes 9m wide;
- Separation between the two corridors;
- Support measures in terms of detection and traffic management.

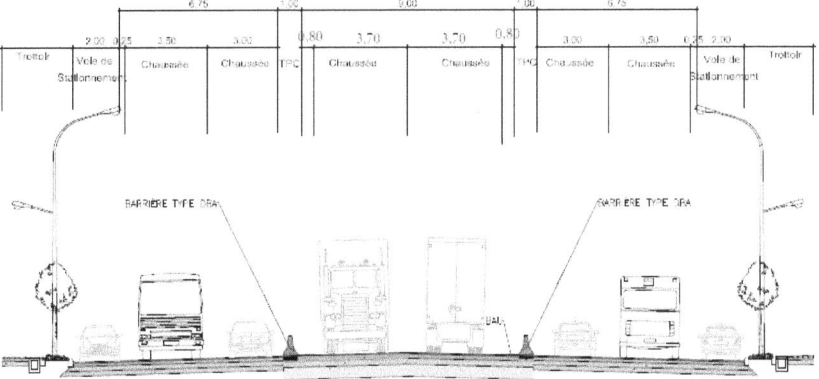

Fig. 1. Cross-Type Profile recommended of sub variant 2.2.

4.2. Checking the feasibility of a planned HGV lane

The figures 2, 3 & 4 show that the proposition of the 9 m width for the two bidirectional lanes dedicated to HGV means the operation in a degraded mode in the event of a truck failure on the lane. Measures can be used in the event of a truck breakdown, such as movable double concrete partitions used to clear HGV through side lanes in the event of a serious accident in the HGV lane.

Fig. 2. HGV road profile.

NOMINAL MODE OPERATION
Vehicles traveling between 50 and 70 km/h

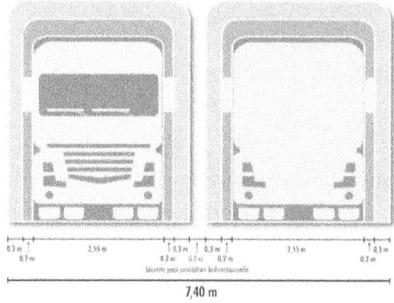

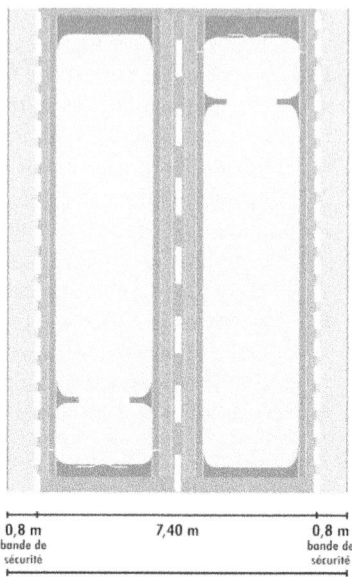

Fig. 3. Checking the gauge of dedicated traffic lanes for HGV in nominal mode operation.

DEGRADED MODE OPERATION
Vehicles traveling between 0 and 20 km/h

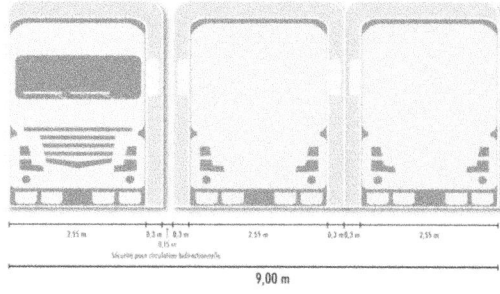

Fig. 4. Checking the gauge of dedicated traffic lanes for HGV in degraded mode operation.

The crossroads. The traffic light intersections are programmed as follows:
- Cycle time of 80 seconds;
- Keeping the main axis green (at least 45 seconds of green to clear the 550 HGV/h per direction);
- Turn left and secondary axes phases on-call (8 seconds of green per phase).

Sensors are placed on the turn left way and on the secondary axes to detect the presence of the car to leave the rest point of the main phase.

The Aimsun software has a "yellow box" function for traffic intersections. When this function is activated on the crossroads, vehicles do not enter if there is a risk of lifts and blocking. Vehicles wait at the light line until the intersection empties. To reproduce the effect of "yellow box, a system of saturation loop and early closing of the light lines is to be expected. (see Fig. 5)

Pedestrians at the crossroads are served during Turn a left phases and secondary axes.

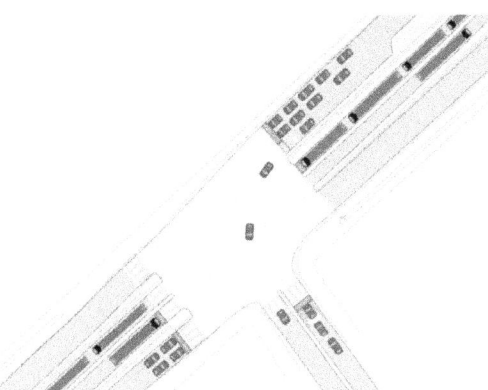

Pedestrians on secondary axes are served during the main phase. Pedestrian detection devices will be used for these crossings to reduce the waiting time for pedestrians by contracting "green cars" if no vehicles are approaching the crossroads.

Fig. 5. Crossroads in "Don't block the box" mode.

Pedestrian crossings in section. To manage pedestrian crossings, a dedicated facility will be created for pedestrian traffic. When a pedestrian is detected, a call is made and the car/HGV phase turns red after 29 seconds so as not to constrain the flow of vehicles. The operation of pedestrian crossings on call does not create a green car wave on the whole road. To optimize the operation of pedestrian crossings transversely, each pedestrian signal opens in the offset to limit the times of red vehicle/HGV.

A "yellow box" is placed downstream of the pedestrian crossing (see Fig. 6). This function is activated in order not to have a vehicle/HGV blocked on the pedestrian

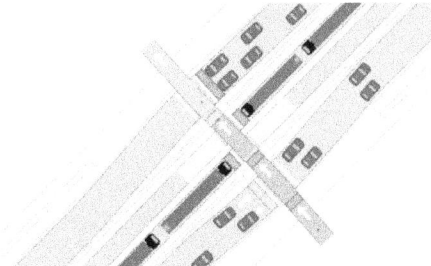

crossing. Trucks do not cross the pedestrian crossing if there is insufficient space for its storage. The blue areas on the pedestrian path are detectors that activate the pedestrian green when it detects a presence.

Fig. 6. Pedestrian crossings in "Don't block the box" mode.

5. Discussion of the results

5.1. Simulation analysis

HGV's traffic on a dedicated central site. All HGV, in simulation, arrive at their destination without too much waiting due to congestion. The average journey time is 12 min over the entire section. Their total downtime is 3min and 40sec. They, therefore, have an average speed of 25 km/h (see Fig. 7).

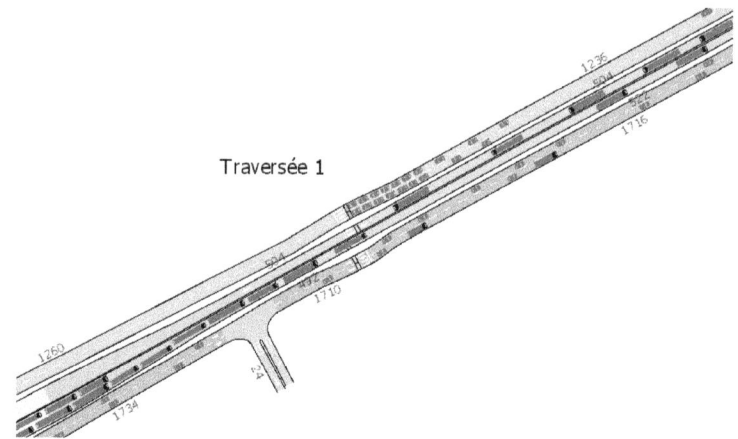

Traversée 1

Fig. 7. Calculation and verification of hourly flow rates in lateral tracks and dedicated lanes.

Vehicle's traffic on the sideways. The crossroads with lights being dimensioned to 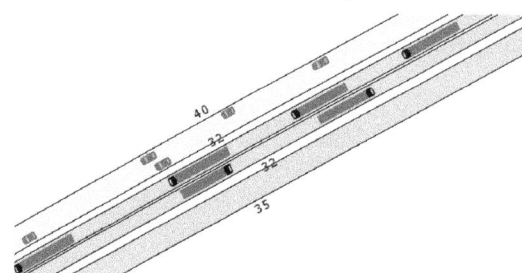 smooth the more important circulation of Heavy Goods Vehicles in the dedicated site (1475 vehicle/lane/direction/hour), the vehicles (825 vehicle/lane/direction/hour) profit from the same times of green HGV. Then, no problem with raising the queue.

Fig. 8. Average speeds in the lateral way and the PL tracks.

The cars cover the 5km in 9min which gives an average speed of 32 km/h (see Fig. 8). Vehicles coming from the secondary axis of an intersection regularly pass to the second cycle. This intersection as a resting point on the main one operates cyclically due to the permanent calls from the secondary axis. Vehicles wait an average of 75 seconds to pass the line of lights (green time being 10 seconds).

Fluidity of pedestrians. The 15 pedestrian crossings in the section are managed to call. Pedestrian is served 29 seconds after its detection to leave a minimum green time for HGV in a dedicated site. On average, a pedestrian takes 54 seconds to cross the entire road (including detection time) with an average speed of 5 km/h.

5.2. Criteria of Intelligent Transport Systems (ITS)

To ensure the optimal simulated operation at the Aimsun Software level, the equipment of the vehicle and pedestrian detection system and ATM must constitute an intelligent transport system and have the following functions and characteristics:

Diagram of the overall functioning of the ATM System. The Diagram below summarizes the overall operation of the integrated traffic operating system allowing ATM with prioritization of HGV flow, detection and securing of pedestrian crossings, intelligent management of traffic lights, information for users at through Variable Message Panels (VMP) and data acquisition and monitoring by WEB. (see Fig. 9)

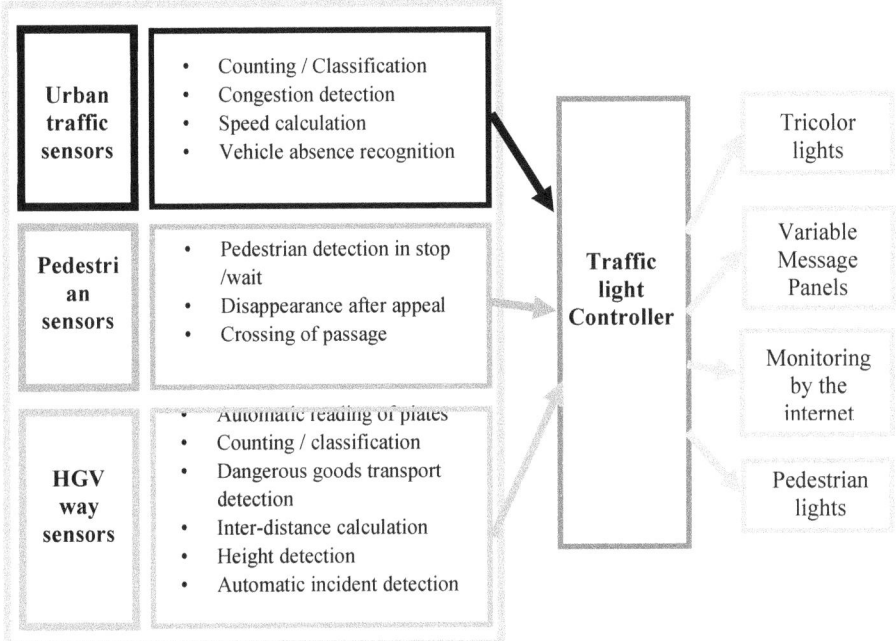

Fig. 9. The overall functioning of the Advanced Traffic Management system.

The adaptive and intelligent traffic light controller. It is an automat dedicated to adaptive management and intelligent regulation of road traffic without necessary a central control having as specific characteristics:

- A history of traffic data, analysis, optimization and evaluation of the effectiveness of the dynamic control system, visualization by the website;
- Recognition of the absence of a vehicle at the intersection to avoid giving unnecessary priority;
- Processing of information from various traffic detectors;
- Compatibility with DIASER and OCIT 2.0 communication protocols;
- Equipped with SIL3 (Safety Integrity Level): redundancy of controls;
- Complies with the requirements of standards EN12675 and EN50556;
- Programmable with third-party software LISA + and VS-PLUS;
- Capable of controlling traffic lights in different voltages (230VAC/110VAC) with bulbs or LEDs, 40VAC for LEDs, 24VDC;
- Possibility of attenuation mode at nightfall with 42VAC dimming module;
- Possibility of policy manual control;
- Priority configuration for firefighters / police / emergencies;

Thermal detection system. It's a thermal imaging camera of vehicles and pedestrians with the following characteristics:
- Detection of saturation of lanes;
- Automatic Incident Detection (AID) on the HGV track for deviation through cross-arrow tanks;
- Detection of pedestrians crossing;
- Dynamic micro-regulation of tricolor lights cycles time;
- Don't block the box flow and saturation control;
- Vehicle counting and classification;
- Over-height detection;
- Reading of Dangerous Goods Transport (DGT) plates;
- Calculation of the speed of the HGV through virtual loops (If V>Vmax: the light turns red);
- Automatic Reading of License Plates (ARLP) by day and by night;
- All-in-one sensor (infrared and CMOS "very high sensitivity");
- 24/7 detection in various weather conditions;
- IP connectivity and configuration via secure Wi-Fi / 3G connection;
- 8 vehicles or pedestrian presence zones;
- Video stream visible in HD (Protocol and RTSP image stream);
- A countdown of the waiting time before going green;
- Management of Variable Message Panels (VMP) and cross-arrow trays.

Polycarbonate signal lanterns: must have the following characteristics:
- Anti-vandalism, slim design, blending in historic urban areas;
- Can be mounted vertically or horizontally;
- Available in ø 100/210/300 mm.

LED modules: must have the following characteristics:
- No visible LED point - central light source;
- Higher anti-ghost performance (class 5);
- Lower energy consumption and brilliant light output;
- Custom masks that can display any symbol;
- Optimized thermal concept, reducing degradation to a minimum;
- Automatic light compensation in case of diode failure;
- Degraded mode functions available in 42V;
- Compliant with DIN VDE 0832 standard.

Pedestrian push button: must have the following characteristics:
- Modular construction allowing adaptation to all types of intersections;
- No moving parts which could be deactivated with toothpicks, gums;
- Laterally tactile symbols appeal to describe the passage for the visually impaired, integrated acoustic units;
- Location of the push button, thanks to the acoustic and optical position signal;
- Meet all the requirements of the directives and regulations in force (RILSA, DIN 32981, DIN VDE 0832, EN 50293).

Environmental sensor: should allow the following measurements:
- Measurement of gases (NO2, O3, CO, CO2, VOC);
- Measurement of polluting fine particles (PM1, PM2.5, PM10);

- Measurement of noise, humidity, temperature, and pressure;

Weighing at the current speed. Weighing-in-motion systems with dynamic weighing sensors help quickly pick up vehicle and axle weights for safer roads and better traffic management.

Conclusion

The installation of 15 secure pedestrian crossings throughout the 5 km of the project in addition to a pedestrian crossing on each of the 5 main intersections has made it possible to reduce HGV speeds, to manage traffic in packages, and to ensure maximum protection of pedestrians. Besides, with the help of ITS, several issues related to traffic regulation and fluidity have been resolved. The dynamic and adaptive management of traffic lights has therefore made it possible to reduce ways dedicated to HGV while minimizing journey time. The use of ITS will allow the registration of traffic data, the collection of information about special events, and the management of system efficiency for real-time. Innovations and intelligent systems deployed which made it possible to bypass the HGV of downtown Casablanca with a significant gain in terms of competitiveness for businesses will have a significant positive impact, also, on the quality of life of citizens and on the city environment. Finally, this research led to the design of a new approach of smart mobility for HGV with introduction of ITS to ensure the optimal use of urban roads in Casablanca.

References

[1] Benevolo, C., Dameri, R. P., & D'Auria, B. (2016) Smart Mobility in Smart City. In T. Torre, A. Braccini, R. Spinelli (Eds.), Empowering Organizations. Lecture Notes in Information Systems and Organization (pp. 13-28).
[2] Mobilise Your City (2019, November 19). MobiliseYourCity in Morroco. Retrieved from https://mobiliseyourcity.net/
[3] SEDD [Secretary of State to the Minister of Energy, Mines and Sustainable Development, in charge of Sustainable Development]. (2017). National Sustainable Development Strategy 2030. Executive summary. 60 p.
[4] METLE [Ministry of Equipment, Transport, Logistics and Water]. (2018). Roadmap for Sustainable Mobility in Morocco. 19 p.
[5] METLE [Ministry of Equipment, Transport, Logistics and Water]. (2017). Synthesis of the National Road Safety Strategy 2017-2026. 27 p.
[6] AMDL [Moroccan Logistics Development Agency]. (2016). Moroccan Green Logistics. Sustainable development at the heart of the logistics dynamic in Morocco. 61 p.
[7] Ministry of the Interior. (2004). Studies of an Urban Travel Plan, a Passenger Transport Account, a Restructuring of the Collective Transport Network, a Traffic Plan on the territory of the Greater Casablanca Region.
[8] DR (Direction of Road), 2019. Road traffic report 2018. 272 p.
[9] ANP (National Ports Agency) (2019). Presentation of the steering committee. 24 p.

Intelligent Environments 2021
E. Bashir and M. Luštrek (Eds.)
© 2021 The authors and IOS Press.
This article is published online with Open Access by IOS Press and distributed under the terms
of the Creative Commons Attribution Non-Commercial License 4.0 (CC BY-NC 4.0).
doi:10.3233/AISE210087

The Smart Chain City

Walter Gaj Tripiano[a,1]
[a] Polo Multimodal Pecem Investimentos, Fortaleza, Brazil

Abstract. The urban, social, industrial and technological evolution in recent years has forced designers, entrepreneurs and public entities to rethink the conceptual and operational logic for the construction of new urban settlements, according to the authentic principles of mutual respect between man and nature, eliminating all thought extremism and purely conceptual theories. In order to make their respective professional and life skills available to future generations, we have set up two international working groups, the ISCW and IS.Smart, in collaboration with the UFC universities of Fortaleza (Brazil), UIR of Rabat (Morocco) and UNITO of Turin (Italy), which for some years now, have been working to transform the notions of smart cities and smart buildings, combined with the new industrial revolution 4.0 and tokenomics, into concrete, multidisciplinary and educational activities. This publication, the result of the work developed in the last three years, intends to be an easy-to-read and application tool for all operators who wish to approach the "smart world" in a professional manner, especially in situations with serious housing shortage and still in the process of social and economic development. By reporting experiences and methodologies already field tested by the author in some working circumstances and that have given rise to case studies, similar to the "Polo Multimodal Pecem" (Brazil), the reader will find useful insights such as the definition of the smart concept and "smart" objectives, the approach to the project, the definition of evaluation parameters, operational examples of management, capital raising and crowdfunding, tokenomics and e-conomy. I do believe that the future of the city will be increasingly linked to these issues, which have now become fundamental and necessary in the projection and planning processes of urban settlements.

Keywords. Smart cities, Block Chain

---

[1] Head Innovation, Polo Multimodal Pecem Investimentos S.A., R. Vicente Linhares, 521 - Sala 2013 Aldeota - Fortaleza (CE. Brazil) www.polomultimodal.com; E-mail: studiogaj@hotmail.com

1. Introduction

The case study of the Polo Multimodal Pecem, located in Brazil, in the State of Ceará, not far from Fortaleza (fig.1), can provide an interesting operational method, showing the steps that have defined the smart choices and procedures in planning the settlement.

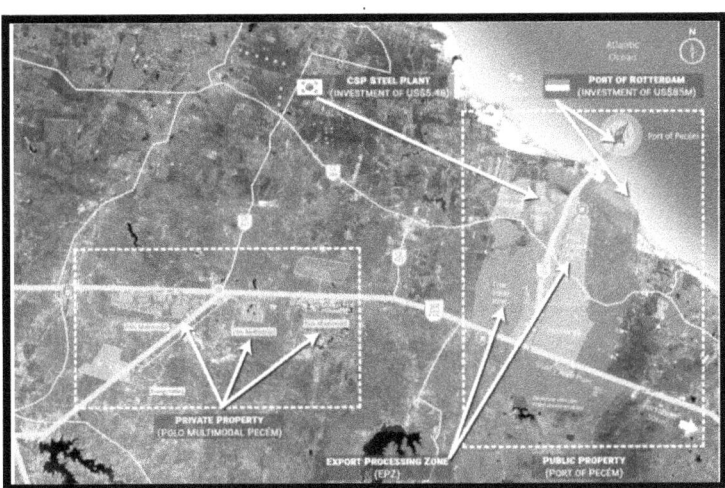

Figure 1. The Polo Multimodal Pecem location : Brazil, in the State of Ceará, not far from Fortaleza

The peculiarity of the settlement is that, in a single intelligent hub, all the different urban activities coexist: residential, health, logistics, production, commercial, development and research, cultural, sports and leisure, accommodation.

The project has been promoted by POLO MULTIMODAL PECEM INVESTIMENTS S.A., based in Fortaleza (Brazil), belonging to the Swiss company POLO SWISS DEVELOPMENT AG, and has been developed by a group of professionals in the area, including myself who, as Smart Concept Manager, was responsible for making both the design choices and the urban and architectural components "smart".

The Polo Multimodal Pecem emerged with the acquisition of the best lands (2000 hectares) located in a strategic point in the hinterland of Pecém (Fortaleza-Ceará), close to the EPZ (Export Processing Zone), a region in strong economic development also thanks to its proximity to the Panama Canal. The company "Port of Rotterdam" has recently acquired a stake in the company "Port of Pecém", to the demonstration that the choice made by the State of Ceará is of strategic importance. The "Smart Chain City" envisages the construction of industrial sheds, logistic, commercial and residential areas, as well as all infrastructure and related activities to meet the demand generated by the port and the large industries settled in the EPZ.The purpose of the project is to provide a competitive solution to the needs of small and medium-sized industries, not only locally, but also internationally. With this in mind, a strategic partnership was signed in Italy with Assoimprese (Association of Small and Medium-sized Industries), which aims to encourage the internationalization of the members in the areas of the Polo Multimodal and EPZ.

2. The project in detail

To create a smart city from its foundations, it is indeed necessary to increase the logistic and production demand, fostering the installation of "smart" and sustainable companies, with the aim of providing all the tools to promote the development of the entire community.

As will be examined in detail below, the innovative process of the Polo Multimodal Pecem is made up of a digital platform called "Polo Digital Platform" which, based conceptually on block chain (smart contracts, smart assets, digital identity) and properly connected and distributed through technology "IoT" (Internet of Things), "Big Data" and "AI" (artificial intelligence), will integrate the control phases in the areas of infrastructure, services, technology and governance. Within this platform there is an exclusive section for companies, called "PoloSuit"; by accessing it companies will be able to take advantage, among other things, of specific services, that help in its simplified, transparent and effective installation, such as rent/purchase the property, company establishment, provision of public services, opening of current account, digital identification and any other necessary services.

The "smart" core of the project consists of an integrated management center called "Polo Chain Lab" (fig.2); an innovative center destined to house laboratories for applied technology in blockchain, advanced robotics, digital information and communication technology and 4.0 systems. This structure, intended for research and innovation, will be advantageous for companies, foundations and institutions, which will benefit from guidance, training, innovation and experimental development services.

Figure 2. The Polo Chain Lab: The "smart" core of the project consists of an integrated management center, an innovative center destined to house laboratories for applied technology in blockchain, advanced robotics, digital information and communication technology and 4.0 systems.

Management through digital systems, with an integrated approach to data sharing, increases the decision-making capacity and its effectiveness. Each service will be more accurate and better tailored to individual needs, while at the same time being less costly to deliver:

- the Polo will combine digital identity with smart assets using intelligent resources registered in the blockchain to verify and identify properties and related rights and obligations;
- IOT-powered AI devices integrated into a secure blockchain infrastructure will facilitate decision making and problem solving.

The project provides for the application of artificial intelligence algorithms to cameras located in the territory, so that an alert is automatically sent to the personnel of the local security operations center when abnormal situations or accidents are detected. The system is completed by specific "totems", also located in green areas, which allow emergency calls and communication with security personnel. In a smart city, where each person produces a large amount of data that must be shared on open platforms (open source), security means above all protecting the systems from "hacking" actions, either "white hat" or "black hat" (of good or bad intentions).When digital and physical systems converge as in smart cities, hacking represents a danger, and the more the city becomes hybrid, the more serious the danger is. The destructive potential for misappropriation of any smart system can be very frightening. Any urban dimension becomes a potential target for hackers: imagine, for example, uncontrolled self-driving cars, overly aggressive reactive architectures or a dangerously asynchronous distribution of energy. Hacking can, however, have an important positive function, both for safety and social reasons. By exploiting system gaps and invading them, white hats are able to uncover vulnerabilities and indicate new patterns, thus bringing innovation to the systems themselves. As paradoxical as it may seem, smart cities can stimulate creativity, if "hackable" to a predefined degree. For this reason, the study of a new computer model is in progress which, respecting the same principles used in the medical field in the analysis of the human body, pursues three objectives: redundancy, that is, a "reserve" of important parts of the system, capable of performing the task should the original part become incapacitated, such as pairs of limbs, eyes, ears in an animal being; segmentation, which allows the replacement of damaged elements before they can affect the healthy ones; proaction, which constantly subjects the system to hypothetical attacks by hackers and leads it to create, consequently, the necessary antivirus.

Encouraging the opening of platforms and data could stimulate new technological and cognitive paths, increase the commitment of residents and users, stimulate confrontations, lead to exchanges and collaborations, with the aim of increasing the level of urban culture. If we consider that open data and platforms have the same value for a smart city as accessible and public spaces have for a traditional city, the processes of urbanization are reduced to political issues that can only be managed through top-down and bottom-up processes. In short, the model applied to the Polo Multimodal Pecem is a new "mixed" type of security system, which provides a "classic" type of protection applied to a part of the data (especially sensitive data) and a protection created thanks to the top-down and bottom-up processes typical of the smart city.

The creation and management of the "PoloCoin" token will facilitate the exchange of goods and services within the smart city, ensuring greater security in business transactions between the companies settled and integrated in the supplychain.

The smart chain city will be powered entirely by clean energy produced by the Polo Energy Valley and will also include a waste recycling center; in addition, the project has been developed for:

- increase the use of green industrial technologies and processes to achieve high levels of savings, both energy and resources used;
- achieve total management of the waste reduction system by promoting reuse and enhancement;
- carry out construction models with the purpose of obtaining specific certifications in the field of greenbuilding.

Since 2017, the project strategy has been in line with the 17 Sustainable Development Goals (SDOs) to be achieved by 2030, approved worldwide and adopted by the United Nations.

From which fundamental points was the project designed?

Once again we must look to the past to project ourselves into the future, starting from the ideas of a great American inventor, Buckminster Fuller, summarized as follows: "The function of what I call the science of design is to solve problems thanks to the contextual introduction of new artifacts whose availability will lead to their spontaneous use by human beings and therefore, simultaneously, will push man to abandon the behaviors and tools from which the problems originated " [1].

Developing a project in certain contexts or in some parts of the world necessarily leads to considering the process known as "leapfrogging", in which some situations of initial technological disadvantage can lead to considerable final benefits; for example, the lack of traditional data transmission networks, in the smart city area, can generate an advantage because fiber optic networks or new technologies can be installed immediately.

One might think at a first reading that the advent of the internet has cancelled the distances and physical places, giving people the possibility to work and operate in any place as it had been predicted by internet scholars since the nineties. Actually, urban space is expanding all over the world, the urban population is growing at a rate of 250,000 inhabitants per day, World Health Organization data indicate that by 2050 75% of human beings will probably live in cities; this is because the essential element of human experience is the importance of physical interaction between people and the environment. As the Greek philosopher Aristotle wrote in the 4th century B.C. in "Politics", man is a social animal because he tends to aggregate with other individuals and to integrate himself into society, a concept taken up long after by the writer Italo Calvino in "The Invisible Cities": "The man who rides for a long time through wild lands is longing for a city" [2]. It is therefore essential that the design of physical spaces, architecture and urban planning, incorporate the digital network, combining virtual and real places, software and hardware. In other words, the smart city becomes the city where spaces are different from traditional ones because they are permeated by digital systems which transform the way people exchange experiences, socialize, communicate and live.

In the smart city, urban sensors transmit a huge amount of data that is processed by acting on the players responsible for modifying urban spaces: this creates a real test bench, a Big Data computing platform that links digital and physical spaces. The public physical space is thus transformed and renewed by users, who acquire a new sense of knowledge and power. Therefore, coherence in the planning of each part of the city is fundamental, because this will automatically achieve a very high standard of operation.

"The word smart, or intelligent, refers to any digital technology used in a given urban context with the intention of producing new resources, improving existing ones, changing user behavior or ensuring other prospective improvements in terms of flexibility, security and sustainability" [3]. One of the most important features of a smart city is the immediate availability of all information, so that the urban system is not only controllable, but can also be optimized thanks to the advancement of information technology that allows connections, communications and interactions in real time. However, this flow of information, to provide advantages in terms of sustainability at the urban level (from the architecture equipped with people detector to public services developed to save resources), must have three essential elements: the instrumentation, basically a series of sensors that detect all the flows in the city; the analysis, data processing algorithms capable of understanding the present and predict future scenarios; the actuators, devices that can react in real time by making changes in physical space. This is how the "Internet of Things" (IoT) is born, that is, the property of each individual object connect bilaterally with other objects in order to transfer the structure of the Internet network to physical space. If we imagine, for example, applying a device to our home pantry, connected in turn with a sensor placed on the packaging of our favorite cookies, which is able to control the level of the product and, if they are finishing, order directly and exactly the same product to the nearest and most convenient grocery store, we would get a structure of interconnection of objects never existed before and similar to the Internet, capable of having important multiple effects.

This could lead to the creation and diffusion of a new generation of products, the so-called "every-ware", deeply linked to urban space; in other words, each urban and architectural element could be designed to work in a coherent and systematic way in order to achieve maximum efficiency in the use of resources.

However, in order to contain the risk that "too much intelligence of objects" will end up creating a certain anxiety in users, a new way of applying technologies is being introduced at the same time which, starting from the background, that is, the citizens, allows to improve its use thanks to the initiatives of each person. For this reason, new forms of aggregation are emerging among individuals with the aim of proposing different ideas and procedures, capable of transferring themselves from virtual space to the physical.

As a result, the dichotomy "bottom-up" / "top-down" arises, which will be discussed in detail later, a key aspect in the creation of a smart city. The bottom-up approach starts from a single element to create a system of increasing complexity, while the top-down approach starts from an overview that is then fragmented into detail.

For the execution of the project developed in Brazil, some strategies were applied:

- Vision of the smart city

Based on the relevant experiences and examples already existing in the world, both with regard to new smart cities and those that have become the result of smart processes, our vision is the meeting point between a technocentric and a holistic model. The path that leads to success is techno-sociological, and integrates the technological components through digital urban development projects shared on collaborative platforms by all involved. Although the purely technological component, services, digital applications and technological infrastructures, represent a significant part in the smart city model, about 60% of the total activities, we must consider that the flow of actions that feed urban growth is of non-technological, ecosystem and governmental nature, in order to have a smart, inclusive, sustainable and lasting city.

- Type of smart city development

The holistic component presupposes a balance between a bottom-up type of development and a top-down model. The goal of the administrative body of a smart city should not be to centralize the decision-making process, but to make it open, inclusive, collaborative and strategically sound. Therefore, administrations participate both in projects related to the technological component and those belonging to the non-technological component; for the latter it is necessary to provide the city with plans, procedures, standards, guidelines, programmatic documents, instructions and measures that make up a general strategic framework (vision, objectives, expected results, priority fields of action, etc.), and create specific management bodies to encourage the construction of a collaborative platform, which constantly involves inhabitants, companies, research centers and government agencies, through the organization of events (fairs, exhibitions, moments of discussion and confrontation, specific training courses, etc.) that make it possible to promote the smart city initiative, fostering awareness, sensitivity and training.

- Smart city collaborative model

Research, industry and government are the driving force behind the process of transformation of urban contexts in the smart city environment; to achieve a better performing model, civil society must also be included. This is what, in jargon, is called the "four drive collaborative model", where nodes, identified by groups of actors, connect to each other according to the type and amount of activities performed (open innovation and user-driven innovation).

- Smart city growth

The development of the smart city involves a close collaboration between organizations of different types and civil society, embedded in an inclusive, heterogeneous and highly participatory ecosystem. The presence of a collective intelligence stimulates the activation of new collaborations and partnerships, enriching the knowledge needed to foster the transformation of projects and making available to the system, sharing them, the creativity and skills of individuals (smart community). From the encounter between individual and collective contribution emerges the intelligence of society, which allows the transformations in an intelligent key and the progressive integration of new digital services and technological infrastructures within the territory.

- Smart city concept

It is necessary to reason in a multidimensional way. On the one hand, the process of intelligent transformation is fed mainly by digital urban development projects that promote energy efficiency and a better use of natural resources, acting mainly in buildings, energy networks and transportation systems; on the other hand, it is necessary to work within a broader and more transversal field of action, which allows the extension of the use of ICT to other sectors where the problems to be addressed are not only related to energy and environmental sustainability, but also to the financial, business, commercial, health, cultural and sports aspects.

What are the operational elements that a team of designers should consider, and from which they should start, when creating a smart city? Or, more specifically, what are the main thematic areas on which architects and engineers should focus their work to propose new models compatible with the smart city concept?

- Mobility

For the Polo Multimodal Pecem project a smart mobility was conceived, carried out both through structural investments and through low cost initiatives that act in social innovation and in the awareness of users/inhabitants.
Structural investments include:
Differentiated road network design according to the use destinations of the areas and the type of transport used;
Versatility of the road network for future transformations, as the smart city is dynamic due to technological innovations and the evolution programs of its inhabitants;
Creation of "park and ride", areas designed to improve the quality of life of the entire urban area, increasing the use of more sustainable means of transport and decreasing noise levels in storage/production areas.

- Energy efficiency

In the Polo Multimodal Pecem project the use of passive energy saving systems has been planned, aimed at ensuring comfort levels without the need for artificially produced energy, thanks only to the characteristics of the building's coating (shape, architecture, exposure, orientation, thermal insulation and thermal mass, protection from solar radiation, materials) and the way heat is transported inside the rooms or outside the building (air, sun, sky). The passive conception, better defined in the building environment with the name of bioclimatic architecture, produces design solutions in which architecture and natural context are closely related to each other, starting from a detailed analysis of the site to determine the climatic agents capable of bringing advantages or against which it is necessary to protect oneself.

- Environmental sustainability

Use of "green energy" and reduction of CO_2 emissions. The Polo Multimodal Pecem project developed green solutions mainly in two directions, the first, of a general nature, considering the construction "on the ground" of a photovoltaic park with large production capacity at the service of the hub itself, the second, of a punctual type, going to define " smart " systems of energy management of buildings and built environments and implementing photovoltaic / wind power plants with lower production capacity connected to the main network of energy distribution in "in/out" mode.

- Smart living

In the smart city, smart living allows urban centers to find, through specific innovative solutions, an important "market launch"; for example, it is possible to advertise an international call to collect ideas that help to identify specific territorial marketing, or to reshape the city's image and design its logo. Web and 2.0 communication are central when it comes to smart living; the Polo Multimodal Pecem project has developed some specific applications designed to improve and enhance the use and appreciation of local resources, as well as increase the opportunities offered to spend free time ("social networking").

- Smart People

Participation, involvement, dialogue, interaction, listening: there can be no smart city if the foundations for good coexistence, open dialogue between citizens and administrators are missing. In the Polo Multimodal Pecem project, through the specific platform called "PoloDigital" (fig.3), and in order to remain in constant contact with the administration and develop common projects, an operational manual is available to users that illustrates specific tools such as the general survey, particular surveys, focus group, surveys about the place where the service is provided, online surveys, collection of demonstrations, suggestions and complaints, petitions and signature campaign. Through PoloDigital, a crowdfunding platform will be created that will improve the collaboration between the administration, the citizen and the companies, aiming at identifying resources and financing the agreed social projects.

- Smart Economy

The Polo Multimodal Pecem project envisages the creation of a specific center called "PoloChain Lab" that deals with investments in the "Knowledge Economy", focusing on research and innovation, encouraging processes of internationalization, sharing knowledge and enhancing creative talent. Investing in the Knowledge Economy means managing the transformation process of the smart city, promoting a synergistic system in which private companies, public agencies, research institutes, universities, science parks, collaborate to raise the technological level, creating a stimulating environment for everyone.

- Smart Governance

The Polo Multimodal Pecem offers, through the "Integrated Management Center" interconnected with the platforms "PoloChain Lab" and "PoloSuit", a "governance", a transparent and friendly promoter to share through open data the flows of information that you will receive daily from users, with the aim of simplifying the administration, digitizing processes and procedures to reach e-government. A smart city must have a smart administration, capable of defining its own strategy, be in harmony with the active citizenship, to aspire to a better and long-term future. In fact, it is through smart governance that smart service projects are developed, from those in the educational-cultural, business, commercial and even mobility, leisure and sports fields [4].

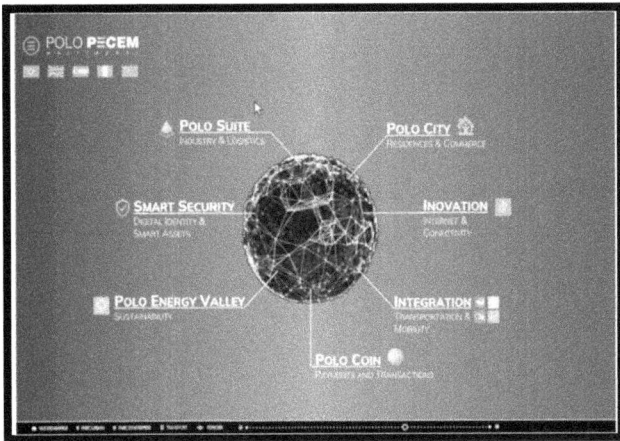

Figure 3. The Polo Digital platform : Participation, involvement, dialogue, interaction, listening: there can be no smart city if the foundations for good coexistence, open dialogue between citizens and administrators are missing. In the Polo Multimodal Pecem project, through the specific platform.

References

[1] Richard Buckminster Fuller and Kiyoshi Kuromiya, Cosmography. A Posthumous Scenario for the Future of Humanity, Macmillan, New York 1992, p.8.
[2]   Italo Calvino, Le città invisibili, Giulio Einaudi editore s.p.a., Turin 1972, p.16.
[3]   Francesca Bria and Evgeny Morozov, Ripensare la smart city, Codice Edizioni,Turin 2018, p.9.
[4] Walter Gaj Tripiano. Smart City Designing the smart city The case study of the Polo Multimodal Pecem Brazil. Fuoril[u]ogo. 2020 Sep

Intelligent Environments 2021
E. Bashir and M. Luštrek (Eds.)
© 2021 The authors and IOS Press.
doi:10.3233/AISE210088

Co-Integration of Oil and Commodity Prices: A Comprehensive Approach

Mohcine Bakhat [a,b,1] and Klaas Würzburg [c]

[a] *Business economics department, University Sultan Moulay Slimane (Morocco)*
E-mail: mohcine.bakhat@usms.ma
[b] *Research fellow, Economics for Energy, Vigo (Spain)*
E-mail: mohcine@uvigo.es
[c] *IESIDE, Business Institute (SPAIN)*
E-mail: klaas.wurzburg@ieside.edu

Abstract. Past research has mainly applied linear cointegration analysis to study the relationship of crude oil prices with the prices of other commodities. However, recent methodological innovations in cointegration analysis allow for a more thorough analysis of the co-movement of commodity prices and detect asymmetric and thresholds co-movements. Following Enders and Siklos [1] and Hansen and Seo [2], we apply threshold cointegration analysis, detecting co-movements that earlier studies based on linear cointegration analysis could not detect. We find that adjustments to positive and negative deviations from the long-run equilibrium are asymmetric for copper, food and agricultural raw materials in the short-run. Moreover, the adjustments for aluminum and nickel are symmetric. The price Granger causalities behave as expected for metals and agricultural raw material prices. Food prices, however, behave differently. In sum, the results of this paper underscore the importance of consistently testing nonlinear cointegration and point out the complex interactions that take place between the markets of oil and other commodities.

Keywords. linear, threshold, cointegration, metals, raw materials, agriculture

1. Introduction

There is remarkable interest in the relationship between the prices of oil and other commodities because there is an urgent need to understand the key characteristics and determinants of long-term commodity price movements, especially with respect to the conditions under which the recent boom occurred. The oil market has experienced high price levels and volatility since the first oil crises back in the 1970s. In the last few years record oil prices and climate-change-related interest in biofuels have intensified the quest for answers in this area, especially due to concerns on growing food prices [3,4,5]. High commodity prices, whether or not related to oil prices, have obvious effects on purchasing power and economic growth [6,7]. In this paper we look at the behavior of commodity prices by using a consistent cointegration analysis.

Since the seminal paper of Balke and Forby [8] on nonlinear cointegration, many empirical studies have demonstrated nonlinear and asymmetric adjustments to a long-run equilibrium in many economic time series [9,10]. The statistical concept of linear cointegration, as originally defined, refers solely to linear combinations of variables linked through a long-run equilibrium relationship. However, this standard cointegration

---

[1] Corresponding Author, E-mail: mohcine.bakhat@usms.ma

technique fails to capture real world economic phenomena such as the possible impact of market frictions, asymmetric information and transaction costs on the adjustment to the long-run equilibrium. Threshold cointegration, as proposed by Balke and Forby [8] includes the discrete adjustment to long-run equilibrium. In their model, the cointegrating relation between two variables is inactive within a certain threshold and because of this, the variables do not adjust to deviations from the equilibrium. Adjustment only takes place when deviations become large and exceed the threshold. Omitting the presence of nonlinear components, like threshold effects in long-run equilibrium, can lead to misinterpretations of equilibrium relationships because the cointegrating vector will no longer be consistently estimated [11].

In dealing with nonlinear long-run equilibrium models, two issues are interwoven. One is whether or not nonlinearity exists, and the other is whether there is a long-run relationship. The approach considered by Balke and Fomby [8] and later by Hansen and Seo [2] is based on a two-step analysis in which the linear no cointegration null hypothesis is first examined against the linear cointegration alternative. Then the linear cointegration null hypothesis is tested against the threshold cointegration alternative. However, in the presence of nonlinearity the linear test in the first step in this approach tends to fail in rejecting the null of no cointegration. Furthermore, the inability to reject the null in a linear cointegration test does not necessarily imply the absence of a long-run relationship since the possibility of nonlinear cointegration still remains [12]. The second step of the procedure testing for nonlinearity can also pose a problem when cointegration does not hold and the cointegrating vector is unknown. In this regard, the Enders and Siklos [1] approach provides additional crucial information because it formally tests for the joint hypothesis of the absence of both nonlinearity and cointegration.

This paper adopts the discrete two-regime threshold cointegration approach and uses the approaches developed by Hansen and Seo [2] and Enders and Siklos [1] jointly to test for threshold behavior between oil and commodity markets. Hence, unlike most previous studies that retained the usual linear cointegration framework or omitted the importance of the power of the tests in the analysis of threshold cointegration, this paper aims to study oil-commodity linkages within linear and nonlinear frameworks. To this end, it uses a battery of tests that provides more accurate information than most previous research. The paper examines the threshold cointegration relationship between crude oil and a set of different commodities. Given that the nexus of prices of crude oil and other commodities extends to different markets that operate under different constraints (especially the food and agricultural raw material markets) the study explicitly considers aluminum, nickel, copper and natural gas. Furthermore, and for the sake of methodology's illustration the respective aggregated price indexes of food, beverages and agricultural raw materials1 are also considered.

This paper is organized as follows. After this introduction, Section 2 presents the literature review. Section 3 discusses the methodology, data description and the empirical results and, finally, Section 4 provides the main discussions and conclusions.

2. Literature review

Several techniques were used in the economic literature to study the link between crude oil price and commodity prices so far. Most previous research on price transmission analysis has applied linear cointegration analysis based on Johansen, Breitung and ARDL approaches including Granger causality. In addition, the studies that apply these methods focus only on certain categories of commodities (like food or metals).

In the food category Abdel and Arshad [13], based on Johansen cointegration and Granger Causality test, identify a cointegration relationship between crude oil and all the four vegetable oils studied. Using the same technique Zhang et al. [14] look at the price relationship of three different fuels with five standard food commodities. They do not find a cointegrating relationship between energy and food commodities. Yu et al. [15] found a similar result for four major traded edible oil prices. However, based on Johansen cointegration and Granger causality tests, they also find that edible oil markets are well interlinked in the contemporaneous time. In the same line, Esmaeili and Shkoohi [16] construct a principal component of prices of different food commodities. They study the Granger causality between the food component and the oil price, among others, and they do not find a direct relationship between the oil price and the food price component. Other authors use the ARDL approach to study cointegration relations between commodities. For example, Sari et al. [17] investigate the energy-grain nexus (crude oil, gasoline, ethanol, corn, soybeans and sugar) focusing particularly on future prices. They identify dependencies that only partially comply with the general view that causal relationships within the energy grain nexus flow from the oil price to the price of gasoline, ethanol and corn. Chen et al. [18] find that the change in the price of each crop price is significantly influenced by changes in crude oil prices and other grain prices during the period extending from early 2005 to mid 2008.

Research on the relationship between oil and natural gas prices is usually carried out within the context of market integration. Cointegration tests are generally applied within a framework of Johansen and Breitung. Brown and Yücel [19] and Villar and Joutz [20] find the prices of natural gas and crude to be cointegrated, and also point out that the price of natural gas reacts to the price of crude oil. Hartley et al. [21] detect an indirect relationship between crude oil and natural gas through the price of residual fuel oil. Panagiotidis and Rutledge [22] study whether oil and gas "decoupled" during the post market deregulation period (1996-2003). They find that both prices are cointegrated before and after liberalization efforts in the UK gas market. In related work, Asche et al. (2006) find cointegration between natural gas and crude oil prices in the UK market after natural gas deregulation, with crude oil prices leading those for natural gas. Relying on daily ICE futures prices of gas and Brent for five contracts, Westgaard et al. [23] find that a long-term relationship exists between prices depending on the length of the contracts.

Oil prices are also suspected to influence commodities other than just food and energy. Chaudhuri [24] argues that oil prices potentially influence the price of other commodities as long as oil is used in the production process and points out that oil price drives an index composed of different commodity prices (including food, metals, and other consumption goods). Moreover, oil price changes affect real exchange rates and the industrial production of (developed) countries that, in turn, affect the world demand for commodities [24].

However, on the analysis of price linkages between oil and commodities most of the literature concentrates on agricultural commodities or on a reduced group of commodities, and only few studies use threshold cointegration in their analysis. Peri and Baldi [5] apply cointegration analysis based on Hansen and Seo [2] on a group of food commodities and find that the cointegration relation of rapeseed and diesel prices is a case of threshold integration. Sunflower oil and soybean oil prices are found to have no cointegration relation with diesel, although Peri and Baldi [5] do not apply the Ender and Siklos test to check whether these two series do feature threshold cointegration. Natanelov et al. [4] use threshold analysis based on Hansen and Seo [2] to investigate the price relationship of future contracts of crude oil, gold and eight food commodities. They find that only cocoa, wheat and gold move together with crude oil in the long-.run over the entire sample period. Hammoudeh et al. [25] investigate the cointegration of future and spot prices for the same commodity. They apply both the method of Hansen and Seo and that of Enders and Siklos but, since they only study cointegration between future and spot prices of the same commodities, they do not report results on co-movements of different commodities.

In contrast to earlier research, this paper will look at a wide range of commodities following a comprehensive procedure that covers linear and nonlinear (Threshold) cointegration. The combination of different cointegration tests, applied in this study and described thoroughly in the next section, yields the maximum detail about the co-moving dynamics of data series that contemporary cointegration analysis can provide. We have no knowledge of a study that has ever applied a system similar to that of commodity price pairs.

3. Methods, data and tests results

Four hypotheses are possible in threshold cointegration models: linear no cointegration, threshold no cointegration, linear cointegration, and threshold cointegration. The methodology applied in this paper starts by testing linear cointegration. If linear cointegration is found between two price series, then the results are contrasted with the Hansen and Seo test, which infers whether this cointegration relationship is indeed a linear or an asymmetric/threshold. In case the initial linear cointegration tests finds 'no cointegration', the Ender and Siklos cointegration test can be applied to test the null of linear no cointegration against the alternative of threshold cointegration. Hence, in our analysis two cointegration tests are always applied to each pair of price series. The linear cointegration test is always performed to start with, followed by an asymmetric cointegration test that, depending on the finding of the initial linear test is either based on Hansen and Seo [2] or Enders and Siklos [1].

3.1. Methods

Linear cointegration
In order to test the presence of linear cointegration between oil price and commodities prices we employ the technique of Johansen [26]. The first requirement of cointegration

is that time series variables must be integrated of the same order. Therefore, several unit roots tests are performed for each of the commodity prices. The cointegration requires pre-testing the order of integration of the variables through the unit root tests. For this purpose, Augmented Dickey Fuller (ADF), Phillips and Perron (PP) and the Breitung test were applied, with the latter test being consistent to structural breaks. Then we follow the procedure proposed by Johansen [26] to test for linear cointegration.[2]

Threshold cointegration

Two methods are applied to study the existence of threshold cointegration or long-run relationships among the variables in the presence of asymmetry in the adjustment process. The first method is the one proposed by Hanson and Seo [2]. It is applied when there is evidence of linear cointegration from the Johansen and Breitung tests. However, the Enders and Siklos [1] method is applied when the Johansen and Breitung tests indicate the absence of linear cointegration. The two methods have different specifications. The first method fits only the long-run adjustments while the second method accommodates for both short- and long-run adjustments. The second method furthermore permits us to examine both threshold autoregressive TAR and momentum threshold autoregressive M-TAR, while the first is appropriate only for TAR. Furthermore, the null hypotheses to be tested are different for each method. We test the null of no cointegration against the alternative of threshold cointegration using the Ender-Siklos method. On the other hand, we use the Hansen and Seo LM statistic to test the null of linear cointegration versus the threshold cointegration. If cointegration exists, the two methods provide us with information regarding the adjustment to long-run equilibrium, that is, we explore the co-movements between the oil price and the commodity prices over time, while we allow for asymmetric adjustments toward the long-run equilibrium.[3]

3.2. Data

This methodology is applied to the commodity price indexes available in the International Monetary Fund database. Prices of the following commodities were taken and included in the analysis: crude oil, aluminum, nickel, copper and natural gas and the aggregated price indexes of food, beverages and agricultural raw materials, respectively. Previous research has shown that the pattern of commodity prices has changed with the new millennium. Therefore, if the investigated time span covers the period prior to the year 2000, this may negatively affect the quality of the estimation results. Our series consists of monthly data from January 2000 to April 2011, to avoid possible distortions from these different data patterns (This gives 114 observations).

---

[2] For more details of the methodology refer to the article of Johansen [26]

[3] Details of the methodology are reported in the articles of Hansen and Seo [2], Ender and Siklos [1]

3.3. Tests results

The combined results of the three tests suggest that the commodities prices time series are, all in all, integrated of order one I(1). Therefore, we go on and conduct the cointegration analysis.[4]

Linear cointegration

The results from the Johansen test and Breitung test reveal that linear cointegration exists, at least a 5% significance level, for the pairs natural gas-crude oil and copper-crude oil. In addition, the results suggest the absence of cointegration between crude oil price and the following commodity prices: nickel, aluminum, agricultural raw materials, food and beverages, respectively.[5]

Threshold cointegration and the asymmetry tests

The presence of threshold cointegration is analyzed using the Ender and Siklos [1] and the Hansen and Seo [2] methods. Starting with the Ender-Siklos method first, we find that there is no evidence of the presence of cointegration between the crude oil price and the beverages price. In addition, the Φ_μ statistics indicate the existence of a pair-wise threshold cointegration between oil price and nickel, aluminum, food, and materials at 1%, and 5% significance level, respectively. We then assess whether the adjustment to the long-run equilibrium is symmetric or asymmetric. In other words, we test if the co-movement towards the long-run equilibrium between the oil price and each of the commodity prices occur at different speeds relative to their being below or above the threshold. The test results suggest asymmetric adjustments for food and agricultural raw materials, whereas symmetric adjustments are shown for nickel and aluminum.

For the crude oil-aluminum pair, the point estimates for the price adjustment show that deviations from the long-run equilibrium would be eliminated at 10.7% in a month, regardless of the sign of the shocks. In other words, it would take 9 months to eliminate deviation from the long-run equilibrium for either positive or negative deviations. Similarly, about the same amount of time is required to fully eliminate long-run deviations for the crude oil-nickel pair.

Applying Hansen-Seo method, which uses the LM threshold test, a threshold cointegrating relationship is found between the oil price and the copper price at a 10% significance level, while no threshold cointegration is detected for natural gas. Hence, the oil prices and gas prices are linearly cointegrated and co-move together towards the long-run equilibrium. Similarly, the Hansen-Seo method allows us to test if the adjustments towards long-run equilibrium are symmetric or asymmetric. Subsequently, the pair-wise combination of the oil price with copper co-moves towards the long-run equilibrium at different speeds that correspond either to being below or above the threshold. In other words, the adjustment towards the respective long-run equilibrium is asymmetric.

---

[4] Results are available upon request from the authors
[5] The lag order of the VAR specification is obtained by using Akaike (AIC), Schwarz (SC) and Hannan and Quinn (HQ) information criteria. We should also recall that, so far, these results are preliminary and have to be corroborated with threshold cointegration tests.

So far, we can summarize the results as follows: First, no cointegration relationship is found between the crude oil price and the price of beverages, which means that no direct effect exists between the oil market and the beverage markets. Second, the crude oil price is linearly cointegrated with the price of natural gas and this relation is featured with a high speed of adjustment to long-run equilibrium. Third, each of aluminum and nickel prices are threshold cointegrated with crude oil prices, and the price responses to either positive or negative shocks are symmetric. In contrast, the copper price shows a different pattern due to the presence of the asymmetric threshold cointegration, characterized with a faster speed of adjustment when deviations are above the threshold than below it. Finally, asymmetric threshold cointegration is also found between the crude oil price and the prices of food and agricultural raw materials price, respectively. The speeds of adjustment to long-run equilibrium in each case are subject to the sign of the shocks.

Results of the error correction model

We first start with the oil and natural gas pair, which exhibits a linear cointegrating relationship. The vector error-correction model yields a high parameter estimate (-1.36). This indicates a strong relationship between oil and natural gas prices. Besides, the statistical significance of the speed of the adjustment coefficient shows that the adjustment towards the long-run equilibrium occurs at the rate of 12.5% per month. The Granger causality outcomes show that crude oil and natural gas prices move together in the long-run with a significant bi-directional causality effect. In the short run, crude oil price leads the prices of natural gas pointing to a close integration between these markets.

Based on the results of the Hansen and Seo [2] threshold VEC, models are specified to assess the asymmetric dynamic behavior in the pairs that are threshold cointegrating. Each of the TVECM is estimated using the negative log-likelihood estimator and the selection of the lag order is determined by AIC and BIC. The parameter estimates are estimated for each regime and their t-statistics are reported in parentheses, where the heteroskedasticity- consistent (Eicker-White) standard errors are considered.

For crude oil, when the error correction term exceeds the threshold value, we see the flat near-zero error-correction effect. Whilst in the first regime ($\omega_{t-1} \leq -1.576$), the response to crude oil price changes is much faster. For copper equations, the adjustment to the long run equilibrium is faster when the error correction term is below the threshold value ($\omega_{t-1} \leq -1.576$), that is, when copper prices decrease or crude oil prices increase. The response of the error-correction effects to copper is significantly larger than the response of crude oil price changes in the second regime ($\omega_{t-1} > -1.576$).

Based on the Ender-Siklos model, the asymmetric correction model with threshold cointegration is estimated for the pairs: crude oil- aluminum, crude oil- copper, crude oil-food and crude oil-agricultural raw materials. AIC and BIC were used to determine the optimal lag length of the model and the residual autocorrelation was analyzed with the Ljung-Box Q statistic.

The results suggest that there is momentum equilibrium adjustment asymmetry for the cases of food and agricultural raw materials. In addition, results show that both food and oil prices respond to deviations from the long-run equilibrium below the threshold in the short term; the food price adjustment is very slow in this regime. In other words, food prices respond to the negative deviations by 3.7% in a month, or alternatively negative deviations take about 27 months to be fully eliminated. Similarly, in the short-run, it is the price of agricultural raw materials and not the crude oil price, which responds to deviations from the long-run equilibrium below the threshold with 14% of the deviations eliminated each month (this results from a decrease in agricultural raw material prices, or equivalently, an increase in crude oil prices). Furthermore, the F statistics show that the cumulative effects in aluminum; nickel and agricultural raw material equations, respectively, are asymmetric.

The hypotheses of Granger causality between crude oil and each one of the commodities were examined with F-tests. Results show a significant Granger causality at 10% level between crude oil and aluminum and copper, respectively, and at 5% level between crude oil and agricultural raw material, and at 1% level between crude oil and food (crude oil → aluminum; crude oil →copper; food →crude oil; crude oil→agricultural raw materials). Yet, no Granger causality is detected between crude oil and beverages. Thus, in the short-term it seems that the prices of metals (aluminum and nickel) and agricultural raw material have depended on the crude oil price. On the other hand, food appears to be leading the price movement of crude oil, with oil prices adjusting to deviations from long-run equilibrium.

4. Conclusions

In this paper we investigate price relationships of crude oil and different types of other commodities applying a system of different cointegration tests. This combination of different cointegration analyses confirms the argument that non-linear and threshold cointegration techniques better represent real markets where frictions, asymmetric information, transaction costs cause non-linear outcomes [10,4,5]. For the price pairs investigated in this paper linear cointegration tests fail to detect any relation between crude oil price and most prices of the commodities. However, when non-linear cointegration analysis as proposed by Enders and Siklos [1] and Hansen and Seo [2] is applied, we find that the estimated thresholds in the case of aluminum, nickel, copper, food and agricultural raw materials are not zeros. We also identify asymmetric cointegration between the crude oil price and the prices of copper, and food and agricultural raw materials, respectively.

Results for the metals category show that each of the aluminum and nickel prices are threshold cointegrated with crude oil prices. The Granger causality tests indicate that the crude oil price leads the prices of aluminum and nickel. Hence, global investors can predict the prices of aluminum and nickel by following fluctuations in the oil prices. In addition, we find that price responses are symmetric in the sense that a shock to crude oil prices of a given magnitude would give rise to the same response in aluminum and nickel respectively, regardless of whether the shock reflected a price increase or price decrease, and deviations after the shocks are digested in roughly 10 months. Copper shows a different asymmetric adjustment; it yields a faster speed of adjustment when deviations are above the threshold ($\omega_{t-1} > -1.576$) rather than below it ($\omega_{t-1} \leq -1.576$). Asymmetry price transmission revealed that pair-wise adjustments of the oil and copper prices are faster when crude oil price decreases than when it increases.

For the relation between crude oil and natural gas we find that the crude oil price is linearly cointegrated with the natural gas price. The main result from the VECM approach is the relatively high magnitude of the parameter estimate (-1.36), which means that a strong relationship exists between crude oil and natural gas prices. On the other hand, the speed of the adjustment coefficient shows that about 12.5% of the adjustment towards long-run equilibrium would take place in each month. The Granger tests reveal that crude oil prices and natural gas prices have been tied up in the long run; short-run shocks can be transferred from the crude oil market to the gas market. These findings do not support the hypothesis that recent developments in natural gas technologies (shale gas, reduced transport costs) would unlink natural gas and crude oil prices. It seems that the link is still strong, much in line with the theoretical substitution relationship between the two fuel types. Previous research [19,21] extensively discussed the possible reasons for this finding.

Results also show that food prices are threshold cointegrated with the crude oil price, underlining the strong price interdependence of food and crude oil for the period under study. Moreover, the different outcomes of the Granger causality test and the adjustment speed asymmetries in this category highlight the complex relationship between food and crude oil prices: The Granger causality showed that the price of food led the crude oil price. That is, fluctuations of food prices are likely to be transmitted to the crude oil market. The results from the asymmetric speed adjustments show persistence of food prices when the crude oil price increases (deviations are eliminated at a rate of 3.7% per month). This particular interaction is interesting and some authors credit it to the increasing role of the biofuel market during the last decade [4,5]. Similarly, agricultural raw material prices show the presence of momentum equilibrium adjustment asymmetry and adjust faster when deviations are below threshold. Results also show a Granger causality running from agricultural raw material prices to crude oil prices, which recommends further detailed research into the linkages between the prices of each agricultural raw material and crude oil prices.

Our work provides a detailed analysis about non-linear cointegration relationships between the crude oil price and prices of a group of different commodities. The different outcomes can provide insight on price movements and their complex interdependences. However, it is relevant to mention that other factors as regulatory interventions, general economic conditions (crises) and other exogenous impacts contribute to the uncertainty and volatility of commodities markets, and increase the complexity of price dynamics between crude oil and other commodities. Therefore, advanced analysis that represents real conditions on markets in the most appropriate way are required to better understand these price dynamics and help policy-makers and other actors in these markets to take the correct decisions.

References

[1] W., Enders, P. L., Siklos, Cointegration and Threshold Adjustment, Journal of Business & Economic Statistics, 19(2), (2001), 166-76.

[2] B.E., Hansen, B., Seo, Testing for two-regime threshold cointegration in vector error-correction models. Journal of Econometrics, 110, (2002) 293–318.

[3] K., Balcombe, A., Bailey, J., Brooks, Threshold Effects in Price Transmission: The Case of Brazilian Wheat, Maize, and Soya Prices, American Journal of Agricultural Economics, 89(2),(2007), p. 308 - 323.

[4] V., Natanelov, M.J., Alam, A.M., McKenzie, G., Van Huylenbroeck, Is there co-movement of agricultural commodities futures prices and crude oil? Energy Policy,39, (2011), 4971-4984.

[5] M., Peri, L., Baldi, Vegetable oil market and biofuel policy: an asymmetric cointegration approach, Energy Economics, 32 (3), (2010), 687–693.

[6] K., Chaudhuri, Long-run prices of primary commodities and oil prices, Applied Economics, 33, (2001), 531-538.

[7] Z., Zhang, L., Lohr, C., Escalante, M., Wetzstein, Food versus fuel: what do prices tell us? Energy Policy 38, (2010), p. 445-451.

[8] N.S., Balke, T.B., Fomby, Threshold cointegration. International Economic Review 38, (1997), p. 627–645.

[9] M.C., Lo, E., Zivot, Threshold Cointegration and Nonlinear Adjustment to the Law of One Price. Macroeconomic Dynamics 5(4),(2001), p. 533–576.

[10] C.C., Douglas, Do gasoline prices exhibit asymmetry? Not usually! Energy Economics 32(4), (2010), p.918–925.

[11] J., Gonzalo, J.Y., Pitarakis, Threshold effects in cointegrating relationships. Oxford Bulletin of Economics and Statistics 68, (2006), p. 813–833.

[12] M., Seo, Bootstrap testing for the null of no cointegration in a threshold vector error correction model. Journal of Econometrics 134, (2006), p. 129–50.

[13] H.A., Abdel, F.M., Arshad, The impact of petroleum prices on vegetable oils prices: evidence from cointegration tests. Oil Palm Industry Economic Journal 9(2), (2009), p. 31-40.

[14] Z., Zhang, L., Lohr, C., Escalante, M., Wetzstein, Food versus fuel: what do prices tell us? Energy Policy 38, (2010), p. 445-451.

[15] T.H., Yu, D.A., Bessler, S. Fuller, Cointegration and Causality Analysis of World Vegetable Oil and Crude Oil Prices. Selected Paper prepared for presentation at the American Agricultural Economics Association Annual Meeting, Long Beach, CA July 23-26 (2006).

[16] A., Esmaeili, Z., Shokoohi, Assessing the effect of oil price on world food prices: Application of principal component analysis, Energy Policy 39, (2011), p. 1022–1025.

[17] R., Sari, S., Hammoudeh, C.L., Chang, M., McAleer, Causality between market liquidity and depth for energy and grains, WP SSRN, (2011).

[18] Chen, S.T., Kuo. H.I., Chen, C.C., 2010. Modeling the relationship between the oil price and global food prices, Applied Energy 87, p. 2517–2525.

[19] S., Brown, M. Yücel, What drives U.S. Natural Gas Prices? presented at the USAEE 26th Annual Conference. Ann Arbor, Mi, (2006).

[20] J., Villar, F., Joutz, The Relationship Between Crude Oil and Natural Gas Prices. EIA manuscript, October 2006, (2006).

[21] P.R., Hartley, K.B., Medlock, J.E., Rosthal, The relationship of natural gas to oil prices, Energy Journal 29, (2008), p. 47–66.

[22] T., Panagiotidis, E., Rutledge, Oil and gas markets in the UK: evidence from a cointegrating approach. Energy Economics 29, (2007), 329–347.

[23] S., Westgaard, M., Estenstad, M., Steim, S., Frydenberg, Co-integration of ICE gas oil and crude oil futures. Energy Economics 33, (2011), p. 311–320.

[24] K., Chaudhuri, Long-run prices of primary commodities and oil prices, Applied Economics 33,(2001), p. 531–538.

[25] S., Hammoudeh, L.H., Chen, B., Fattouh, Asymmetric adjustments in oil and metals markets, Energy Journal 31(4), (2010), p. 183–202.

[26] S., Johansen, Likelihood Based Inference in Cointegrated Vector Error Correction Models, Oxford University Press, Oxford, (1995).

Intelligent Environments 2021
E. Bashir and M. Luštrek (Eds.)
© 2021 The authors and IOS Press.
This article is published online with Open Access by IOS Press and distributed under the terms
of the Creative Commons Attribution Non-Commercial License 4.0 (CC BY-NC 4.0).
doi:10.3233/AISE210089

BESTest for Integrated Outdoor-Indoor Energy Balance Modelling

Mohammed BAKKALI[a,b,1] and Yasunobu ASHIE[c]

[a] The Bartlett School, University College London, UK

[b] The International University of Rabat, Morocco

[c] Building Research Institute, Japan

Abstract. In our growing cities, climate change and energy related uncertainties are of great concern. The impact of the Urban Heat Island on comfort, health and the way we use energy still requires further clarification. The outdoor-indoor energy balance model (3D-City Irradiance) presented in this article was developed so as to address these issues. The effects of view factors between urban surfaces on three-dimensional radiation and the effects of fully integrated outdoor-indoor energy balance schemes on heat islands and building indoor thermal loads could be included within different building blocks at a resolution of several metres. The model operated under the 'stand alone' mode. It was tested using the Building Energy Simulation Test (BESTest) which demonstrated good levels of agreement for diurnal and seasonal simulations.

Keywords
Urban Heat Island, three-dimensional radiation, outdoor-indoor energy balance schemes, building blocks, diurnal and seasonal simulations, local climates

1. Introduction

The Urban Heat Island (UHI) is a significant physical process that involves a changing built environment including buildings, land cover and human activity. Exhaustive definition of the pros and cons of UHI effects still requires further research and effective mitigation strategies will have to be implemented in order to maintain the positive contributions and remove the undesirable effects [1-5]. The thermal interaction between indoor and outdoor environments through conduction, radiation and convection is essential for a better implementation of micro-climatic boundary conditions. Outdoor-indoor energy budget models enable the assessment of the impact of outdoor adjacent air/surface temperatures on building indoor air/surface temperatures by taking into account reflected and emitted radiation from surrounding urban surfaces besides other micro-scale effects. The effect of indoor conditions on outdoor environments can also be determined e.g., via quantifying conductive transient heat, energy demand from heating and cooling loads and related sensible and latent waste heat. Such models can

---

[1] Corresponding Author: The Bartlett, UCL Institute for Environmental Design and Engineering, Central House, 14 Upper Woburn Place, WC1H 0NN London, UK; E-mail: mohammed.bakkali.10@ucl.ac.uk.

undoubtedly broaden our understanding of the physics behind of the UHI through depicting the patterns of different outdoor turbulent heat fluxes, three-dimensional radiation, waste heat from AC systems, building envelopes and other urban surfaces. In meso-scale meteorology, outdoor-indoor energy budget schemes were recently developed. WRF-Urban was implemented with a multi-scaled coupled BEM and a multi-layer urban canopy model (Building Effect Parameterization (BEP)) [6-8]. The three-dimensional Meso-scale meteorological Model (MM) was similarly coupled with a one-dimensional urban Canopy Model (CM), and a BEM [9, 10]. Likewise, an iterative scheme was developed for coupling EnergyPlus and the Town Energy Balance (TEB) [11]. In micro-scale meteorology, few outdoor-indoor energy balance models have been developed to-date. From urban block to neighbourhood scale, they enable the calculation of heat exchange between indoor and outdoor environments. The effect of thermal radiation and conduction from buildings and ground surfaces were taken into account by some models.

Initially, the Temperature of Urban Facets Indoor–Outdoor Building Energy Simulator (TUF-IOBES) was applied to an idealised domain [12]. Computational Fluid Dynamics (CFD) model (Fluent) was coupled with a thermo-radiative simulation tool (Solene) with a typical application on an urban fragment [13]. A three-dimensional (3D) Computer Aided Design (CAD) based simulation tool was developed for predicting the effect of outdoor thermal environment on building thermal performance in an urban block with detailed morphology [14, 15]. A CFD model was coupled with three-dimensional radiation and one-dimensional heat conduction schemes in addition to air conditioning heat loads [16-19].

2. Model Description

The surface temperature of any building depends essentially on the colour of its surface and that of the surrounding surfaces, its orientation and location, solar radiation, air temperature both outside and inside the buildings and essential meteorological conditions such as air pressure, humidity, wind speed and direction, and finally the thermal properties of different construction materials used to build the walls and roofs of buildings. These parameters led us to study surface temperatures, and to predict them in various urban locations and environmental conditions [1].

2.1 Inputs and Outputs

Input data includes weather components, water temperature, building data, construction materials and their thermo physical properties in addition to indoor conditions. Output data encompasses radiation levels, surface temperatures, building thermal loads, energy usage profiles so on and so forth (see, figure 1).

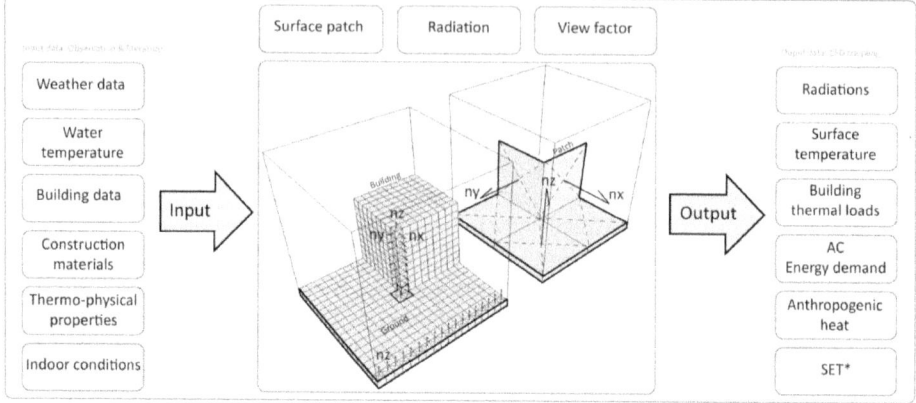

Figure 1. Inputs and Outputs of the heat balance model (3D-City Irradiance)

2.2 Indoor and Outdoor energy budget

2.2.1 Outdoor 3D Irradiation

The temperature of outdoor urban surfaces is mainly determined by the calculation of atmospheric and solar radiation and air temperature, which is assumed to be hourly uniform, and it is inputted in the boundary conditions. The configuration of shaded/sunlit surface distributions is obtained by using the Ray-Tracing function where a sunlit surface patch has a value of one and a shaded solid surface patch has a value of zero, i.e. no incident solar radiation and it therefore not counted for the calculation of adjacent shortwave length radiation. Mutual reflected short wavelength radiation, emitted and reflected long wavelength radiation between surrounding buildings; urban infrastructures and ground are considered here in the energy budget of the urban canopy. The Monte Carlo Method was used in order to compute view factor values of complicated urban geometries at every solid surface patch. Surface facets were assumed to be grey diffuse surfaces and radiation values were determined through the Radiosity Method, which is more suitable for radiation calculations over larger localities [20].

The governing equations are set as follow:

- Absorbed solar and short wave length radiation from the sun, the atmosphere and the surroundings urban surfaces at every surface patch (eq. 1 -2):

$$S_i = A_i \alpha_i I_i \qquad \text{Where} \quad I_i = \beta_i I_{direct} \cos \theta_i + F_{Si} I_{diffuse} \quad (1)$$

$$R_{Si} = A_i \alpha_i \sum_{j=1}^{m} G_{Sj} F_{ji} \quad \text{Where} \quad G_{Sj} = (1 - \alpha_j) \sum_{i=1}^{n} G_{Si} F_{ij} \quad (2)$$

- Net long wave length radiation at different urban surfaces (eq. 3 – 4):

$$L_i = A_i F_{Si} \varepsilon_i \sigma T_a^4 (a + b\sqrt{e}) \qquad (3)$$

$$R_{Li} = A_i \varepsilon_i (\sum_{j=1}^{m} G_{Lj} F_{ji} - \sigma T_i^4) \quad \text{Where} \quad G_{Lj} = \varepsilon_j \sigma T_j^4 + (1 - \varepsilon_j) \sum_{i=1}^{n} G_{Li} F_{ij} \quad (4)$$

2.2.2 Outdoor Energy Budget

Ambient air temperature and wind velocity are uniformly distributed and forced from site measurements or through general circulation and/or regional numerical modelling. Weather conditions are more relevant to the model when wind velocities are lower. Surface temperature at each solid surface patch designated as i is calculated through its energy budget (eq. 5). Latent heat can be considered, for instance, if the solid surface patch i contains water. Conductive transient heat through construction layers in building roofs, walls and ground is calculated by means of the unsteady state one-dimensional thermal conduction equation applied at the normal direction across each solid surface patch i and its constituent construction layers (eq. 6 - 7). The effect of building thermal bridges is not considered here. G_i is used for outdoor boundary conditions of solid surface patch i. The Finite Difference Method (FDM) is used to fully discrete transient heat conduction in each solid surface patch i and time step t. The implicit Euler Method is used for time discretisation. The Newton Method is used to solve the latter equation especially for long wave radiation. Indoor building surface temperature such as walls, ceilings and floors were then determined accordingly.

- Outdoor energy balance equation at each solid surface patch i:

$$S_i + R_{Si} + L_i + R_{Li} + G_i + H_i + E_i = 0 \quad (5)$$

$$G_i = A_i k_i \frac{\partial T_i}{\partial x}\bigg|_{x=0} \quad (6)$$

- Unsteady state one-dimensional thermal conduction equation:

$$\rho_i c_i \frac{\partial T_i}{\partial t} = \frac{\partial}{\partial x}\left(k_i \frac{\partial T_i}{\partial x}\right) \quad (7)$$

- Sensible heat flux at solid surface patch i:
$$H_i = A_i h(T_{adjacent} - T_i) \quad (8)$$

- Latent heat flux at solid surface patch i:
$$E_i = A_i L \beta h_q (q_a - q_i) \quad (9)$$

2.2.3 Building Thermal Loads

In this case, the model determined indoor air temperature through the calculation of building indoor thermal loads via the indoor energy budget. Outdoor and indoor energy balance schemes were dynamically used as mutual boundary conditions (eq. 6 and 11).

By adding conductive transient heat loads (eq. 11), radiation across windows and other transparent surfaces (eq. 12), ventilation heat loads (eq. 13), indoor generated heat loads from electrical appliances for instance and radiation and convection heat loads from indoor floors (eq. 14), the total building indoor thermal loads was determined (eq. 10). The model simulates indoor generated heat, which is mainly affected by occupant behaviour along with sensible and latent heat from ventilation.

- Building thermal loads are shown in eq. (10) - (14):

$$Q = H_t + H_s + H_{is} + H_{i\ell} + H_{vs} + H_{v\ell} + H_f \ (10)$$

- Transient heat loads at indoor solid surface patch i:

$$H_t = \sum_{i=1}^{n} A_i h_i (T_i - T_r) \ (11)$$

- Solar heat loads at indoor solid surface patch i:

$$H_s = \sum_{i=1}^{n} A_i \eta_i S_i \tau_i \qquad \tau_i = e^{-k_i \frac{d_i}{\cos\theta_i}} \ (12)$$

- Heat loads from ventilation:

$$H_{vs} = c_a \rho_a V \eta (T_a - T_r) \qquad\qquad H_{v\ell} = L c_a \rho_a V \eta (q_a - q_r) \ (13)$$

- Reflective and convective heat loads from the floor:

$$H_f = \sum_i A_{fi} [\alpha_{fi} S_i + h_{fi}(T_{fi} - T_r)] \ (14)$$

3. Building Energy Simulation Test (BESTest)

Output results from 3D-City Irradiance were compared with other building energy simulation programs. Common boundary conditions were used. They were obtained from the International Energy Agency Building Energy Simulation Test and Diagnostic Method [21]. This procedure aims to methodically test thermal and building energy simulation models. It diagnoses the level of disagreements through a comparative assessment framework based on the features of the building fabric. The Case 600 was used for the testing of annual and daily heating and cooling loads. This case is designated for the assessment of the performance of lightweight building envelope. The building is of a rectangular shape. The widths of south and north elevations are set equally to eight metres. The width of west and east elevations is set to six metres. The height of the building is 2.7 metres. There are no interior partitions inside. The south elevation has 12 m2 of double-glazed windows. The mechanical system is 100% efficient with no duct losses, infinite capacity and no latent heat extraction. A dual set point thermostat was set within a dead band between 20°C and 27°C. The weather data was extracted from the

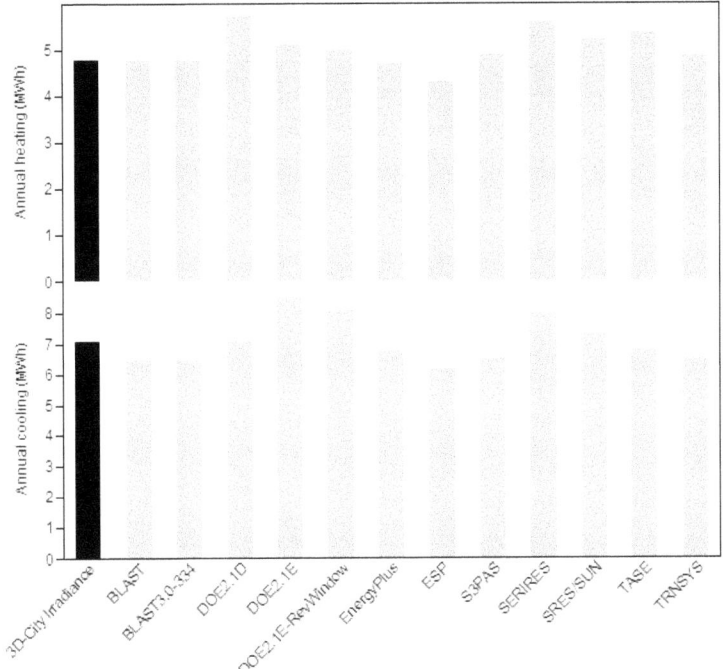

Figure 2. Inter-comparative study between simulated annual heating loads and from different models using BESTest.

Typical Meteorological Year (TMY) in Denver, Colorado. The latitude of the site is 39.8 north, longitude is 104.9 west and altitude is 1609m. The minimum and maximum annual dry bulb temperature is -24.39°C and 35°C respectively. The ground temperature is 10°C. Detailed information on building construction characteristics and input data can be found in the literature [21]. A comparison of annual and daily heating and cooling loads is presented here. These models cannot reflect building energy use truthfully and the goal is rather to observe if predicted thermal loads fall close to or within the range of results from other programs.

3.1. Annual heating and cooling loads

This study compares output results from heating and cooling loads between 13 different models assessed by means of the BESTest (see, figure 2). 3D-City Irradiance calculated a total of 4.77 MWh for annual heating loads. Ten models estimated higher annual heating loads than this value. From higher to lower annual heating loads, these models are DOE2.1D, SERIRES, TASE, SRES/SUN, DOE2.1E, DOE2.1E-RevWindow, S3PAS, TRNSYS, BLAST and BLAST3.0-334 respectively. However, two models have estimated lower annual heating loads than this value, EnergyPlus and ESP.

On the other hand, 3D-City Irradiance calculated a total of 7.09 MWh for annual cooling loads. Four models estimated higher annual cooling loads than this value. From higher to lower annual cooling loads, we find DOE2.1E, DOE2.1E-RevWindow, SERIRES and SRES/SUN respectively. Nonetheless, eight models estimated lower annual cooling loads than this value. From higher to lower annual cooling loads, we find DOE2.1D,

EnergyPlus, TASE, S3PAS, TRNSYS, BLAST, BLAST3.0-334 and ESP respectively. ESP estimated minimum heating and cooling loads, 4.296 MWh and 6.137 MWh respectively. For maximum annual heating loads DOE2.1D estimated 5.709 MWh and DOE2.1E estimated 8.5 MWh. These results show that 3D-City Irradiance complies with other Building Energy Models (BEMs) for a lightweight building envelope. This is an encouraging outcome that will lead to further testing in the future.

3.2. Daily heating and cooling loads

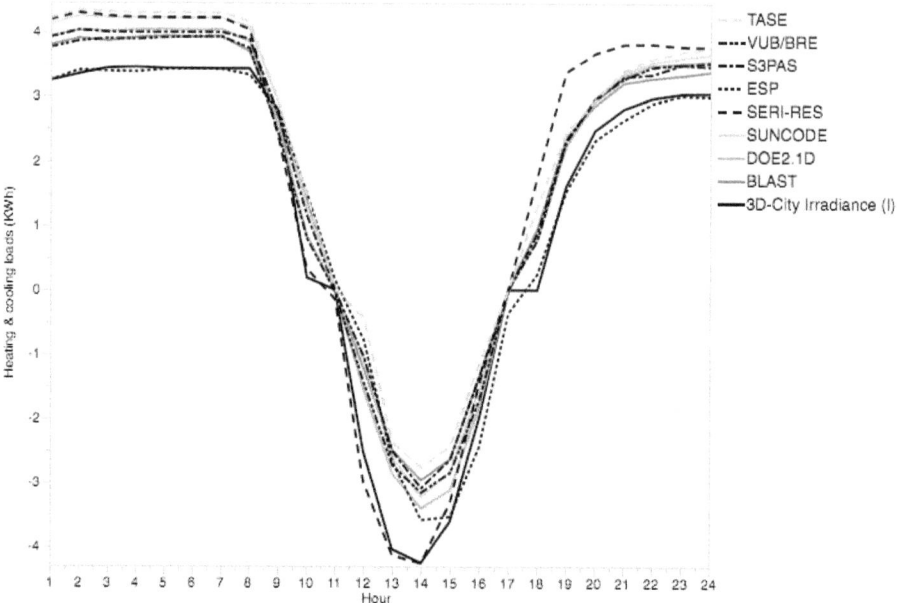

Figure 3. Inter-comparative study between simulated heating and cooling loads from different models using BESTest for one day, 4th January from TMY.

Output results for daily heating and cooling loads from 3D-City Irradiance were compared with eight other models using BESTest on the 4th of January (see, figure 3). In this case, values with (-) represent cooling loads and values with (+) represent heating loads. 3D-City Irradiance estimated daily heating and cooling loads within the range of values estimated by other models. Regarding heating loads, the calculated values by the model were closer to those calculated by ESP (lowest values amid the range) and concerning cooling loads, the model calculated values closer to SERI-RES model (highest values amid the range). In order to allow 3D-City Irradiance model to fulfil other requirements of BESTest, an eventual implementation of detailed outdoor convective and indoor convective and radiative heat transfer coefficients are essential.

Conclusion

The 3D-City Irradiance model is an outdoor-indoor energy balance model where comprehensive radiative and conductive heat loads within the built environment were assessed comprehensively all together. The model was adapted and further developed to meet the demands of BESTest. The results show that 3D-City Irradiance complies with other Building Energy Models (BEMs) for a lightweight building envelope. This is an encouraging outcome that will lead to further testing in the future. The binary outdoor-indoor energy balance approach is vital for the assessment of certain mitigation strategies.

Acknowledgements

We would like to thank the Japanese Society for the Promotion of Science for funding this research and the Japanese National Institute for Land and Infrastructure Management (NILIM) for hosting this collaboration in Tsukuba City.

References

[1] Akbari, H., Konopacki, S., 2005. Calculating energy-saving potentials of heat-island reduction strategies. Energy Policy 33, 721–756.
[2] Rosenfeld, A.H., Akbari, H., Bretz, S., Fishman, B.L., Kurn, D.M., Sailor, D., Taha, H., 1995. Mitigation of urban heat islands: materials, utility programs, updates. Energy Build. 22, 255–265.
[3] Mackey, C.W., Lee, X., Smith, R.B., 2012. Remotely sensing the cooling effects of city scale efforts to reduce urban heat island. Build. Environ. 49, 348–358.
[4] Johnson, G.T., Oke, T.R., Lyons, T.J., Steyn, D.G., Watson, I.D., Voogt, J.A., 1991. Simulation of surface urban heat islands under "IDEAL" conditions at night part 1: Theory and tests against field data. Bound.-Layer Meteorol. 56, 275–294.
[5] Oke, T.R., Johnson, G.T., Steyn, D.G., Watson, I.D., 1991. Simulation of surface urban heat islands under "ideal" conditions at night part 2: Diagnosis of causation. Bound.-Layer Meteorol. 56, 339–358.
[6] Chen, F., Kusaka, H., Bornstein, R., Ching, J., Grimmond, C.S.B., Grossman-Clarke, S., Loridan, T., Manning, K.W., Martilli, A., Miao, S., Sailor, D., Salamanca, F.P., Taha, H., Tewari, M., Wang, X., Wyszogrodzki, A.A., Zhang, C., 2011. The integrated WRF/urban modelling system: development, evaluation, and applications to urban environmental problems. Int. J. Clim. 31, 273–288.
[7] Salamanca, F., Krpo, A., Martilli, A., Clappier, A., 2010. A new building energy model coupled with an urban canopy parameterization for urban climate simulations—part I. formulation, verification, and sensitivity analysis of the model. Theor. Appl. Clim. 99, 331–344.
[8] Salamanca, F., Martilli, A., 2010. A new Building Energy Model coupled with an Urban Canopy Parameterization for urban climate simulations—part II. Validation with one dimension off-line simulations. Theor. Appl. Clim. 99, 345–356.
[9] Kikegawa, Y., Genchi, Y., Yoshikado, H., Kondo, H., 2003. Development of a numerical simulation system toward comprehensive assessments of urban warming countermeasures including their impacts upon the urban buildings' energy-demands. Appl. Energy 76, 449–466.
[10] Kikegawa, Y., Genchi, Y., Kondo, H., Hanaki, K., 2006. Impacts of city-block-scale countermeasures against urban heat-island phenomena upon a building's energy-consumption for air-conditioning. Appl. Energy 83, 649–668.
[11] Bueno, B., Norford, L., Pigeon, G., Britter, R., 2011. Combining a Detailed Building Energy Model with a Physically-Based Urban Canopy Model. Bound.-Layer Meteorol. 140, 471–489.
[12] Yaghoobian, N., Kleissl, J., 2012. An indoor–outdoor building energy simulator to study urban modification effects on building energy use – Model description and validation. Energy Build. 54, 407–417.
[13] Bouyer, J., Inard, C., Musy, M., 2011. Microclimatic coupling as a solution to improve building energy simulation in an urban context. Energy Build. 43, 1549–1559.
[14] Asawa, T., Hoyano, A., Nakaohkubo, K., 2008. Thermal design tool for outdoor spaces based on heat balance simulation using a 3D-CAD system. Build. Environ. 43, 2112–2123.
[15] He, J., Hoyano, A., Asawa, T., 2009. A numerical simulation tool for predicting the impact of outdoor thermal environment on building energy performance. Appl. Energy 86, 1596–1605.

[16] Huang, H., Ooka, R., Kato, S., 2005. Urban thermal environment measurements and numerical simulation for an actual complex urban area covering a large district heating and cooling system in summer. Atmos. Environ. 39, 6362–6375.

[17] Chen, H., Ooka, R., Harayama, K., Kato, S., Li, X., 2004a. Study on outdoor thermal environment of apartment block in Shenzhen, China with coupled simulation of convection, radiation and conduction. Energy Build. 36, 1247–1258.

[18] Chen, H., Ooka, R., Harayama, K., Kato, S., Li, X., 2004b. Study on outdoor thermal environment of apartment block in Shenzhen, China with coupled simulation of convection, radiation and conduction. Energy Build. 36, 1247–1258.

[19] Bakkali M., Ashie Y., Rans Modelling for Local Climates, Energy Use and Comfort Predictions in Cities. *Ambient Intelligence and Smart Environments*, Volume 26: Intelligent Environments, 2019, 76 – 88, doi: 10.3233/AISE190026.

[20] Sparrow, E.M., Cess, R.D., 1978. Radiation heat transfer. Hemisphere Pub. Corp.

[21] Judkoff, R., Neymark, J., 1995. International Energy Agency building energy simulation test (BESTEST) and diagnostic method.

List of Symbols

a, b [-] Constants from Brunt's formula

α_{fi} [-] Reflectivity of floor

α_i [-] Absorption coefficient at solid surface patch i

A_i [m2] Area of solid surface patch i

A_{fi} [m^2] Area of floor

β [-] Evaporation efficiency

β_i [-] Solar view factor at solid surface patch i

c_a [J / kg.K] Specific heat capacity of air

c_i [J/kg.K] Specific heat capacity of solid surface patch i

COP [-] Coefficient of performance of AC systems

e [mmHg] Water vapour pressure

ε_i [-] Emissivity of solid surface patch i

E_i [W/m^2] Latent heat flux at solid surface patch i

F_{ji} [-] View factor at solid surface patch i

F_{Si} [-] Sky view factor at solid surface patch i

G_i [W/m^2] Conductive heat flux at solid surface patch i

G_{Lj} [-] Long wave length radiosity of solid surface patch j

G_{Sj} [-] Total short wave radiosity from solid surface patch j

h [W/m^2. K] Convective heat transfer coefficient

h_{fi} [W/m^2. K] Heat transfer coefficient of floor

h_i [W/m^2. K] Convective heat transfer coefficient of solid surface patch i

h_q [kg/m^2.s (kg/kg)] Mass transfer coefficient

η [-] Ventilation efficiency of the room

η_i [-] Surface ratio of window surface patch i to solid surface patch i

H_f [W/m2] Radiated heat loads from floor per building averaged per m2

H_i [W/m^2] Sensible heat flux at solid surface patch i

$H_{i\ell}$ [W/m2] Latent heat generated indoor per building type then averaged per m2

H_{is} [W/m2] Sensible heat generated indoor per building type then averaged per m2 [W/m2]

H_s [W/m2] Solar heat loads per building then averaged per m2

H_t [W/m2] Transient heat loads through building envelope per building averaged per m2

$H_{v\ell}$ [W/m2] Latent heat loads from ventilation per building averaged per m2

H_{vs} [W/m2] Sensible heat loads from ventilation per building averaged per m2

$I_{diffuse}$ [W/m2] Incident diffuse solar radiation

I_{direct} [W/m2] Incident direct solar radiation

I_i [W/m2] Total direct and diffuse solar radiation at solid surface patch i

k_i [W/m.K] Thermal conductivity

L [J / kg] Latent heat from evaporation

L_i [W/m²] Absorbed long wave length radiation from the atmosphere at solid surface patch i

ρ_a [kg/m³] Density of air

ρ_i [kg / m³] Density of solid surface patch i

q_a [kg/kg] Specific humidity of ambient air

q_i [kg / kg] Specific humidity of solid surface patch i (Mostly ground surface)

q_r [kg/kg] Specific humidity of indoor room.

Q [W/m2] Total average indoor thermal loads per building then averaged per m2

R_{Li} [W/m²] Absorbed long-wave radiation from urban surroundings and total emitted long wave length radiations from solid surface patch i

R_{Si} [W/m²] Absorbed short wave radiation from urban surroundings

S_i [W/m²] Absorbed incident solar radiation from the sun and the sky at solid surface patch i

τ_i [ ] Solar transmittance of i.

$T_{adjacent}$ [K] Temperature of adjacent outdoor air

T_a [K] Temperature of ambient air

T_i [k] Surface temperature at solid surface patch i

T_{fi} [k] Surface temperature of floor

T_r [k] Temperaure of indoor room

T_{ir} [k] Surface temperature of indoor room

V [m³] Volume of room

W_{heat} [W/m2] Total waste heat from AC systems per building then averaged per m2

θ_i [rad] Angle formed by incident direct solar radiation and normal vector of solid surface patch i

σ [W/m².K⁴] Stefan-Boltzmann Constant

1st International Workshop on Self-Learning in Intelligent Environments (SeLIE'21)

Intelligent Environments 2021
E. Bashir and M. Luštrek (Eds.)
© 2021 The authors and IOS Press.
This article is published online with Open Access by IOS Press and distributed under the terms
of the Creative Commons Attribution Non-Commercial License 4.0 (CC BY-NC 4.0).
doi:10.3233/AISE210091

Preface to the Proceedings of the First International Workshop on Self-Learning in Intelligent Environments

Antonio CORONATO [a,1], Giovanna DI MARZO SERUGENDO [b]

[a] Istituto di Calcolo e Reti ad Alte Prestazioni, CNR-ICAR, Italy
[b] University of Geneva, Switzerlanz

Self-learning systems are artificial agents able to acquire and renew knowledge over the time by themselves, without any hard coding. These are adaptive systems whose functions improve by a learning process based -typically- on the method of trial and error. A self-learning system interacts with its users or surrounding environment initially by attempts and observes the changes produced by its actions.

This workshop focuses on the design, implementation and exploitation of self-learning features -either within an Intelligent Environment [2] as a whole, or within some of its components- by means of leading technologies ([8,6,7,5]) and even in critical environments (e.g. healthcare [3,1,4]). The workshop will represent an opportunity for both the academia and industry to debate the state-of-the-art, challenges and open issues.

This first edition of the workshop has accepted for publication and presentation seven papers.

Aslan et al. have proposed an algorithm for learning to move the desired object by humanoid robots. In this algorithm, the semantic segmentation algorithm and Deep Reinforcement Learning (DRL) algorithms are combined. The semantic segmentation algorithm is used to detect and recognize the object be moved. DRL algorithms are used at the walking and grasping steps.

Donnici et al. have presented an intelligent system for supporting patients during their home medical treatment. The system can assist impaired patients in taking medicines in accordance with their treatment plans. The demonstration of the system via mobile app shows promising results and can improve the quality of healthcare at home.

Hayat et al. have proposed a framework that self-learns and automatically classifies any given news headline into its corresponding news category using artificial intelligence methods i.e. text mining and machine learning algorithms.

Ribino and Bonomolo have reported an approach based on reinforcement learning to support the rearrangement of indoor spaces by maximizing the indoor environmental quality index in terms of thermal, acoustic and visual comfort in the new furniture layout scheme.

Shah and Coronato have exploited an IRL method named Max-Margin Algorithm (MMA) to learn the reward function for a robotic navigation problem. The learned reward function reveals the demonstrated policy (expert policy) better than all other poli-

---

[1] Corresponding Author: Antonio Coronato; E-mail: antonio.coronato@icar.cnr.it

cies. Results show that this method has better convergence and learned reward functions through the adopted method represents expert behavior more efficiently.

Shah and De Pietro have surveyed IRL algorithms. The purpose of their paper is to provide an overview and theoretical background of IRL in the field of Machine Learning and Artificial Intelligence.

Aamir et al. have introduced a novel supervised machine learning based approach for breast prediction that embodies Random Forest, Gradient Boosting, Support Vector Machine, Artificial Neural Network and Multilayer Perception methods.

We deeply appreciate the Intelligent Environments 2021 conference organizers for their help on hosting this event.

References

[1] A. Coronato and A. Cuzzocrea. An innovative risk assessment methodology for medical information systems. *IEEE Transactions on Knowledge and Data Engineering*, pages 1–1, 2020.

[2] A. Coronato and G. De Pietro. Tools for the rapid prototyping of provably correct ambient intelligence applications. *IEEE Transactions on Software Engineering*, 38(4):975–991, 2012.

[3] Antonio Coronato, Muddasar Naeem, Giuseppe De Pietro, and Giovanni Paragliola. Reinforcement learning for intelligent healthcare applications: A survey. *Artificial Intelligence in Medicine*, 109:101964, 2020.

[4] Muddasar Naeem, Giovanni Paragliola, and Antonio Coronato. A reinforcement learning and deep learning based intelligent system for the support of impaired patients in home treatment. *Expert Systems with Applications*, page 114285, 2020.

[5] Muddasar Naeem, S Tahir H Rizvi, and Antonio Coronato. A gentle introduction to reinforcement learning and its application in different fields. *IEEE Access*, 2020.

[6] Andrew Y Ng, Stuart J Russell, et al. Algorithms for inverse reinforcement learning. In *Icml*, volume 1, page 2, 2000.

[7] Martin L Puterman. *Markov decision processes: discrete stochastic dynamic programming*. John Wiley & Sons, 2014.

[8] Richard S. Sutton and Andrew G. Barto. *Reinforcement Learning: An Introduction*. The MIT Press, second edition, 2018.

Intelligent Environments 2021
E. Bashir and M. Luštrek (Eds.)
doi:10.3233/AISE210092

Learning to Move an Object by the Humanoid Robots by Using Deep Reinforcement Learning

Simge Nur ASLAN [a,1], Burak TAŞÇI [a,2], Ayşegül UÇAR [a,3], and Cüneyt GÜZELİŞ [b]

[a1,3] Firat University, Department of Mechatronics Engineering, Elazig, Turkey
[a1,2] Firat University, Vocational School of Technical Sciences, Elazig, Turkey
[b] Yaşar University Department of Electrical and Electronics Engineering, Izmir, Turkey

Abstract. This paper proposes an algorithm for learning to move the desired object by humanoid robots. In this algorithm, the semantic segmentation algorithm and Deep Reinforcement Learning (DRL) algorithms are combined. The semantic segmentation algorithm is used to detect and recognize the object be moved. DRL algorithms are used at the walking and grasping steps. Deep Q Network (DQN) is used to walk towards the target object by means of the previously defined actions at the gate manager and the different head positions of the robot. Deep Deterministic Policy Gradient (DDPG) network is used for grasping by means of the continuous actions. The previously defined commands are finally assigned for the robot to stand up, turn left side and move forward together with the object. In the experimental setup, the Robotis-Op3 humanoid robot is used. The obtained results show that the proposed algorithm has successfully worked.

Keywords. Humanoid robots, DQN, DDPG, deep semantic segmentation, object manipulation, locomotion

1. Introduction

The humanoid robots are expected to navigate by themselves in environments such as the house, office, and hospital in a normal daily life and move the desired objects [1-4]. The combination of the locomotion and manipulation skills of the robots is required for the application. Especially, the robots should be capable of object detection and recognition capabilities. Hence, the development of artificial intelligence algorithms for robots to autonomously transport objects is still a challenging research topic.

Deep learning algorithms were used for a lot of applications in smart environments [5-9]. In this paper, we investigate different deep learning algorithms for the walking

---

[1] Simge Nur Aslan, Corresponding author, Department of Mechatronics Engineering, Firat University, 23119, Elazig, Turkey; E-mail: simgeaslan124@gmail.com.

[2] Burak Taşçı, Author, Vocational School of Technical Sciences, Firat University, 23119, Elazig, Turkey; E-mail: btasci@firat.edu.tr.

[3] Ayşegül Uçar, Author, Department of Mechatronics Engineering, Firat University, 23119, Elazig, Turkey; E-mail: agulucar@firat.edu.tr.

4 Cüneyt Güzeliş, Author, Department of Electrical and Electronics Engineering, Yaşar University, Izmir, Turkey; E-mail: cuneyt.guzelis@yasar.edu.tr

task towards the object and the manipulating task it. In this context, the recent attention has focused on Deep Reinforcement Learning (DRL) for solving complex and high dimensional problems [9-18]. In the literature, the locomotion and manipulation tasks were generally taken into consideration in separate scenarios. In [10-15], the DRL algorithms were worked the basic tasks such as pushing, grasping, and pulling were worked on robotic manipulators. In [16], Reinforcement Learning (RL) method was used for reaching task for Baxter humanoid robot arm to a target pose that is controlled by a human. [17] proposed two RL based hierarchies plans to perform the task of manipulating and grasping the objects with uncertain disturbance for a humanoid like mobile robot having the human like upper body includes two robotic arms. In [18-20], some deep learning methods were applied for object detection and manipulation in cluttered environment on the Nao Robot, the REEM-C robot, and Romeo. They did not include DRL. In [21-26], few powerful DRL algorithms were developed for locomotion tasks of both biped and humanoid robots. In our earlier works, we carried out separately the locomotion and grasping tasks for the humanoid robots [27-30]. We applied the demonstration learning by using different deep learning models such as Convolutional Neural Networks (CNNs) and Long Short-Term Memory (LSTM) networks in [27]. In [28], we applied semantic segmentation for grasping tasks. We deployed DRL for different scenarios including walking tasks in [29-30].

In this paper, the locomotion and manipulation tasks are combined by using DRL as different from the works in the literature. The semantic segmentation algorithm is proposed to detect the target object. Pyramid Scene Parsing Network (PSPNet) is used for semantic segmentation [31]. The outputs of the network are imported to the input of the next stage. Two modules for walking and manipulation are planned. Deep Q Network (DQN) is prepared to walk towards the target object by using all combinations of the defined actions using the gait manager and the different head positions of the robot. Deep Deterministic Policy Gradient (DDPG) network is employed for manipulating the object [32-35]. The previously defined commands are finally assigned for the robot to stand up, turn left side and to move forward together with the grasped object. In the experimental setup, the Robotis-Op3 humanoid robot [36] is used. The obtained results show that the proposed algorithm successfully worked.

The rest of this paper is organized as follows. In section 2, the fundamentals of CNNs, PSPNet, DQN, and DDPG are introduced. Section 3 gives the details of the proposed vision based algorithm. The experimental setup and results are presented, respectively in Section 4. The conclusions and future direction are presented in Section 5.

2. Theoretical background

This section briefly reviews the fundamental concepts of CNNs, PSPNet, DQN, and DDPG.

2.1. Convolutional neural networks

CNNs are a kind of artificial neural networks with multiple layers [37-38]. The inputs of the network are the images and/or the signals. They are processed end-to-end. Classification, detection, and recognition tasks are carried out without using a feature extraction method. A CNN structure consists of the combinations of the input layer, the convolutional layers, the Batch Normalization layers (BN), the pooling layers, and the fully-connected layers [39]. The input layer accepts two-dimensional signals or three-dimensional images. If we consider using the images for our aim, the images are then filtered in the convolutional layer by sliding the filter matrices. In the pooling layer, the dimensions of image matrices are reduced by using the functions average, max poling, and min. In other ways, the layer is a feature selection layer. In BN layer, the layer inputs are normalized to provide zero mean and unit variance. Rectified Linear Unit Activation Function (ReLU) activation function is usually used after BN layer [40]. ReLU provides zero output for negative input, and itself input for the others. In a fully-connected layer, each unit is exactly connected to the next layer similar to feed forward neural networks. The data in a vector form are obtained. The final layer of CNN includes a softmax or a linear/nonlinear activation function. The softmax activation function calculates the class probabilities relating to each output [38-39].

2.2. Pyramid scene parsing network

PSPNet is a network used for semantic segmentation. The network consists of the encoder and decoder parts. The structure is shown in Fig.1 [31]. Fig. 1 consists of the encoder part and the decoder part. The encoder part can be generated by a vanilla CNN constructed with (BN), ReLU activation, max pooling, and zero padding or famous CNN models such as ResNet, VGG16, MobileNet, and FCNs [41-44]. The decoder part is a single standard structure. The decoder part is depicted by the blocks called Pool, Conv, Upsample, Concat, and Conv. Pool, Conv, and Upsample blocks are generated from four layers of average pooling, convolution, BN, ReLU activation, and resize. The feature map dimensions are [64, 128, 256, 256, 256] and the polling outputs are [1x1, 2x2, 3x3, 6x6]. The Concat block means the concentration as a single tensor of the inputs at the final. The final Conv block includes multiple layers such as the convolution, BN, ReLU, convolution, resize, reshape, and Softmax activation layers, respectively.

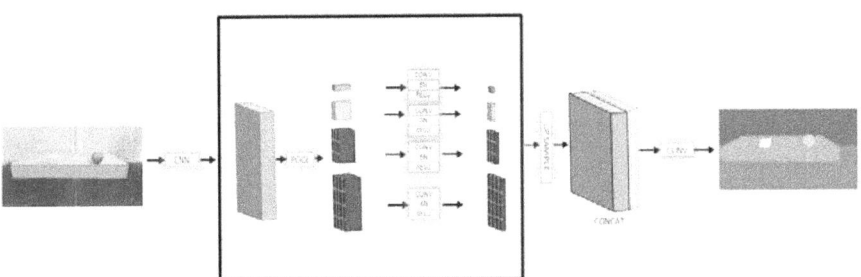

Figure 1. PSPNet architecture [31].

2.3. Deep Q network

There are three learning methods classified as supervised, unsupervised, and RL. RL methods use a reward and penalty values. The correct action provides a reward and the wrong action provides a penalty. The agent interacts with the environment, receives the values, and then applies the appropriate action [32-33]. Thus, the best policy is provided. AlphaGo game is the best known example of RL. The computer algorithm being powered with the RL has won the game against the best Go player [34].

Fig. 2 shows a kind of DRL algorithm called DQN. In Fig. 2, the agent is a CNN. It decides to an action $a_t \in A$ from action space with respect to the state S_t by interacting with the environment at t∈[0, T] and obtains a reward or penalty R_t with respect to the selected action and pass to a new state s_{t+1}. It generates the policy $a_t = \pi(s_t)$.

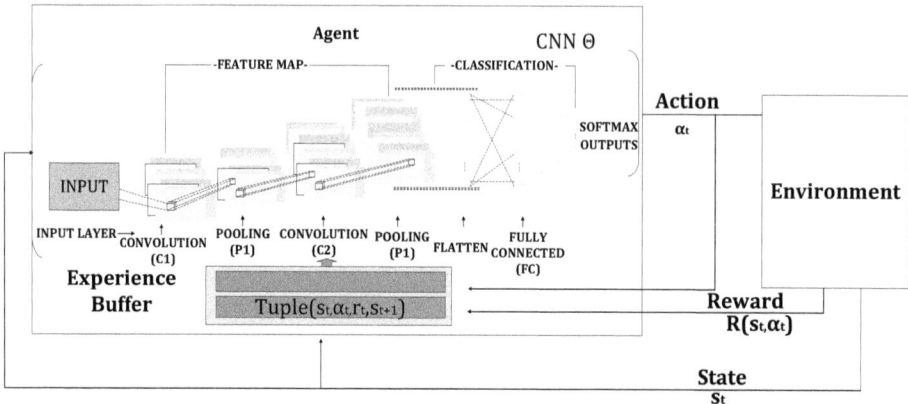

Figure 2. Structure of deep Q network.

In DQN, the cumulative reward is maximized as in

$$R_t = \sum_{i=t}^{T} \gamma^{i-t} r_i \tag{1}$$

where γ is a discount factor in the range of [0,1]. Being given a policy, the optimal Q-value function is obtained by using Belmann equality

$$Q^\pi(s_t, a_t) = \mathrm{E}[R_t | s_t, a_t, \pi], \tag{2}$$

$$Q^\pi(s_t, a_t) = \mathrm{E}[r_t + \gamma \mathrm{E}[Q^\pi(s_{t+1}, a_{t+1}) | s_t, a_t, \pi]], \tag{3}$$

$$Q^*(s_t, a_t) = \mathrm{E}_{s_{t+1}}[r + \gamma max_{a_{t+1}} Q^*(s_{t+1}, a_{t+1}) | s_t, a_t]. \tag{4}$$

Q-learning is realized by:

$$Q(s_t, a_t) \leftarrow Q(s_t, a_t) + \alpha_t[r_{t+1} + \gamma max_{a_{t+1}} Q(s_{t+1}, a_{t+1}) - Q(s_t, a_t)]. \tag{5}$$

DDPG in Fig.3 integrates both the actor-critic and DQN methods to learn the policies in the continuous domain. It provides the continuous actions. In DDPG, $Q(s, a)$, Q-value network represents a critic function estimating the value of state-action pairs. The actor function called the policy network is updated by

$$\nabla_{\theta^\mu} \approx \mathrm{E}_{s_t \sim \rho_\pi} \left[\nabla_a Q(s, a | \theta^Q) |_{s=s_t, a=\mu(s_t)} \nabla_{\theta_\mu} \mu(s | \theta^\mu) |_{s_t} \right] \tag{6}$$

where ρ_π means the transitions obtained from the π stochastic behavior policy. The policy is generally a Gaussian distribution with the center of $\mu(s | \theta^\mu)$ depending on the parameter θ.

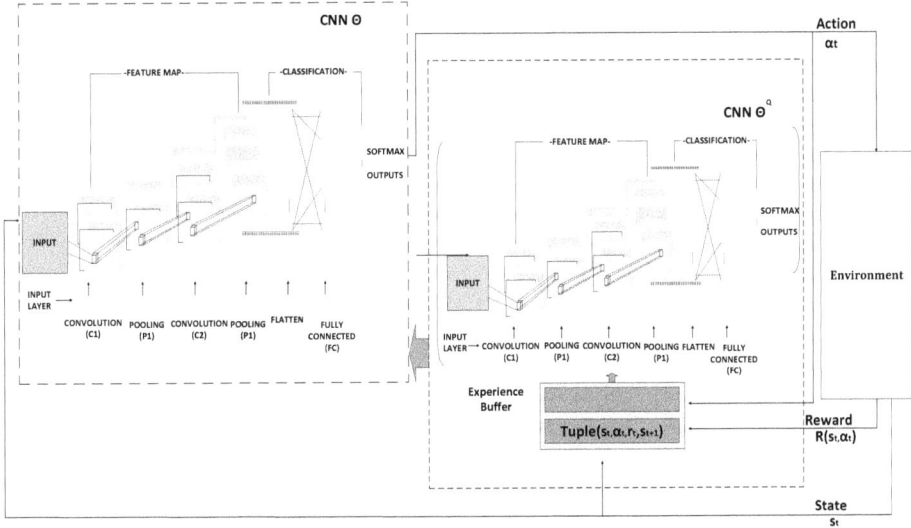

Figure 3. The structure of deep deterministic policy gradient network.

3. Vision-based walking and object manipulation

In this paper, a vision based experimental procedure is designed to walk to the target and to manipulate the object. The flowchart in Fig. 4 depicts the experimental plan for the pilot scenario. In this plan, Robotis-Op3 is controlled using a keyboard and the images are collected by using its camera as moving the robot. Three deep learning algorithms are then executed. One of them is the deep semantic segmentation algorithm and the others are DRL algorithms. PSPNet semantic model is applied to segment the objects in the environment and find the target object. To use at the training of the PSPNet model, all images are manually labeled in Matlab. DQN is applied for the robot to walk to the target. DDPG is applied for the robot to manipulate the target object. Finally, the networks are transferred to the real robot by means of Robot Operating System (ROS) and the pre-defined actions are applied to the robot in order.

4. Experimental results

In the study, we used the Robotis-Op3 humanoid robot and experimental environment consisting of the room of (2m x 4m) in Fig. 5 [36]. We aimed that the robot walks to the target object and to manipulate it. Robotis-Op3 consists of 20 axes and has a Logitech C920 HD Pro camera, 9 degrees of freedom inertial measurement unit, Intel NUC i3, Linux operating system, Dynamixel SDK, and ROS. In the experiments, we worked on a workstation having Nvidia Titan XP.

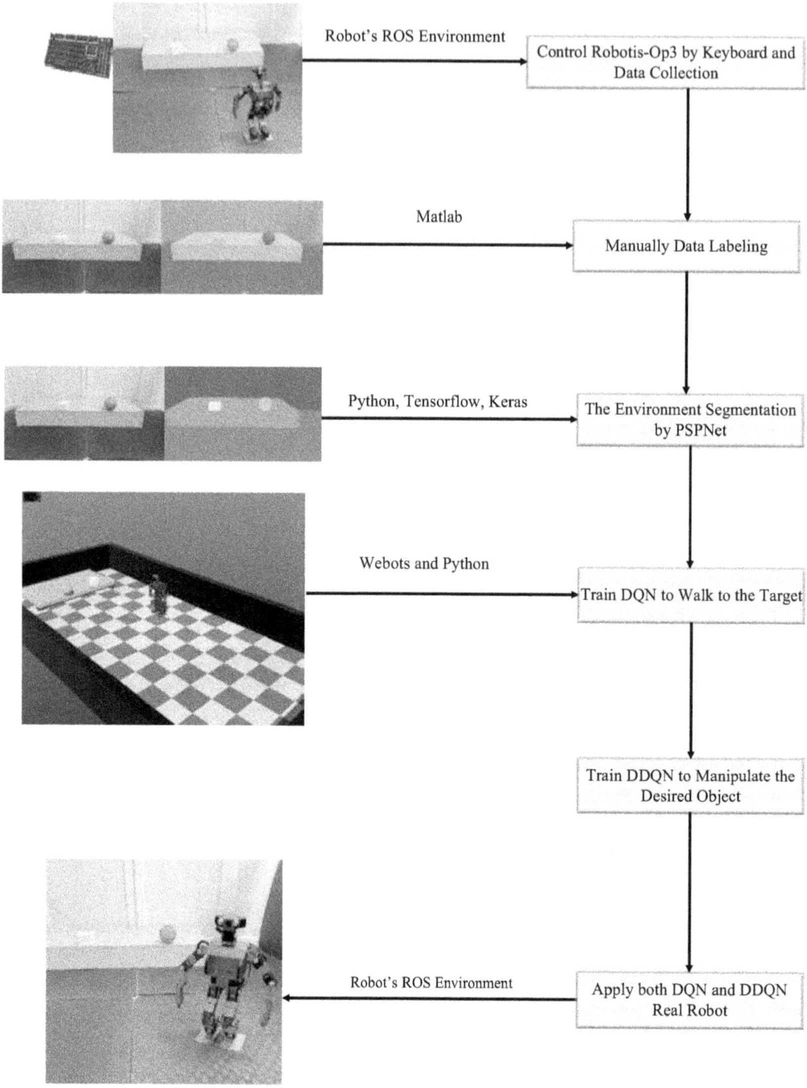

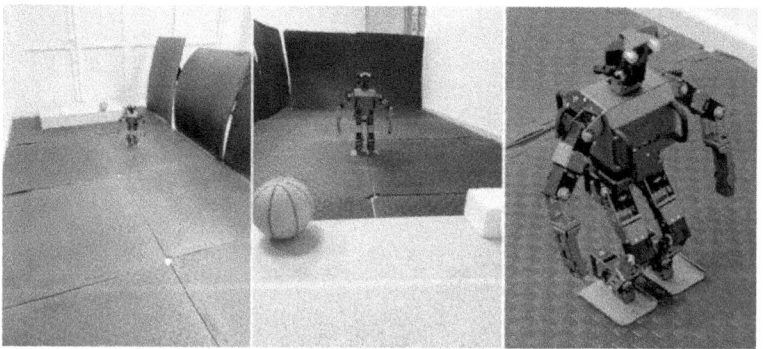

Robot's ROS Environment → Control Robotis-Op3 by Keyboard and Data Collection

Matlab → Manually Data Labeling

Python, Tensorflow, Keras → The Environment Segmentation by PSPNet

Webots and Python → Train DQN to Walk to the Target

Train DDQN to Manipulate the Desired Object

Robot's ROS Environment ← Apply both DQN and DDQN Real Robot

Figure 4. The flowchart of the proposed system.

Figure 5. Robotis-Op3 humanoid robot and the experimental environment.

Figure 6. Some images used in the training data set.

To walk towards the desired object, we used the gait pattern generator that is available in Robotis-Op3 source code [36]. The pattern generator deploys three coupled oscillators, for the right foot, the left foot and the center of mass with sinusoidal trajectory in a synchronized way with respect to the movement. Moreover, it has closed loop control to balance using the gyro data to get rid of falling [45]. In our experiment, firstly, we controlled the robot by a keyboard. We captured a total of 3211 images consisting of 2269 training images and 942 validation images from the real Robotis-Op3 camera by using ROS to walk to target.

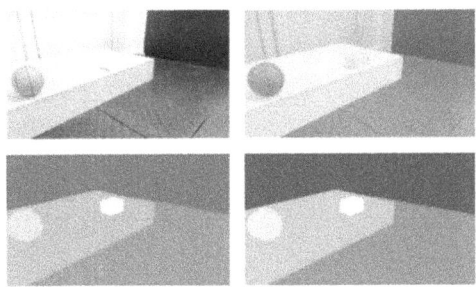

Figure 7. The RGB image used in the training data set and its labelled version (top), the segmented image and its depth version (bottom).

Fig. 6 shows the sample images in the training data set. We labeled five objects consisting of the ball, cube, platform, floor, and wall. Each object painted with a different color. Fig. 7 shows the captured real RGB image, the labeled images, the segmented images, and the depth images. We used completely real images for the training. We tried different models such as ResNet50-PSPNet, VGG16-UNet, VGG16-FCN32, VGG16-PSPNet, and VGG16-Segnet [41-44, 46-47]. We trained all models by the Adedelta optimization algorithm for 40 epochs, 2 batch size, and 100 steps [48]. We evaluated the network by the accuracy, the Dice coefficient, and the Intersection Over Union (IOU) coefficient to measure the segmentation performance [49-51]. The values near one of the accuracy, Dice, and IOU exhibit perfect object boundaries [49-50]. We used ResNet50-PSPNet in our scenario since it achieved better performance than the other networks. The results of ResNet50-PSPNet relating to the accuracy, Dice, and IOU are shown in Fig. 8a-c for each epoch, respectively. As can be seen from Fig. 8 and Table 1, the value of {0.9983, 0.9988, 0.9945} were obtained for accuracy, Dice, and IOU in the training stage, respectively by ResNet50-PSPNet. In the validation stage, we obtained the values of

{0.9971, 0.9978, 0.9928}. The results reveal that the ResNet50-PSPNet can successfully segment the experimental environment.

Table 1. Segmentation performance of state-of-art algorithms.

Model	Train			Validation		
Encoder-Decoder	Accuracy	Dice	IOU	Accuracy	Dice	IOU
ResNet50-PSPNet	0.9983	0.9988	0.9945	0.9971	0.9978	0.9928
VGG16-PSPNet	0.9719	0.9730	0.9709	0.9663	0.9676	0.9705
VGG16-UNet	0.9844	0.9865	0.9740	0.9713	0.9726	0.9457
VGG16-FCN32	0.9834	0.9832	0.9810	0.9738	0.9663	0.9386
VGG16-Segnet	0.9811	0.9856	0.9534	0.9768	0.9767	0.9402

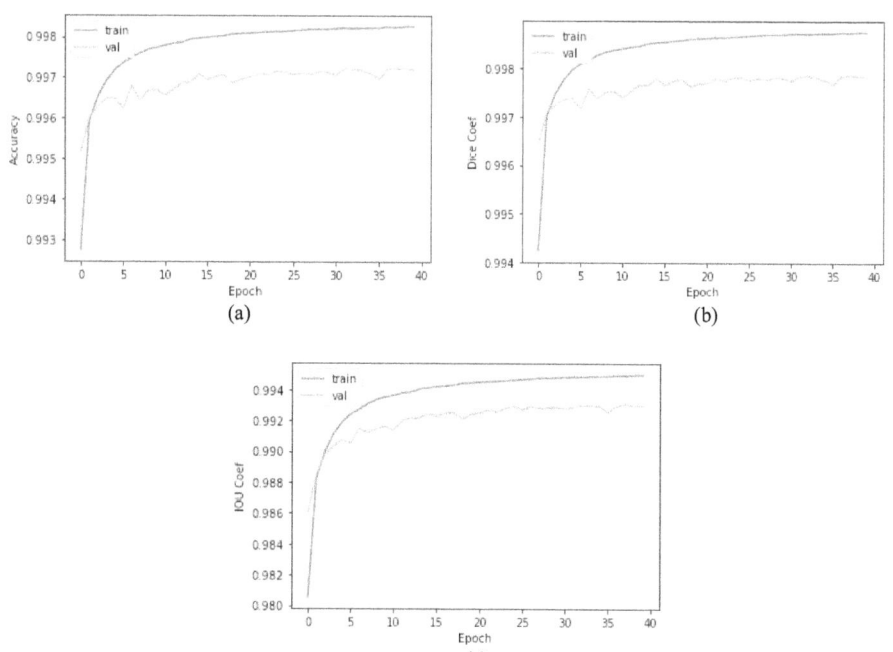

Figure 8. The training and validation results for the accuracy, Dice, and IOU.

Secondly, we constructed our experimental environment in the Webots simulation platform for the simulation works [52]. We trained our robot to learn to walk towards the target object by using DQN in Webots. We defined the combinations of discrete actions consisting of forward, back, left, and right of the movement and right, left, straight, down on the head and applied DQN structure in [29]. We used both the depth image and segmented image as the DQN inputs. Depth image corrupted with random noise and blur was used for achieving the robustness to the differences of illumination and appearance between the real and simulation environments. The segmented image was used to determine the target. We centered the object on the image taken from the robot camera in respect to the neck and head positions [36,52-53]. We calculated the distance and the orientation to the target using the head tilt angle and the neck pan angle, respectively [36,52-53]. We selected the eight actions such as forward, backward, turn left, turn right, and stop. Table 2 shows the selected discrete actions in respect to the robot velocities and orientation. We determined the reward with respect to the distance to the target. We randomly selected the label of the target object in the environment by using the segmented image at the beginning of each episode. We fixed the distance between the objects since the robot is able easily to grasp one object, but we randomly changed the initial locations of both objects on the table at the beginning of each episode. We randomly re-initialized the location of the robot at each episode. When the robot fell down and the robot walked more than 8 minutes in the room of 2mx4m, we terminated and reset, respectively. The obtained mean reward from the used DQN at the walking task towards the desired object is given in Fig. 9. The results reveal that our agents can successfully learn to walk to the target object no matter what the size, shape, and location of it are.

Table 2. The actions to robot velocities and orientation.

Action	Linear (cm/s), Angle (degree)
Forward	[6, ±-15] and [1-3, ±[0-15]]
Backward	[1,7],[1,-15] [8,-7.5]
Noop	-

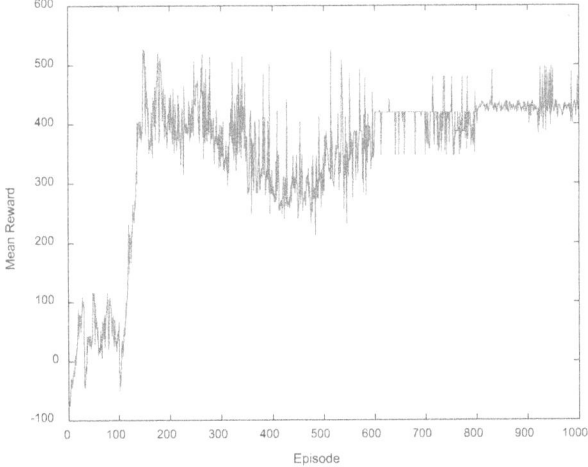

Figure 9. Mean reward with respect to training episodes for DQN.

To test the performance of DQN, we measured the success rate of the reaching to the target when moving the robot from four different initial positions and the rotation. We determined the middle point of the environment as (0,0) point. Each position was evaluated 20 times, resulting in the success percentages reported in Table 3. The results show that our agent is successful in both the real-world and simulation scenarios.

Table 3. Comparison of and simulation and real world experiments to walk to the target.

Initial Value (x,y, angle)	(0,0,35°)	(0,0,0°)	(0.6,0.6,35°)	(-0.5,-0.5,-90°)
Real-World	86.24 %	98.12 %	83.87 %	69.45 %
Simulation	92.03 %	100 %	93.45 %	88 .67%

Thirdly, in order to manipulate object, we then put the robot to the sitting position and run DDPG for grasping as an autonomous step in the algorithm. We controlled the position of right shoulder and left shoulder, right arm upper and left arm upper, right arm lower and left arm lower by using the continuous actions thanks to DDPG. In DDPG having 4 layered actor and critic networks, we selected critic learning rate, actor learning rate, discount factor, critic l2 regularization as 1e-3, 1e-4, 0.9, 1e-2, respectively. We trained the robot in Webots. We received exactly the simulated RGB images as the inputs. We limited the robot arm's workspace and started a new episode and penalized the robot when it digresses off limits. We evaluated the relative location of the object and robot arm in the reward function. We determined the positive and negative reward values as the object moves up and down, respectively. The obtained mean reward is given in Fig. 10. As can be seen from Fig. 10, the models learn perfectly all steps to manipulate the object after 300 episodes.

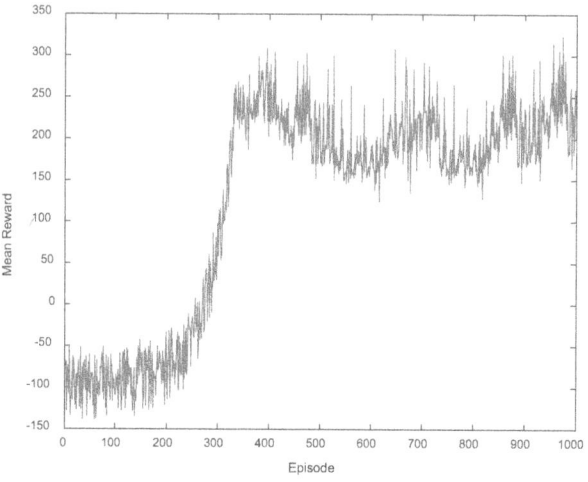

Figure 10. Mean reward with respect to training episodes for DDPG.

We applied all steps of the algorithm and embedded the DQN and DDPG networks into the robot in the ROS environment. To test the performance of DDPG, we measured the success rate of the object manipulation. We repeated 20 times the manipulation task for the object in both real world and simulation environments. As can be seen from Table 4, our agent provides the good results for both real world and simulation. Failure of manipulation trials are due to the robot hand with clamp shape. It is generally not seen as the failure of the agent. The robot catches the object, but it could not hold the object without dropping it due to the hand's clamp shape.

Table 4. Comparison of and simulation and real world experiments to manipulate the object

Model	Real-World	Simulation
DDPG	67 %	79 %

After carrying out the manipulation task, we finally applied the commands of standing up, turning left side, and straight walking, respectively to move the object.

5. Conclusions

In this paper, we presented new algorithm steps for the humanoid robots to learn to move an object by using DRL. The algorithm included three tasks, as object detection, locomotion, and manipulation. The first task was carried out by a deep semantic segmentation network, the second and third tasks deployed DQN and DDPG respectively.

All models were trained in the simulation environment and then the obtained networks were embedded into the robot. All algorithm steps were successfully performed on the real robot. The algorithm can be applied to walk towards one from previously labelled objects in a cluttered environment and to grasp it without re-training the networks. Moreover, the algorithm can be extended for the obstacle avoidance task with the appropriate reward and actions. Future studies will go ahead to work in the clutter environments with the same algorithm.

Acknowledgment

This work was supported by the Scientific and Technological Research Council of Turkey (TUBITAK) grant numbers 117E589. In addition, GTX Titan X Pascal GPU in this research was donated by the NVIDIA Corporation

References

[1] M. Grey, S. Joo, and M. Zucker, Planning heavy lifts for humanoid robots, in: 2014 IEEE-RAS International Conference on Humanoid Robots, IEEE, 2014: pp. 640–645.

[2] S.A.A. Moosavian, A. Janati, and M.H. Ghazikhani, Object manipulation by two humanoid robots using MTJ control, in: 2011 IEEE International Conference on Mechatronics and Automation, IEEE, 2011: pp. 1286–1290.

[3] M.-H. Wu, A. Konno, S. Ogawa, and S. Komizunai, Symmetry cooperative object transportation by multiple humanoid robots, in: 2014 IEEE International Conference on Robotics and Automation (ICRA), IEEE, 2014: pp. 3446–3451.

[4] A. Rioux, and W. Suleiman, Autonomous SLAM based humanoid navigation in a cluttered environment while transporting a heavy load, *Robotics and Autonomous Systems*. **99** (2018) 50–62.

[5] S.I. Muzaffar, K. Shahzad, K. Malik, and K. Mahmood, Intention mining: A deep learning-based approach for smart devices, *Journal of Ambient Intelligence and Smart Environments*. (2020) 1–13.

[6] M.A. Guillén, A. Llanes, B. Imbernón, R. Martínez-España, A. Bueno-Crespo, J.-C. Cano, and J.M. Cecilia, Performance evaluation of edge-computing platforms for the prediction of low temperatures in agriculture using deep learning, *The Journal of Supercomputing*. **77** (2021) 818–840.

[7] J. Kar, M.V. Cohen, S.P. McQuiston, and C.M. Malozzi, A deep-learning semantic segmentation approach to fully automated MRI-based left-ventricular deformation analysis in cardiotoxicity, *Magnetic Resonance Imaging*. **78** (2021) 127–139.

[8] Z.-W. Hong, C. Yu-Ming, S.-Y. Su, T.-Y. Shann, Y.-H. Chang, H.-K. Yang, B.H.-L. Ho, C.-C. Tu, Y.-C. Chang, and T.-C. Hsiao, Virtual-to-real: Learning to control in visual semantic segmentation, *ArXiv Preprint ArXiv:1802.00285*. (2018).

[9] S. Kumra, S. Josh, and F. Sahin, Learning Robotic Manipulation Tasks through Visual Planning, *ArXiv Preprint ArXiv:2103.01434*. (2021).

[10] S. Gu, E. Holly, T. Lillicrap, and S. Levine, Deep reinforcement learning for robotic manipulation with asynchronous off-policy updates, in: 2017 IEEE International Conference on Robotics and Automation (ICRA), IEEE, 2017: pp. 3389–3396.

[11] T. Haarnoja, V. Pong, A. Zhou, M. Dalal, P. Abbeel, and S. Levine, Composable deep reinforcement learning for robotic manipulation, in: 2018 IEEE International Conference on Robotics and Automation (ICRA), IEEE, 2018: pp. 6244–6251.

[12] A. Hundt, B. Killeen, N. Greene, H. Wu, H. Kwon, C. Paxton, and G.D. Hager, "Good Robot!": Efficient Reinforcement Learning for Multi-Step Visual Tasks with Sim to Real Transfer, *IEEE Robotics and Automation Letters*. **5** (2020) 6724–6731.

[13] A. Rajeswaran, V. Kumar, A. Gupta, G. Vezzani, J. Schulman, E. Todorov, and S. Levine, Learning complex dexterous manipulation with deep reinforcement learning and demonstrations, *ArXiv Preprint ArXiv:1709.10087*. (2017).

[14] S. Joshi, S. Kumra, and F. Sahin, Robotic grasping using deep reinforcement learning, in: 2020 IEEE 16th International Conference on Automation Science and Engineering (CASE), IEEE, 2020: pp. 1461–1466.

[15] P. Florence, L. Manuelli, and R. Tedrake, Self-supervised correspondence in visuomotor policy learning, *IEEE Robotics and Automation Letters*. **5** (2019) 492–499.

[16] R. Silva, M. Faria, F.S. Melo, and M. Veloso, Adaptive indirect control through communication in collaborative human-robot interaction, in: 2017 IEEE/RSJ International Conference on Intelligent Robots and Systems (IROS), IEEE, 2017: pp. 3617–3622.

[17] Z. Li, T. Zhao, F. Chen, Y. Hu, C.-Y. Su, and T. Fukuda, Reinforcement learning of manipulation and grasping using dynamical movement primitives for a humanoidlike mobile manipulator, *IEEE/ASME Transactions on Mechatronics*. **23** (2017) 121–131.

[18] A. Hornung, S. Böttcher, J. Schlagenhauf, C. Dornhege, A. Hertle, and M. Bennewitz, Mobile manipulation in cluttered environments with humanoids: Integrated perception, task planning, and action execution, in: 2014 IEEE-RAS International Conference on Humanoid Robots, IEEE, 2014: pp. 773–778.

[19] P. Regier, A. Milioto, C. Stachniss, and M. Bennewitz, Classifying obstacles and exploiting class information for humanoid navigation through cluttered environments, *International Journal of Humanoid Robotics*. **17** (2020) 2050013.

[20] G. Claudio, F. Spindler, and F. Chaumette, Vision-based manipulation with the humanoid robot Romeo, in: 2016 IEEE-RAS 16th International Conference on Humanoid Robots (Humanoids), IEEE, 2016: pp. 286–293.

[21] D.R. Song, C. Yang, C. McGreavy, and Z. Li, Recurrent deterministic policy gradient method for bipedal locomotion on rough terrain challenge, in: 2018 15th International Conference on Control, Automation, Robotics and Vision (ICARCV), IEEE, 2018: pp. 311–318.

[22] X.B. Peng, G. Berseth, and M. Van de Panne, Terrain-adaptive locomotion skills using deep reinforcement learning, *ACM Transactions on Graphics (TOG)*. **35** (2016) 1–12.

[23] X.B. Peng, G. Berseth, K. Yin, and M. Van De Panne, Deeploco: Dynamic locomotion skills using hierarchical deep reinforcement learning, *ACM Transactions on Graphics (TOG)*. **36** (2017) 1–13.

[24] K. Lobos-Tsunekawa, F. Leiva, and J. Ruiz-del-Solar, Visual navigation for biped humanoid robots using deep reinforcement learning, *IEEE Robotics and Automation Letters*. **3** (2018) 3247–3254.

[25] S. Khatibi, M. Teimouri, and M. Rezaei, Real-time Active Vision for a Humanoid Soccer Robot Using Deep Reinforcement Learning, *ArXiv Preprint ArXiv:2011.13851*. (2020).

[26] Q. Shi, W. Ying, L. Lv, and J. Xie, Deep reinforcement learning-based attitude motion control for humanoid robots with stability constraints, *Industrial Robot: The International Journal of Robotics Research and Application*. (2020).

[27] S.N. Aslan, R. Ozalp, A. Uçar, and C. Güzeliş, End-To-End Learning from Demonstation for Object Manipulation of Robotis-Op3 Humanoid Robot, in: 2020 International Conference on INnovations in Intelligent SysTems and Applications (INISTA), IEEE, 2020: pp. 1–6.

[28] S.N. Aslan, A. Uçar, and C. Güzeliş, Semantic Segmentation for Object Detection and Grasping with Humanoid Robots, in: 2020 Innovations in Intelligent Systems and Applications Conference (ASYU), IEEE, 2020: pp. 1–6.

[29] R. Özalp, C. Kaymak, Ö. Yildirim, A. Ucar, Y. Demir, and C. Güzeliş, An implementation of vision based deep reinforcement learning for humanoid robot locomotion, in: 2019 IEEE International Symposium on INnovations in Intelligent SysTems and Applications (INISTA), IEEE, 2019: pp. 1–5.

[30] S.N. Aslan, A. Uçar, and C. Güzeliş, Development of Deep Learning Algorithm for Humanoid Robots to Walk to the Target Using Semantic Segmentation and Deep Q Network, in: 2020 Innovations in Intelligent Systems and Applications Conference (ASYU), IEEE, 2020: pp. 1–6.

[31] H. Zhao, J. Shi, X. Qi, X. Wang, and J. Jia, Pyramid scene parsing network, in: Proceedings of the IEEE Conference on Computer Vision and Pattern Recognition, 2017: pp. 2881–2890.

[32] R.S. Sutton, and A.G. Barto, Reinforcement learning: An introduction, MIT press, 2018.

[33] V. Mnih, K. Kavukcuoglu, D. Silver, A.A. Rusu, J. Veness, M.G. Bellemare, A. Graves, M. Riedmiller, A.K. Fidjeland, and G. Ostrovski, Human-level control through deep reinforcement learning, *Nature.* **518** (2015) 529–533.

[34] D. Silver, A. Huang, C.J. Maddison, A. Guez, L. Sifre, G. Van Den Driessche, J. Schrittwieser, I. Antonoglou, V. Panneershelvam, and M. Lanctot, Mastering the game of Go with deep neural networks and tree search, *Nature.* **529** (2016) 484–489.

[35] T.P. Lillicrap, J.J. Hunt, A. Pritzel, N. Heess, T. Erez, Y. Tassa, D. Silver, and D. Wierstra, Continuous control with deep reinforcement learning, *ArXiv Preprint ArXiv:1509.02971.* (2015).

[36] Robotis-Op3, http://emanual.robotis.com/docs/en/platform/op3/introduction/ Accessed 5 October 2020.

[37] A. Uçar, Y. Demir, and C. Güzeliş, Object recognition and detection with deep learning for autonomous driving applications, *Simulation.* **93** (2017) 759–769.

[38] Y. LeCun, Y. Bengio, and G. Hinton, Deep learning. nature 521 (7553), 436-444, *Google Scholar Google Scholar Cross Ref Cross Ref.* (2015).

[39] S. Ioffe, and C. Szegedy, Batch normalization: Accelerating deep network training by reducing internal covariate shift, in: International Conference on Machine Learning, PMLR, 2015: pp. 448–456.

[40] V. Nair, and G.E. Hinton, Rectified linear units improve restricted boltzmann machines, in: Icml, 2010.

[41] K. He, X. Zhang, S. Ren, and J. Sun, Deep residual learning for image recognition, in: Proceedings of the IEEE Conference on Computer Vision and Pattern Recognition, 2016: pp. 770–778.

[42] K. Simonyan, and A. Zisserman, Very deep convolutional networks for large-scale image recognition, *ArXiv Preprint ArXiv:1409.1556.* (2014).

[43] J. Long, E. Shelhamer, and T. Darrell, Fully convolutional networks for semantic segmentation, in: Proceedings of the IEEE Conference on Computer Vision and Pattern Recognition, 2015: pp. 3431–3440.

[44] A.G. Howard, M. Zhu, B. Chen, D. Kalenichenko, W. Wang, T. Weyand, M. Andreetto, and H. Adam, Mobilenets: Efficient convolutional neural networks for mobile vision applications, *ArXiv Preprint ArXiv:1704.04861.* (2017).

[45] I. Ha, Y. Tamura, and H. Asama, Gait pattern generation and stabilization for humanoid robot based on coupled oscillators, in: 2011 IEEE/RSJ International Conference on Intelligent Robots and Systems, IEEE, 2011: pp. 3207–3212.

[46] O. Ronneberger, P. Fischer, and T. Brox, U-net: Convolutional networks for biomedical image segmentation, in: International Conference on Medical Image Computing and Computer-Assisted Intervention, Springer, 2015: pp. 234–241.

[47] V. Badrinarayanan, A. Kendall, and R. Cipolla, Segnet: A deep convolutional encoder-decoder architecture for image segmentation, *IEEE Transactions on Pattern Analysis and Machine Intelligence.* **39** (2017) 2481–2495.

[48] M.D. Zeiler, Adadelta: an adaptive learning rate method, *ArXiv Preprint ArXiv:1212.5701.* (2012).

[49] A. Popovic, M. De la Fuente, M. Engelhardt, and K. Radermacher, Statistical validation metric for accuracy assessment in medical image segmentation, *International Journal of Computer Assisted Radiology and Surgery.* **2** (2007) 169–181.

[50] L.R. Dice, Measures of the amount of ecologic association between species, *Ecology.* **26** (1945) 297–302.

[51] S.V. des S. Naturelles, Bulletin de la Société vaudoise des sciences naturelles, F. Rouge, 1864.

[52] O. Michel, Cyberbotics Ltd. Webots™: professional mobile robot simulation, *International Journal of Advanced Robotic Systems.* **1** (2004) 5.

[53] W. Budiharto, B. Kanigoro, and V. Noviantri, Ball Distance Estimation and Tracking System of Humanoid Soccer Robot, in: Information and Communication Technology-EurAsia Conference, Springer, 2014: pp. 170–178.

Intelligent Environments 2021
E. Bashir and M. Luštrek (Eds.)
© *2021 The authors and IOS Press.*
This article is published online with Open Access by IOS Press and distributed under the terms
of the Creative Commons Attribution Non-Commercial License 4.0 (CC BY-NC 4.0).
doi:10.3233/AISE210093

A Self-Learning Autonomous and Intelligent System for the Reduction of Medication Errors in Home Treatments

Rosamaria DONNICI [a], Antonio CORONATO [a] and Muddasar NAEEM [a,1]

[a] *ICAR-CNR Napoli, Italy*

Abstract. The treatment process at home after hospitalization may become challenging for elders and people having any physical or cognitive disability. Such patients can, nowadays, be supported by Autonomous and Intelligent Monitoring Systems (AIMSs) that may get new levels of functionalities thanks to technologies like Reinforcement Learning, Deep Learning and Internet of Things.

We present an AIMS that can assist impaired patients in taking medicines in accordance with their treatment plans. The demonstration of the AIMS via mobile app shows promising results and can improve the quality of healthcare at home.

Keywords. Artificial intelligence, Reinforcement Learning, Autonomous Systems, Deep Learning, Internet of Things, Healthcare.

1. Introduction

Clinical factors and daily activities of impaired persons if left unnoticed, can cause issues and hazardous situations for the safety of the patient. AIMS based on Artificial Intelligence (AI) (e.g. Reinforcement Learning (RL) and Deep Learning (DL)) and Internet of Things (IoT) is a novel solution for continuous and autonomous monitoring.

Devices used for recording, communicating, sensing and displaying can play an important role in monitoring the routine activities and health parameters.

The interconnection of IoT devices has allowed the deployment of large-scale applications such as, for example, smart cities [27], [47], large-scale smart networks and radios [1] and smart campus systems [7], [1]. Many applications of IoT devices and sensors, improved by self-learning technologies such as RL, have been proposed in healthcare [18].

Medication errors (i.e., "any preventable event that may cause or lead to inappropriate medication use or patient harm while the medication is in the control of the health care professional, patient, or consumer" [45]) have been estimated by the United States Institute of Medicine to harm 1.5 million people and kill 7,000 patients annually in the USA alone. The situation is not very different in Europe as well [20].

The costs due to medication errors have been estimated at US $42 billion annually or almost 1% of total global health expenditure [46]. In Europe, according to the Euro-

---

[1] Corresponding Author: Muddasar Naeem ICAR-CNR
Email: muddasar.naeem@icar.cnr.it

pean Medicines Agency, the annual cost of medication errors is estimated between €4.5 billion and €21.8 billion [20].

Medication errors may be located either in hospital settings or in home care settings [6]. The World Health Organisation (WHO) classified different medication errors such as wrong dosage, wrong frequency, wrong medication or omission [45] and emphasised that "the elderly population may also encounter special issues related to medication errors".

This paper presents a set of Intelligent Services specifically devised to support fragile patients in their domestic healthcare treatments. The main functions are offered through a Mobile App that interacts with the patient with the aim of i) self-assessing the patient's skills in terms of audio, visual and cognitive capabilities to customize the way to communicate with the patient; ii) reminding promptly the patient to take his medication accordingly with his treatment plan; iii) checking the correctness of the medication that patient is going to take by identifying the medication box while handled by the patient.

The remaining part of the paper is structured as follows. Next section presents an overview of the related work. Section 3 introduces the fundamentals of RL, DL and IoT. Next, Section 4 presents our system model. Section 5 shows some results. Finally, section 6 reports conclusion and future work.

2. Related Work

Anomaly detection is being performed on health records in the project MavHome to investigate to drifts and outliers in smart homes [23]. Some other projects related to smart home have been started in order to provide elderly people and patients more comfort at home with more safety. Among the most relevant projects there are those under development by Microsoft Research [10], Intel Research [12], University of Colorado [13] and the Georgia Tech [31].

Relevant Health Monitoring System (HMS) taht exploits IoT and AI are proposed in [3] and [35].

MIT House and the Consortium and Technology processing company (TIAX, LLC) have developed the place lab [24], where they are researching techniques to validate performance of the daily life activities and biometric monitoring. The place lab is using rich sensing infrastructures to develop methods to recognize patterns of recreation, socializing, eating, sleeping, and etc. In particular, for elder people, variations in basic Activities of Daily Life (ADL) are considered early critical indicators of possible health issues. In addition, their work on the recognition of ADL using ubiquitous sensors in a domestic scenario has potentialities for monitoring the general health status of a person, but could be improved by a more widespread exploitation of AI technologies.

An overview of the literature on the use of IoT in HMS is presented in [2] and recent developments and trends of HMS in terms of security issues, wireless communications, frameworks and health parameters are discussed with advantages and limitations.

Ambient assisted living techniques may improve the quality of life of elderly with cognitive and physical disabilities, by solving assistive tasks such as medication management [19]. Social humanoid robot [9] can help to monitor indoor environmental quality [37].

A project on the "Improvement of the Elderly Quality of Life and Care through Smart Emotion Regulation" is proposed in [22] to investigate solutions for the improvement in the care and quality of life of elderly people through the use of sensors, cameras and emotion regulation methods. A distributed fuzzy system able to infer in real-time critical situations by analysing data gathered from user's smart-phones about the environment and the individual is presented in [38].

A data analytic technique, which exploits ML and smartphone's inertial sensors to recognize human activities, is given in [11]. A method for norm compliance in the context of open and goal-directed intelligent systems working in dynamic normative environments is discussed in [39].

A screening framework for clinical care of disabled and elderly people is presented in [26]. A novel vision enhancing method for low vision impairments is presented in [30]. Monitoring is realized by a platform that verifies some clinical conditions accordingly to pre-determined thresholds.

An intelligent environment for the identification of critical situation for fragile patients has been presented in [17]. A risk management system at nuclear medicial department is presented in [36].

It is worth noting that there are just few works that focus on the use self-learning technologies for the assistance of patients in their home treatment [18]. In a previous one, we have presented a preliminary version of the services for the checking of the medication via deep learning [34].

3. Background

This section presents a brief introduction to reinforcement learning, deep learning and internet of things.

3.1. Reinforcement Learning

Markov Decision Process (MDP) is the central concept of all RL problems [32]. The goal of the RL algorithms is to find the solution to an MDP. An MDP model has the following components:

-Set of states: $S = s_1, s_2, s_3, \ldots s_n$
-Set of actions: $A = a_1, a_2, a_3, \ldots a_n$
-Transition model: $T(s_t, a, s_{t+1})$
-Reward R

Reward and transition model depend upon on current state s_t, selected action a_t and resulting state s_{t+1}.

The target of RL algorithms is to interact with a given world either with some prior knowledge i.e reward and transition model (model-based RL e.g Dynamic Programming which includes value iteration and policy iteration) or without any prior knowledge (model-free RL). Examples of model free algorithms are Monte Carlo and Temporal difference (TD) learning. Q-learning and SARSA are widely used as TD algorithms [40]. After many repetitive interactions with the environment, the agent learns the characteristics of the environment. The target of RL agent is to find an optimal action out of available actions in each state. Optimal action returns best desired numerical reward to the agent. An agent choose an action in each state which results in a policy π. An optimal policy maximizes the aggregated future reward for a specific problem. The working framework of RL is shown in figure 1.

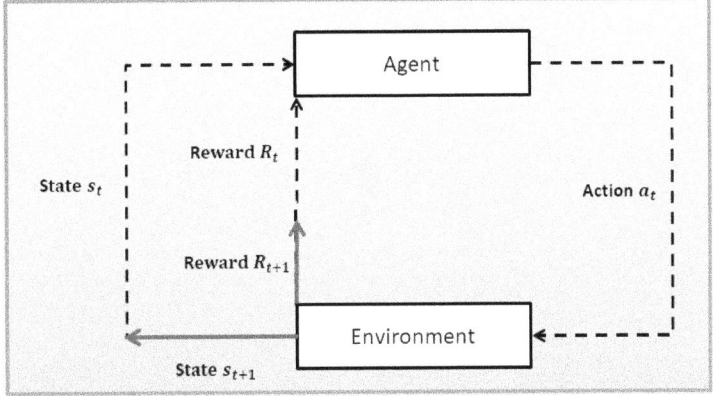

Figure 1. The Reinforcement Learning Problem.

3.2. Deep Learning

Deep Learning (DL) is a sub-field of ML concerned with methods inspired by the structure and function of the brain called Artificial Neural Networks (ANN). ''Deep learning algorithms aim at learning feature hierarchies with features from higher levels of the hierarchy formed by the composition of lower level features. Automatically learning features at multiple levels of abstraction allow a system to learn complex functions mapping the input to the output directly from data, without depending completely on human-crafted features" [8].

Typical ANN challenges are: training time, overfitting and initialization of the parameters. These issues are solvable now due to various methods. We can use layer normalization [5], Batch normalization [28], weight normalization [41] and normalization propagation [4] to accelerate the training process of ANNs. Moreover, Dropouts [42] is helpful in minimizing over fitting.

DL, especially Convolutional Neural Networks (CNN), powers the major progress in computer vision and object detection e.g. face recognition [43], pedestrian and gesture detection [44], drug identification [33], [29] and self-driving cars [21].

3.3. Internet of Things

IoT is a system of interconnected smart machines and physical devices having unique identifiers [25]. IoT devices consist of sensors, electronics, radios and software. IoT devices create digital representation of the surrounding world. IoT devices collect data from the real world and then transmit the collected data to the platform for continuous monitoring.

Sensors such as thermometers, accelerometers, cameras and microphones convert a physical process into an electrical signal. A smartphone is probably a good example of interconected device that embeds many heterogeneous sensors including magnetometers, minimum two cameras, microphone arrays and accelerometers. First generation mobile phones typically included six sensors while now a days a Galaxy S5, for example, has 26 sensors including humidity, cameras, gyro, IR, proximity, microphones, pressure, accelerometer, magnetometer etc.

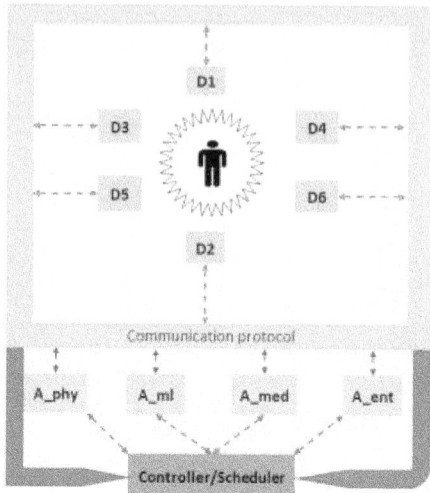

Figure 2. The proposed model.

4. System Model

This section presents the proposed system model which consists of multiple RL agents, DL agent and IoT system. Figure 2 shows the model which consist of multiple IoT devices connected to multiple RL agents, DL agent and scheduler through a communication link. Next we briefly introduce each component of the AIMS.

RL agent.1

The role of this agent is to monitor physical activities of the persons such as: sleeping, awaking, dressing, cleaning etc and will be referred as A_{phy} in the rest of the document. As an example, we have added the demonstration of one of the RL agent i.e. A_{phy} as shown in figure 3. The is dependent on three IoT devices and is being controlled by A_{main}

RL agent.2

The role of RL agent.2 is to monitor eating and drinking activities of a person e.g. preparing and eating breakfast/lunch/dinner; preparing and drinking beverage. RL agent.2 will be referred as A_{ml} afterwords.

RL agent.3

The duty of the RL agent.3 is to manage the entertainment for the person i.e. listening music, watching TV and will be referred as A_{ent}.

RL agent.4 The task of fourth agent is to take care of the treatment schedule of the person if any and to remind him/her about medication time. We will use term A_{med} for this agent onwards. We have implemented this agent on mobile app and its demonstration will be presented in result section. Some of the salient features of agent A_{med} are given next:

- **Omissions** - The A_{med} enable the AIMS to remind the patient about taking a certain pill through the mobile app;
- **Bad timing** - First the A_{med} reminds on time the medication and the then verifies that it is taken on time;

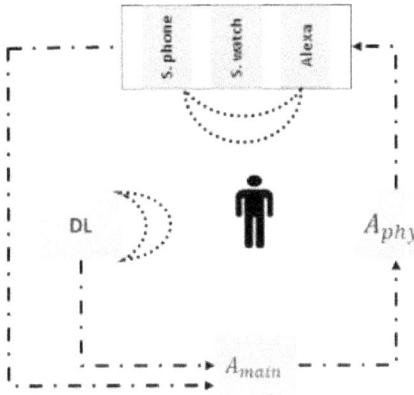

Figure 3. The work flow of A_{phy}

- **Wrong medication** - The A_{med} also verifies through a camera, that the patient is going to take the right pill;
- **Incorrect or omitted therapy annotation** - The A_{med} automatically annotates the medication that the patient has really taken.

Controller/ Scheduler

This is the central agent which control all other agents and act as controller/scheduler. It is referred as A_{main} and its goal is to schedule all agents for their respective jobs in an optimal way. A_{main} is a DL agent that is assisted with CCTV cameras to monitor the persons at different time intervals and give the status of the person to the A_{main} as input. The scheduler then activates another agent for respective task. Next we will discuss the training of DL agent.

First step to train a deep network is the selection of framework. Some of most popular frameworks deep learning models are Tensorflow, PyTorch, Theano and Caffe. We selected Tensorflow framework by considering different criteria like: 1). library management, 2). optimization on CPU, 3). open source and adoption level, 4). debugging and 5). graph visualization.

Moreover, final step before training a model is choice of network or topology. Depending on the projecting requirements including the final deployment on an edge device one should consider important factors like size, time to train, accuracy and inference speed. Some of the famous network are:

-ResNet-50
-VGG16 and VGG19
-Inception v3
-MobileNet.

However, there is no one size fits all methods for these parameters and one has to find an optimal solution by trial and error. Communication protocol Communication protocol acts as a bridge between IoT devices and multiple agents.

Most widely used IoT communication protocols are WiFi, ZigBee, Bluetooth, Z-Wave, long range wide area network and near field communication. Nowadays, WiFi and Bluetooth are frequently used but near field communication is progressing enormously. To choose which communication protocol is best for IoT devices is difficult, but the one

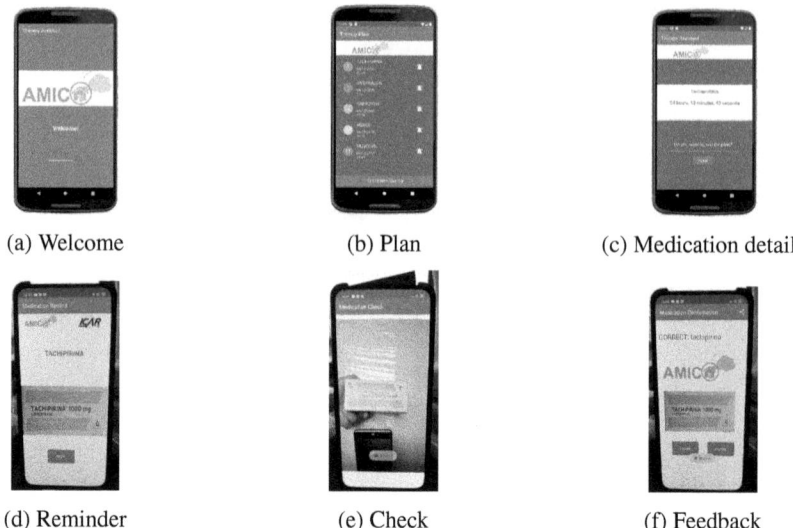

(a) Welcome   (b) Plan   (c) Medication details

(d) Reminder   (e) Check   (f) Feedback

Figure 4. Demonstration of A_{med}

that is easily available in most of the newly manufactured IoT devices and smart phones at a reasonable price.

5. Results and Discussion

This section presents the demonstration of the AIMS, in particular, we have presented results of RL based A_{med} agent. IoT devices continuously monitor the state of the person and send the collected data to the trained DL agent which predict the state of the person. Then main agent i.e the controller activate the appropriate agent according to the predicted state/condition of the person. The AIMS is also able to predict level of risk e.g. normal, serious or emergency and in case of emergency can send automatic message to user relative or doctor.

Figure 4 presents the demonstration of the mobile app i.e A_{med} agent. Left most image in the top row of figure 4 shows the welcome interface that is displayed at the launching of the mobile App. Middle image in the top row of figure 4 indicates the treatment plan that the A_{med} agent has received from the cloud service. Medicine information may be shown by clicking on one of the drugs of the list (as an example, the figure 4c).

A reminder is sent to patient as shown in figure 4d via audio, textual or visual message (in this case reminder is being sent via the combination of scientific name of the medicine and image of the same medicine). An acoustic alert that terminates when the patient pushes the "Stop" button. Immediately after pushing the button, the App starts the camera and extracts images from the scene.

In Figure 4e the image is taken when the patient shows the pill-box. All images extracted by the camera are sent to the DL agent that verifies whether the patient is going to take the same medicine as suggested. When a medication is recognised, it is compared with the one planned for the patient. If the medication is not the one supposed to be taken by the patient, a negative feedback is returned to the App otherwise a positive feedback is

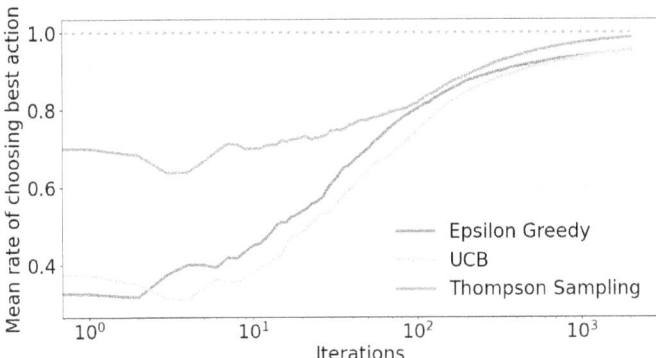

Figure 5. Learning curve of the RL agent

appeared on App as shown in figure 4f to confirm that the taken medicine was correct. We have experimented different Bayesian RL algorithms and found that Thompson sampling shows comparatively better performance. Figure 5 shows learning curves of Thompson sampling, Upper Confidence Bound (UCB) and Epsilon decreasing algorithm. We can evaluate that the Thompson sampling algorithm takes less number of trials to learn the best action.

6. Conclusion

We have presented an Autonomous and Intelligent Monitoring System consisted of RL agents, DL agents and IoT devices to assist patients and elderly at home. We have shown the feasibility of the system through mobile application. In future work, we will assess the application with respect to the risks for the patient [14] that is needed by the regulatory environment in order to approach the market. The mitigation strategies will be designed accordingly with a methodology [16] and adopting rapid prototyping tools such as [15].

Acknowledgement
This work is supported by the AMICO project, which has received funding from the National Programs (PON) of the Italian Ministry of Education, Universities and Research (MIUR): code ARS0100900 (Decree n.1989, 26 July 2018).

List of Acronyms

AIMS Autonomous and Intelligent Monitoring System
MDP Markov Decision Process
RL Reinforcement Learning
AI Artificial Intelligence
DL Deep Learning
IoT Internet of Things
HMS Health Monitoring System
ADL Activities of Daily Life

References

[1] Attila Adamkó, Tamãs Kãdek, and Mãrk Kósa. Intelligent and adaptive services for a smart campus. In *2014 5th IEEE Conference on Cognitive Infocommunications (CogInfoCom)*, pages 505–509. IEEE, 2014.

[2] Mobyen Uddin Ahmed, Mats Björkman, Aida Čaušević, Hossein Fotouhi, and Maria Lindén. An overview on the internet of things for health monitoring systems. In *International Internet of Things Summit*, pages 429–436. Springer, 2015.

[3] Amaya Arcelus, Megan Howell Jones, Rafik Goubran, and Frank Knoefel. Integration of smart home technologies in a health monitoring system for the elderly. In *21st International Conference on Advanced Information Networking and Applications Workshops (AINAW'07)*, volume 2, pages 820–825. IEEE, 2007.

[4] Devansh Arpit, Yingbo Zhou, Bhargava Kota, and Venu Govindaraju. Normalization propagation: A parametric technique for removing internal covariate shift in deep networks. In *International Conference on Machine Learning*, pages 1168–1176. PMLR, 2016.

[5] Jimmy Lei Ba, Jamie Ryan Kiros, and Geoffrey E Hinton. Layer normalization. *arXiv preprint arXiv:1607.06450*, 2016.

[6] Nick Barber, David Alldred, David Raynor, Rebecca Dickinson, Sara Garfield, B Jesson, Rosemary Lim, I Savage, C Standage, Peter Buckle, J Carpenter, Bryony Franklin, Maria Woloshynowych, and Arnold Zermansky. Care homes' use of medicines study: Prevalence, causes and potential harm of medication errors in care homes for older people. *Quality & safety in health care*, 18:341–6, 10 2009.

[7] Paolo Bellavista, Giuseppe Cardone, Antonio Corradi, and Luca Foschini. Convergence of manet and wsn in iot urban scenarios. *IEEE Sensors Journal*, 13(10):3558–3567, 2013.

[8] Yoshua Bengio. *Learning deep architectures for AI*. Now Publishers Inc, 2009.

[9] Marina Bonomolo, Patrizia Ribino, and Gianpaolo Vitale. Explainable post-occupancy evaluation using a humanoid robot. *Applied Sciences*, 10(21):7906, 2020.

[10] Barry Brumitt and Jonathan J Cadiz. " let there be light!": Comparing interfaces for homes of the future. 2000.

[11] Girija Chetty, Matthew White, and Farnaz Akther. Smart phone based data mining for human activity recognition. *Procedia Computer Science*, 46:1181–1187, 2015.

[12] Sunny Consolvo, Peter Roessler, Brett E Shelton, Anthony LaMarca, Bill Schilit, and Sara Bly. Technology for care networks of elders. *IEEE pervasive computing*, 3(2):22–29, 2004.

[13] Diane Cook and Sajal Kumar Das. *Smart environments: technology, protocols, and applications*, volume 43. John Wiley & Sons, 2004.

[14] A. Coronato and A. Cuzzocrea. An innovative risk assessment methodology for medical information systems. *IEEE Transactions on Knowledge and Data Engineering*, pages 1–1, 2020.

[15] A. Coronato and G. De Pietro. Tools for the rapid prototyping of provably correct ambient intelligence applications. *IEEE Transactions on Software Engineering*, 38(4):975–991, 2012.

[16] A. Coronato and G. D. Pietro. Formal design of ambient intelligence applications. *Computer*, 43(12):60–68, 2010.

[17] Antonio Coronato, Giuseppe De Pietro, and Giovanni Paragliola. A situation-aware system for the detection of motion disorders of patients with autism spectrum disorders. *Expert Systems with Applications*, 41(17):7868–7877, 2014.

[18] Antonio Coronato, Muddasar Naeem, Giuseppe De Pietro, and Giovanni Paragliola. Reinforcement learning for intelligent healthcare applications: A survey. *Artificial Intelligence in Medicine*, 109:101964, 2020.

[19] Claudia Di Napoli, Patrizia Ribino, and Luca Serino. Customisable assistive plans as dynamic composition of services with normed-qos. *Journal of Ambient Intelligence and Humanized Computing*, pages 1–26, 2021.

[20] European Medicines Agency. Streamlining EMA public communication on medication errors. 2015.

[21] Clément Farabet, Camille Couprie, Laurent Najman, and Yann LeCun. Scene parsing with multiscale feature learning, purity trees, and optimal covers. *arXiv preprint arXiv:1202.2160*, 2012.

[22] Antonio Fernández-Caballero, José Miguel Latorre, José Manuel Pastor, and Alicia Fernández-Sotos. Improvement of the elderly quality of life and care through smart emotion regulation. In *International Workshop on Ambient Assisted Living*, pages 348–355. Springer, 2014.

[23] JAIN Gaurav and J Daine. Monitoring health by detecting drifts and outliers for a smart environment inhabitant. In *Proceedings of the 4th International Conference on Smart Homes and Health Telematics*,

volume 19, page 114, 2006.

[24] JAIN Gaurav and J Daine. Monitoring health by detecting drifts and outliers for a smart environment inhabitant. In *Proceedings of the 4th International Conference on Smart Homes and Health Telematics*, volume 19, page 114, 2006.

[25] Jayavardhana Gubbi, Rajkumar Buyya, Slaven Marusic, and Marimuthu Palaniswami. Internet of things (iot): A vision, architectural elements, and future directions. *Future generation computer systems*, 29(7):1645–1660, 2013.

[26] Aamir Hussain, Rao Wenbi, Aristides Lopes da Silva, Muhammad Nadher, and Muhammad Mudhish. Health and emergency-care platform for the elderly and disabled people in the smart city. *Journal of Systems and Software*, 110:253–263, 2015.

[27] Jong-Sung Hwang. The evolution of smart city in south korea: The smart city winter and the city-as-a-platform. In *Smart Cities in Asia*. Edward Elgar Publishing, 2020.

[28] Sergey Ioffe and Christian Szegedy. Batch normalization: Accelerating deep network training by reducing internal covariate shift. In *International conference on machine learning*, pages 448–456. PMLR, 2015.

[29] Soyeong Lee, Sunhae Jung, and Hyunjoo Song. Cnn-based drug recognition and braille embosser system for the blind. *Journal of Computing Science and Engineering*, 12(4):149–156, 2018.

[30] Carmelo Lodato and Patrizia Ribino. A novel vision-enhancing technology for low-vision impairments. *Journal of medical systems*, 42(12):1–13, 2018.

[31] Elizabeth D Mynatt, Jim Rowan, Sarah Craighill, and Annie Jacobs. Digital family portraits: supporting peace of mind for extended family members. In *Proceedings of the SIGCHI conference on Human factors in computing systems*, pages 333–340, 2001.

[32] M. Naeem, S. T. H. Rizvi, and A. Coronato. A gentle introduction to reinforcement learning and its application in different fields. *IEEE Access*, 8:209320–209344, 2020.

[33] Muddasar Naeem, Giovanni Paragiola, Antonio Coronato, and Giuseppe De Pietro. A cnn based monitoring system to minimize medication errors during treatment process at home. In *Proceedings of the 3rd International Conference on Applications of Intelligent Systems*, pages 1–5, 2020.

[34] Muddasar Naeem, Giovanni Paragliola, and Antonio Coronato. A reinforcement learning and deep learning based intelligent system for the support of impaired patients in home treatment. *Expert Systems with Applications*, 168:114285, 2021.

[35] Suvarna Nandyal, Roopini U Kulkarni, and Rooprani P Metre. Old age people health monitoring system using iot and ml.

[36] Giovanni Paragliola and Muddasar Naeem. Risk management for nuclear medical department using reinforcement learning algorithms. *Journal of Reliable Intelligent Environments*, 5(2):105–113, 2019.

[37] Patrizia Ribino, Marina Bonomolo, Carmelo Lodato, and Gianpaolo Vitale. A humanoid social robot based approach for indoor environment quality monitoring and well-being improvement. *International Journal of Social Robotics*, pages 1–20, 2020.

[38] Patrizia Ribino and Carmelo Lodato. A distributed fuzzy system for dangerous events real-time alerting. *Journal of Ambient Intelligence and Humanized Computing*, 10(11):4263–4282, 2019.

[39] Patrizia Ribino and Carmelo Lodato. A norm compliance approach for open and goal-directed intelligent systems. *Complexity*, 2019, 2019.

[40] Stuart Russell and Peter Norvig. Artificial intelligence: a modern approach. 2002.

[41] Tim Salimans and Diederik P Kingma. Weight normalization: A simple reparameterization to accelerate training of deep neural networks. *arXiv preprint arXiv:1602.07868*, 2016.

[42] Nitish Srivastava, Geoffrey Hinton, Alex Krizhevsky, Ilya Sutskever, and Ruslan Salakhutdinov. Dropout: a simple way to prevent neural networks from overfitting. *The journal of machine learning research*, 15(1):1929–1958, 2014.

[43] Yaniv Taigman, Ming Yang, Marc'Aurelio Ranzato, and Lior Wolf. Deepface: Closing the gap to human-level performance in face verification. In *Proceedings of the IEEE conference on computer vision and pattern recognition*, pages 1701–1708, 2014.

[44] Jonathan Tompson, Ross Goroshin, Arjun Jain, Yann LeCun, and Christoph Bregler. Efficient object localization using convolutional networks. In *Proceedings of the IEEE conference on computer vision and pattern recognition*, pages 648–656, 2015.

[45] World Health Organization. *Medication errors*. Technical Series on Safer Primary Care. World Health Organization, 2016.

[46] World Health Organization. Medication without harm. 2017.

[47] Andrea Zanella, Nicola Bui, Angelo Castellani, Lorenzo Vangelista, and Michele Zorzi. Internet of things for smart cities. *IEEE Internet of Things journal*, 1(1):22–32, 2014.

Intelligent Environments 2021
E. Bashir and M. Luštrek (Eds.)
© *2021 The authors and IOS Press.*
167

doi:10.3233/AISE210094

Self Learning of News Category Using AI Techniques

Zara HAYAT [a], Aqsa RAHIM [a], Sajid BASHIR [b] and Muddasar NAEEM [c]

[a] *National University of Science And Technology*
[b] *National University of Technology*
[c] *ICAR-CNR*

Abstract. Numerous e-news channels publish the daily happenings in the world from different sources. These huge amounts of news articles have lamentably conceived the information overload issue among the users. Hence text mining, which aims in extracting previously unknown information from unstructured text, has been widely used by several researchers to segregate full news articles however, the news headlines categorization is still specifically limited. Therefore, considering this limitation, the current research aims to propose a framework that will self-learn and automatically classify any given news headline into its corresponding news category using artificial intelligence methods i.e. text mining and machine learning algorithms. The proposed framework consists of three stages: Exploratory Data Analysis, Text Pre-processing, and Text Classification. For exploratory data analysis, the top 10 most frequent balanced news categories are chosen so that further processing of data can be done on a more balanced version of the dataset. After exploring the data, text pre-processing techniques are applied to make the data transformed, normalized, and structured. Finally, text classification is carried out with two approaches: unsupervised classification using Mean Shift and K-means algorithms and supervised classification using Logistic Regression with Bag of Words and TF-IDF algorithm. To depict the working of the proposed framework, a case study is presented on a news headlines dataset which accurately performed news headlines classification.

Keywords. Self-learning, Artificial Intelligence, Text Mining, Machine Learning, Exploratory Data Analysis, Text Pre-processing, TF-IDF, Bag of Words, Unsupervised Clustering, Supervised Classification, Logistic Regression

1. Introduction

With the escalating popularity of using the internet and smartphones, a substantial number of internet users have been amplified remarkably. By taking advantage of an enormous cluster of online users and attaining more viewers and readers, a lot of newspaper companies and news providers have started publishing and updating the news editorial articles on their websites and weblogs. Many users regularly get news from these sources [24]. These news sites include news articles aligned with a corresponding news headline. The headlines of the news exemplify the central concept of the accompanying news article in audited textual information and a much-summarized format.

Research on self-learning of news category [11] indicates that computer-aided categorization of news is more accurate and efficient than humans because once the machines are fed with some task, they will perform it faster than humans therefore, a computer-based classification is a better solution to go for. Browsing through categories benefits in boosting the search results by allowing the users to search and filter the outcomes based on the pre-defined news categories i.e. business, weather, social, technology, politics, sports, entertainment, and many more.

Therefore, the news categorization of headlines can save the efforts and time of the readers by lessening the need to search heterogeneous full-text news articles [34]. It is observed from the several works that as much as categorization of the textual data is essential in terms of time and effort, it is also regarded as one of the toughest classification methods in machine learning. During the last decade, numerous manual newspapers and magazine companies shifted to the digital world by developing their websites to update news to online users. Reading important news is quite valuable to users, but on the other hand side, it is also cumbersome as readers have to go through the entire article to find the useful news out of the articles. Therefore, the classification of news into its various categories seems to be essential to acquire the most relevant and useful information out of the long-length articles quickly.

Recently, the development in computing technology and the introduction of new machine learning algorithms e.g. reinforcement learning [22], text mining the goal of Artificial Intelligence (AI) has become a step closer. AI has important application in diverse fields including:healthcare [5]and [21], robotics and autonomous control [26] and [2],dynamic normative environments [28], ambient assisted living techniques for improvement in the quality of life of elder persons [8], drug identification [20], intelligent environments [3], games and self-organized system [4], vision enhancing method for low vision impairments [19] scheduling and management and configuration of resources, distributed fuzzy system for inferring in real-time critical situations [27], risk management [23] and computer vision.

Text mining often referred to as text data mining [31], is the analysis of textual semi-structured or unstructured data. Since the unstructured and fuzzy text is involved in text mining, therefore, it is regarded as more complex than the data mining process [16]. Therefore, the core aim of text mining is to transform the textual data into numerical data to apply data mining algorithms to it. As text mining is a multidisciplinary area of research, the current research will follow the application of text categorization [30] from news headlines.

The rest of the paper is as follows: a literature review is described in section 2, section 3 represents the proposed framework, section 4 represents the case study and section 5 concludes the paper.

2. Literature Review

Over the recent years, plenty of methodologies have been formulated in the domain of news exploration systems and web news mining. There are numerous studies found in the literature on the automatic categorization of textual data [32]. Text Categorization is the automated allocation of textual documents into their pre-defined categories. It follows machine learning classification algorithms to build models. Several works proposed and compared various algorithms for text categorizing. A lot of techniques and algorithms are there to classify the textual data including Naïve Bayes, Support Vector Machines,

Decision Trees, Neural Network, Random forest, and many more. The notion of these classifiers is to automatically predict the incoming news article to some pre-defined class using a trained classifier.

Over the last few years, classification of news headlines has been an area of research including, classification of emotions, classification of financial news [10], headlines classification with N-gram model [18], automated categorization of news headlines [25], classification of news headlines with SVM [7], mining of emotions from headlines of news [14], and Twitter classification of news from short news headlines [9]. It is very important to have a proper news headlines categorization in our lives. Text classification is a method of allocating predefined categories to a text in conformity with its contents [12]. A well-categorized dataset of news has to be utilized for exploration such as prediction of the stock market, news categorization and trading system, and news-oriented stock trend prediction. [29].

[1] concluded that enhancement for lessening the manual effort is attained by providing the name of the category as the only input keyword for text classification. The multiplication of Word-Net and similarity-based Latent Semantic Analysis (LSA) is carried to compute the final similarity score of documents with category names. Reuters-10 corpus revealed improvement in the precision, with additional amendments and variations in lexical references and context model indicated in [1]. A category name as an initial input-based classification scheme is shown in [17]. The work in [32] carried out relative literature on different algorithms comprising of Support Vector Machines (SVM), Linear Least-Squares Fit (LLSF), K-Nearest Neighbor (kNN), Neural Network (Net), Naive Bayes (NB). He concluded that SVM generated the best performance [15].

A comparative literature review on feature selection in text classification is presented in [33]. He stated that one of the major issues in the categorization of text is the feature space's high dimensionality. The feature set for textual documents consists of unique words that appear in all the documents. To build an efficient and effective model, feature selection is a technique applied. In [13] the author worked on the accuracy of full news article classification using neural networks. Authors in [6] order the news to various classes using SVM, then applied to preprocess techniques and feature selection based on TF-IDF. He used two datasets namely BBC news and 20newsgroup respectively.

It is witnessed from the literature that major work has been presented on long-length news categorization in text mining whereas; the research work on news headlines is still specifically limited. In the current research, the core purpose is to conduct news categorization on news headlines instead of complete text news because long-length news classification is noticeably tough, tiresome, and is computationally expensive. In comparison to the news headlines, the chances of misclassification in large descriptive news articles are more conspicuous.

Therefore, the core objective of the current research is to provide a framework for the E-News channels and news portals to automatically categorize the online news headlines into their pre-defined class by applying effective text preprocessing and classification techniques on a rich news categories dataset that will ultimately improve the overall classification process and yield better accuracy and lessen the computational complexity.

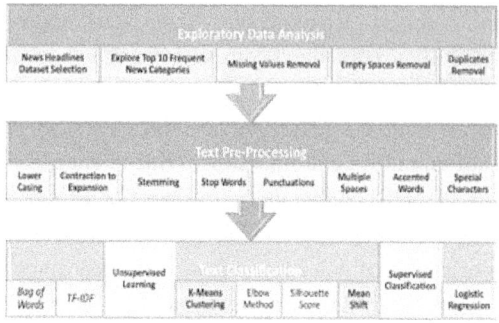

Figure 1. The proposed model.

3. System Model

The current paper provides a framework that can automatically classify news headlines into their pre-defined news categories using machine learning and text mining techniques. Figure 1 represents the proposed framework that follows three steps; exploratory data analysis, text pre-processing, and text classification. As a pre-requisite to news headlines classification, firstly a news headlines dataset is chosen so that the whole proposed process can be applied. After the selection of an appropriate news headlines dataset, the dataset is explored to see if it is an unbalanced dataset or not. Furthermore, all the duplication, empty slots, missing values are removed to make the textual data structured.

The next step is text pre-processing that will make the data clean by applying text pre-processing techniques; lower casing, contraction to expansion, stemming, removal of URL, stop words, multiple spaces, punctuations, accented words, and special characters. The last step is the text classification step which uses two approaches: unsupervised classification with K-means Clustering and Means shift algorithm. The optimal numbers of clusters are automatically predicted in K-means clustering using the Elbow method. Whereas the quality of clusters in both the algorithms i.e., K-means clustering algorithm and Mean shift algorithms, is evaluated using Silhouette score that predicts the Silhouette coefficient which tells if the clusters are defined well or not. The other approach of news headlines classification is a supervised classification which uses Logistic Regression Classifier to yield the performance of the news headlines classification efficiently.

4. Case Study

This section depicts a case study that uses the proposed framework to classify news headlines into its news categories. Each step of the proposed framework is discussed in detail below.

4.1. Dataset

For the current work, News Category Dataset is taken from the Kaggle dataset repository, which contained 200,000 news headlines from 2012 to 2018. Each news headlines have a corresponding news category. The dataset contained six columns; category, headline, author, link, short description, and date.

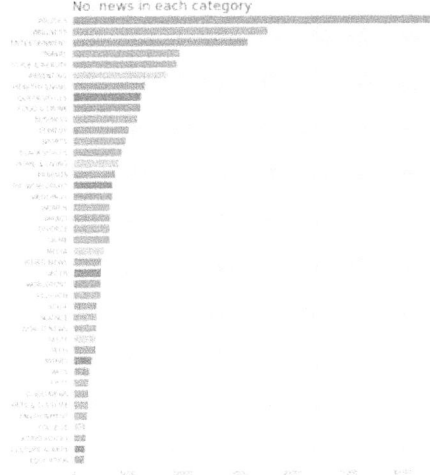

Figure 2. Total amount of news headlines in each category.

4.2. Exploratory Data Analysis

Exploratory data analysis is applied to the News Category Dataset. There were 200,583 news article values in the dataset. Figure 2 represents the 41 categories in the dataset. The exploration of data stated that the dataset was unbalanced as the three categories i.e., politics, wellness, and entertainment were more frequent than the other categories. An unbalanced dataset can highly affect the overall accuracy of the model, therefore; only those news categories were focused on that had enough news articles so that the model can be trained.

Consequently, for data exploratory only the top 10 most frequent balanced categories of news articles were explored. Subsequently, the textual redundancies, missing values, and faulty points from the sample were filtered. From a total of 200,583 news articles, only 120,008 rows were left after exploratory data analysis.

4.3. Text Pre-Processing

Next text pre-processing is applied which normalized, segmented and cleaned the data to make the data predictable by the machine learning algorithm. Firstly, a random sample of 7 rows was taken from the dataset consequently, the unique values, duplicates, empty values were checked. Authors, link and date columns from the dataset were removed as they were not giving any information. Following are the different techniques applied to pre-process, structure and clean the data.

4.3.1. Lower Casing

Firstly lower casing is applied on the text of three columns i.e., category, headline, short description to resolve the sparsity issue and get better outcomes. Making them in lower case helps in the consistency of the output.

	category	headline	short_description	headline_lemmatized_tokenized
77200	wellness	gps guide sandy c newbiggings simple steps cla...	stress strain constantly connected life wellbe...	[gps, guide, sandy, c, newbiggings, simple, st...
79743	food & drink	secrets skinny chef	combat getting weight control thinking balance...	[secret, skinny, chef]
9540	entertainment	jon stewarts war bs continues jordan klepper	opposition trying accomplish jon stewarts fina...	[jon, stewart, war, b, continues, jordan, klep...
33909	healthy living	common myths sleep aids debunked	myth 2 sleep aids improve quality sleep	[common, myth, sleep, aid, debunked]
40945	politics	conservative group urges republicans embrace e...	gop direct money medical research education in...	[conservative, group, urge, republican, embrac...

Figure 3. Data sample after Stemming.

4.3.2. Contraction to Expansion

Contraction to expansion method expanded all the abbreviated words in the data sample. The removal of contraction contributed to text standardization.

4.3.3. Uniform Resource Locators (URL)

Next step was to find the Uniform Resource Locators (URL), which are the textual references to a location on web. As they were not adding any information therefore, they were removed.

4.3.4. Stop Words

Furthermore, stop words were removed which are the most commonly used words in English , removing them means removing the meaningless words from the dataset and only concentrate on the highly meaningful words which can aid in the classification.

4.3.5. Stemming

Next stemming is applied on the data that uses a crude heuristic procedure to cut off the ends of words in the hope of properly transforming them into their root words. It helps in improving classification accuracy which saves required time and space. Figure 3 represents the data obtained after stemming. Likewise, the punctuations, multiple spaces, special characters and accented words were removed from the data as they were not adding any value to the data. Removing them made the data more standardized and usable for extracting meaningful insights.Figure 4 represents the unbalanced dataset with a huge difference between the first categories; politics as 32736 or 26.7% news headlines and the tenth category as 5937 or 4.63% news headlines. To train the machine learning model the news category was normalized with 5937 news headlines occurrences as this was the lowest value of the tenth row.

4.4. Feature Extraction

Feature extraction is the prerequisite for training a model, which transforms the textual data into numerical data. For the current research two approaches of feature extraction are used i.e., Bag of Words model and TF-IDF algorithm.

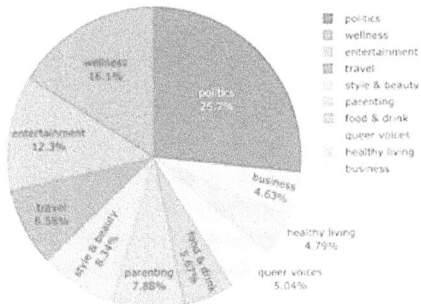

Figure 4. Top 10 categories represents it is an unbalanced dataset.

4.4.1. Bag of Words (BoW

As machine learning algorithms cannot work with raw text directly so the textual data was converted into numerical vectors of numbers using BoW. A list of unique words from the textual data obtained from the preprocessed top 10 categories were identified to design the vocabulary and then on the basis of their presence, these words were scored in each document to form a vector which can be used as an input to the machine learning model. It disregarded the order of words and all the grammatical details in the document and thus easily the textual data is represented into its equivalent vector of numbers.

4.4.2.

The next step was to apply Term Frequency-Inverse Document Frequency which spotted all the important words from the 10 categories. Words with less tfidf scores were considered as they had high importance in the document matrix. Following is the calculation of TF-IDF score for the word t in document d from the document set D:

$$\text{tf idf}\,(t, d, D) = \text{if}\,(t, d).\ \text{idf}(t, D) \tag{1}$$

$$\text{tf}\,(t, d) = \log(I + \text{freq}(t, d)) \tag{2}$$

N is the total number of documents in our text corpus, therefore, IDF Inverse Document Frequency is:

$$\text{idf}\,(t, d) = \log\left(\frac{N}{\text{count}d \in D; t \in d}\right) \tag{3}$$

After the text was converted into numerical data, the next step was to apply text classification using two approaches of machine learning classification, supervised classification and unsupervised classification.

4.5. Text Classification

The aim of text classification is to label the textual data into its relevant categories. In the current case study, the news headlines are divided into two data sets; train data set

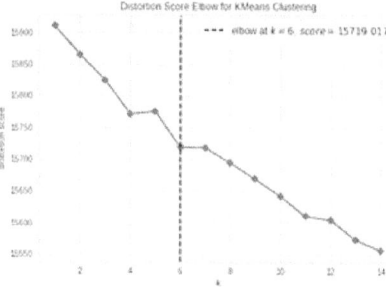

Figure 5. Distortion Score Elbow for K-Means Clustering

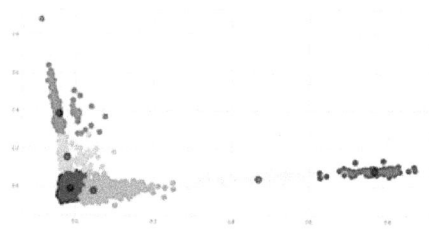

Figure 6. Visual representation of 6 clusters obtained from K-Means

which comprises of 80% of the news articles and the remaining 20% of news articles for testing. 10-fold cross validation is also applied to let the model learn from the test data. For the text classification only 20,000 news headlines samples are chosen because of memory issues.

4.5.1. Unsupervised Learning

Unsupervised learning is used to train a machine learning model using unlabeled data and allows the model to discover the unknown information, hidden structure, patterns in data and find features that can be useful for categorization without any prior training. Two algorithms are used to classify news headlines categories.

    -K-Means clustering

    K-Means clustering also known as distance-based algorithm, is the most efficient method to cluster data. It is applied on the preprocessed news text sample that was first converted into numerical data using TF-IDF and Bag of Words model. Each datapoint in the data sample was assigned to its closest centroid to form a cluster which minimized the distance of the data points within the cluster. To choose the right number of clusters, elbow method is used.

Elbow method automatically selected the optimal number of clusters with KElbowVisualizer method for K-Means algorithm by fitting the model with a range of values for K. Figure 5 shows the distortion score Elbow for K-means clustering, where distortion is the sum of squared distances from each point to its assigned center point. Elbow method predicted 6 clusters for news categories. Figure 6 shows the visual representation of the 6 clusters obtained from Elbow method.

The next step was to check the quality of the clusters obtained from K-means algorithm and elbow method using **Silhouette** score with higher Silhouette coefficient which proved that the model contained well defined clusters score. After K-means clustering

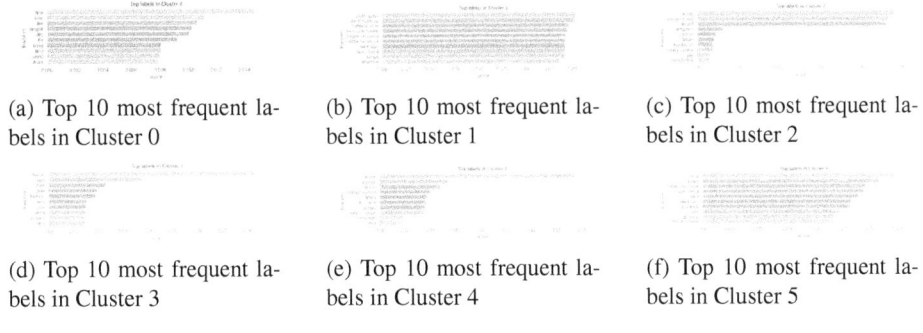

(a) Top 10 most frequent labels in Cluster 0

(b) Top 10 most frequent labels in Cluster 1

(c) Top 10 most frequent labels in Cluster 2

(d) Top 10 most frequent labels in Cluster 3

(e) Top 10 most frequent labels in Cluster 4

(f) Top 10 most frequent labels in Cluster 5

Figure 7. Top 10 most frequent labels in Cluster

Figure 8. 27 Clusters obtained from Mean Shift Algorithm

is done, top 10 most frequent words in each 6 predicted clusters are analyzed. Figure 7 represents the results of most frequent words and their scores in each predicted cluster.

-**Mean Shift**

Next step was to use mean shift algorithm to find clusters from the data sample. Mean Shift is also referred to as centroid based algorithm or mode seeking algorithm. In a given region, it shifts the candidate data points to the closest centroids of the clusters iteratively to be the mean of the points. Then these data points are filtered in a later stage to remove the redundant points to form final set of centroids. Unlike K-means clustering, it does not requires specifying the number of clusters as the number of clusters is automatically determined by the algorithm with respect to the data. Figure 8 represents the 27 clusters obtained by Mean Shift algorithm. The quality of clusters obtained from Mean shift algorithm using Silhouette score is checked. Similarly, top 10 words from the predicted 27 clusters were obtained.

4.5.2. Supervised Classification

The last step is the supervised learning which is training the machine learning model with sorted and labeled data. As the data sample was text so it was first converted into numerical data with TF-IDF algorithm and then Logistic regression algorithm on the sample data was applied. Logistic regression algorithm is most used machine learning predictive analysis algorithm which predicts the categorical dependent variable using a set of given independent variables to assess the probability of an event's success or failure. It is very fast and efficient to train and discover the classification of the unknown records.

The experimental results obtained by logistic regression proved that prediction of news headlines into their pre-defined categories can be best gained by the use of logistic re-

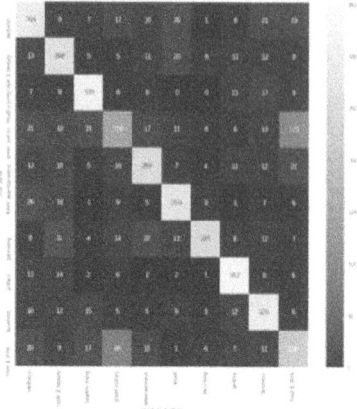

Figure 9. Heatmap representing prediction by Logistic Regression.

gression as it gave 74% accuracy in classifying the news categories. Figure 9 represents the heatmap obtained by Logistic Regression having higher values on diagonal verifies that the model is predicting the headlines to their pre-defined categories accurately. Therefore, the case study proved that with the use of the proposed framework, online news channels and news portals can easily increase their speed and efficiency of news headlines classification and reduce their computational complexity.

5. Conclusion and Future Work

We have presented a framework that self learn the category of a news. The system in the first step, exploratory data analysis on the news dataset is applied, and then text pre-processing techniques are applied to remove the unwanted and useless information from the data to get meaningful insights. Lastly, text classification is carried out using two approaches of classification, unsupervised clustering with K- means clustering and mean shift algorithm, and supervised classification with logistic regression. The proposed framework is persistent and cost-effective which will help the e-news channels and news portals to automatically classify the huge set of scattered news headlines into their pre-defined news categories efficiently and effectively. Thus, the automatic news headlines categorization will reduce the computational complexity.

In the future, the proposed framework can be extended to analyze the sentiments of users on news articles using sentiment analysis. Currently, this framework is applied to news headlines classification however, in the future; this framework methodology can be applied to other datasets such as movies or talk shows datasets to automatically categorize the movies and talk shows into their pre-defined categories efficiently and hence the accuracy can be improved. Similarly, developing a recommender system that recommends news categories to the users based on their user profiles remains a milestone to be achieved in the future.

References

[1] Libby Barak, Ido Dagan, and Eyal Shnarch. Text categorization from category name via lexical reference. In *Proceedings of Human Language Technologies: The 2009 Annual Conference of the North American Chapter of the Association for Computational Linguistics, Companion Volume: Short Papers*, pages 33–36, 2009.

[2] Marina Bonomolo, Patrizia Ribino, and Gianpaolo Vitale. Explainable post-occupancy evaluation using a humanoid robot. *Applied Sciences*, 10(21):7906, 2020.

[3] A. Coronato and G. De Pietro. Tools for the rapid prototyping of provably correct ambient intelligence applications. *IEEE Transactions on Software Engineering*, 38(4):975–991, 2012.

[4] A. Coronato and G. D. Pietro. Formal design of ambient intelligence applications. *Computer*, 43(12):60–68, 2010.

[5] Antonio Coronato, Muddasar Naeem, Giuseppe De Pietro, and Giovanni Paragliola. Reinforcement learning for intelligent healthcare applications: A survey. *Artificial Intelligence in Medicine*, 109:101964, 2020.

[6] Seyyed Mohammad Hossein Dadgar, Mohammad Shirzad Araghi, and Morteza Mastery Farahani. A novel text mining approach based on tf-idf and support vector machine for news classification. In *2016 IEEE International Conference on Engineering and Technology (ICETECH)*, pages 112–116. IEEE, 2016.

[7] RR Deshmukh and DK Kirange. Classifying news headlines for providing user centered e-newspaper using svm. *International Journal of Emerging Trends & Technology in Computer Science (IJETTCS)*, 2(3):157–160, 2013.

[8] Claudia Di Napoli, Patrizia Ribino, and Luca Serino. Customisable assistive plans as dynamic composition of services with normed-qos. *Journal of Ambient Intelligence and Humanized Computing*, pages 1–26, 2021.

[9] Inoshika Dilrukshi, Kasun De Zoysa, and Amitha Caldera. Twitter news classification using svm. In *2013 8th International Conference on Computer Science & Education*, pages 287–291. IEEE, 2013.

[10] Brett Drury, Luis Torgo, and JJ Almeida. Classifying news stories to estimate the direction of a stock market index. In *6th Iberian Conference on Information Systems and Technologies (CISTI 2011)*, pages 1–4. IEEE, 2011.

[11] Ari Aulia Hakim, Alva Erwin, Kho I Eng, Maulahikmah Galinium, and Wahyu Muliady. Automated document classification for news article in bahasa indonesia based on term frequency inverse document frequency (tf-idf) approach. In *2014 6th international conference on information technology and electrical engineering (ICITEE)*, pages 1–4. IEEE, 2014.

[12] Rajni Jindal, Ruchika Malhotra, and Abha Jain. Techniques for text classification: Literature review and current trends. *webology*, 12(2), 2015.

[13] Sandeep Kaur and Navdeep Kaur Khiva. Online news classification using deep learning technique. *International Research Journal of Engineering and Technology (IRJET)*, 3(10):558–563, 2016.

[14] DK Kirange and RR Deshmukh. Emotion classification of news headlines using svm. *Asian Journal of Computer Science and Information Technology*, 5(2):104–106, 2012.

[15] David D Lewis, Yiming Yang, Tony Russell-Rose, and Fan Li. Rcv1: A new benchmark collection for text categorization research. *Journal of machine learning research*, 5(Apr):361–397, 2004.

[16] Shu-Hsien Liao, Pei-Hui Chu, and Pei-Yuan Hsiao. Data mining techniques and applications–a decade review from 2000 to 2011. *Expert systems with applications*, 39(12):11303–11311, 2012.

[17] Chaya Liebeskind, Lili Kotlerman, and Ido Dagan. Text categorization from category name in an industry-motivated scenario. *Language resources and evaluation*, 49(2):227–261, 2015.

[18] Xin Liu, Gao Rujia, and Song Liufu. Internet news headlines classification method based on the n-gram language model. In *2012 International Conference on Computer Science and Information Processing (CSIP)*, pages 826–828. IEEE, 2012.

[19] Carmelo Lodato and Patrizia Ribino. A novel vision-enhancing technology for low-vision impairments. *Journal of medical systems*, 42(12):1–13, 2018.

[20] Muddasar Naeem, Giovanni Paragiola, Antonio Coronato, and Giuseppe De Pietro. A cnn based monitoring system to minimize medication errors during treatment process at home. In *Proceedings of the 3rd International Conference on Applications of Intelligent Systems*, pages 1–5, 2020.

[21] Muddasar Naeem, Giovanni Paragliola, and Antonio Coronato. A reinforcement learning and deep learning based intelligent system for the support of impaired patients in home treatment. *Expert Systems with Applications*, page 114285, 2020.

[22]   Muddasar Naeem, S Tahir H Rizvi, and Antonio Coronato. A gentle introduction to reinforcement learning and its application in different fields. *IEEE Access*, 2020.

[23]   Giovanni Paragliola and Muddasar Naeem. Risk management for nuclear medical department using reinforcement learning algorithms. *Journal of Reliable Intelligent Environments*, 5(2):105–113, 2019.

[24]   [1] Pew Research Center U.S. Politics Policy. Americans spending more time following the news. Available at: ¡https://www.pewresearch.org/politics/2010/09/12/americans-spending-more-time-following-the-news, 2021.

[25]   Mark W Pope. Automatic classification of online news headlines. 2007.

[26]   Patrizia Ribino, Marina Bonomolo, Carmelo Lodato, and Gianpaolo Vitale. A humanoid social robot based approach for indoor environment quality monitoring and well-being improvement. *International Journal of Social Robotics*, pages 1–20, 2020.

[27]   Patrizia Ribino and Carmelo Lodato. A distributed fuzzy system for dangerous events real-time alerting. *Journal of Ambient Intelligence and Humanized Computing*, 10(11):4263–4282, 2019.

[28]   Patrizia Ribino and Carmelo Lodato. A norm compliance approach for open and goal-directed intelligent systems. *Complexity*, 2019, 2019.

[29]   Satoru Takahashi, Masakazu Takahashi, Hiroshi Takahashi, and Kazuhiko Tsuda. Analysis of the relation between stock price returns and headline news using text categorization. In *International Conference on Knowledge-Based and Intelligent Information and Engineering Systems*, pages 1339–1345. Springer, 2007.

[30]   Ah-Hwee Tan et al. Text mining: The state of the art and the challenges. In *Proceedings of the pakdd 1999 workshop on knowledge disocovery from advanced databases*, volume 8, pages 65–70. Citeseer, 1999.

[31]   Sholom M Weiss, Nitin Indurkhya, Tong Zhang, and Fred Damerau. *Text mining: predictive methods for analyzing unstructured information*. Springer Science & Business Media, 2010.

[32]   Yiming Yang and Xin Liu. A re-examination of text categorization methods. In *Proceedings of the 22nd annual international ACM SIGIR conference on Research and development in information retrieval*, pages 42–49, 1999.

[33]   Yiming Yang and Jan O Pedersen. A comparative study on feature selection in text categorization. In *Icml*, volume 97, page 35. Nashville, TN, USA, 1997.

[34]   Yingwu Zhu. Measurement and analysis of an online content voting network: a case study of digg. In *Proceedings of the 19th international conference on World wide web*, pages 1039–1048, 2010.

Intelligent Environments 2021
E. Bashir and M. Luštrek (Eds.)
© *2021 The authors and IOS Press.*
This article is published online with Open Access by IOS Press and distributed under the terms
of the Creative Commons Attribution Non-Commercial License 4.0 (CC BY-NC 4.0).
doi:10.3233/AISE210095

An RL-Based Approach for IEQ Optimization in Reorganizing Interior Spaces for Home-Working

Patrizia RIBINO [a,1], and Marina BONOMOLO [b]

[a] *Istituto di Calcolo e Reti ad Alte prestazioni (ICAR), National Research Council, Italy*
[b] *Dipartimento di Ingegneria, Universita degli Studi di Palermo*

Abstract. Indoor Environmental Quality (IEQ) is a very important aspect in the design of spaces. It depends on fundamental requirements, and it aims to improve living quality and ensure high well-being for the occupants. Poor IEQ has significant health consequences as people spend a considerable portion of their time indoors. The restrictions imposed by the current SARS-CoV-2 pandemic have worsened this situation by requiring people to spend more time at home and adopt home-working as a primary work model. Such a situation requires individuals to reconfigure their interior home spaces for adapting to new emerging needs. This paper proposes an approach based on reinforcement learning to support the rearrangement of indoor spaces by maximizing the indoor environmental quality index in terms of thermal, acoustic and visual comfort in the new furniture layout scheme.

Keywords. Indoor Environmental Quality, Reinforcement Learning, Interior Layout Design

1. Introduction

People spend most of their time in buildings [1]. It is estimated that individuals spend more than 80% of the day indoors. The maintenance of indoor environmental quality (IEQ) is therefore significant for improving occupants' feeling of comfort, health, working efficiency, and productivity[2]. Nowadays, this requirement is becoming more relevant due to the pandemic emergency caused by the SARS-CoV-2. The extremely contagious nature of the virus and the severity of the Covid-19 disease have resulted in strict stay-in-home and social distancing policies in several countries across the world. In particular, to contain the spread of the coronavirus and curb contagion, home working [2] is becoming the most common model of work. Such a new situation leads people to rearrange some home spaces and to reconfigure new furniture layout schemes to be comfortable, as much as possible, in their home-work. Thermal, visual and acoustic conditions are the key elements that jointly influence indoor comfort [2]. Many standards [3] and

---

[1] Corresponding Author: Istituto di Calcolo e Reti ad Alte prestazioni (ICAR), via Ugo La Mafa 113, Palermo, Italy; E-mail: patrizia.ribino@icar.cnr.it.

[2] In this work, we prefer to use the term *home-working* rather than the widely used term of 'smart working' because this latter is not constrained to be performed strictly at home as the Covid-19 restrictions impose.

studies propose methods and indexes to evaluate the quality of the environment. They indicate methods and criteria for designing indoor environments related to the internal air quality, illuminance level, and thermal and acoustic parameters. In general, such standards are based on measurable values directly collected from the environment employing appropriate sensors. Although several existing works deal with the interior automatic layout design, they do not consider IEQ issues. To cite a few, Yang et al. [4] present a learning-based solution for an automatic interior layout that considers user-specified furniture to be placed in appropriate functional areas. Liu et al. [5] reconstructed 3D scenes from photos capturing different orientations. Yu et al. [6] present a method based on energy function for the overall layout by considering relations between furniture and the room function, introducing layout constraints.

Moreover, in recent years, extensive research efforts have been spent on the development of artificial intelligence (AI) techniques, particularly in contexts where individual well-being is a key element, such as ambient assisted living (AAL) [7, 8], health-care [9, 10] and intelligent environments [11]. Among AI techniques, reinforcement learning [12] plays an increasingly important role in achieving specific goals in unknown environments by adopting a trial-and-error paradigm.

The contribution of this paper is a novel method for generating optimal layout configurations in terms of environmental comfort conditions. In particular, we propose the use of Reinforcement Learning techniques for an automatic and intelligent furniture layout generation of a given room with office end-use that maximizes the environmental comfort of an individual in performing his/her working activities. In particular, we formulated the interior layout problem as a Markov decision process by defining states, actions, and reward function according to the constraints given by the IEQ standards.

A pilot study about a home office configuration is presented. Preliminary results give evidence that the proposed IEQ-based RL approach could be a valuable tool for an automatic indoor design that evaluates the conformity to the standard levels of environmental comfort during the design phase. Indeed, a common method called *Post Occupancy Evaluation* [13, 14] is used to evaluate an occupant's comfort conditions (after (s)he lived in that space) and make some successive modifications to reach desired occupant comfort conditions. The proposed approach allows evaluating a priori the best comfortable conditions that can be reached by an environment endowed with specific equipment. The current version of the software prototype is realized for finding optimal configurations of room with office end-use. However, it can be extended as a generalized tool for dealing with different functional environments and furniture.

The rest of the paper is organized as follows. Section 2 introduces a theoretical background of IEQ and RL. Section 3 presents the RL approach for generating comfortable indoor layouts. Section 4 shows a pilot study about a rearrangement of home space for working destination. Finally, in Section 5, conclusions and future works are shown.

2. Theoretical Background

2.1. Indoor comfort conditions

The primary contribution to IEQ satisfaction is linked to Thermal, Visual and Acoustics Comfort and Air Quality [2]. In this paper, we consider the first three.

Thermal Comfort –   The first model of thermal comfort was established by Fanger in the 1970s [15]. Predicted Mean Vote (PMV), as a result of Fanger's comfort equation, predicts the mean response of a larger group of people according to the ASHRAE thermal sense scale [15]. PMV depends on a combination of environmental and individual variables $PMV = f(t_a, t_{mr}, v_a, p_v, M, I_{cl})$ where t_a is the air temperature ($^\circ C$), t_{mr} is the mean radiant temperature ($^\circ C$), v_a is the relative air velocity (m/s), p_v is the water vapor partial pressure (Pa), M is the metabolic rate (W/m^2), and I_{cl} is the clothing insulation ($m^2 K/W$). PMV can be expressed in a concise form as:

$$PMV = (0.303 \times exp^{-0.036M} + 0.028) \times L \qquad (1)$$

where L is the thermal load on the body. It is the difference between the rate of metabolic heat generation and the calculated heat loss from the body to the actual environmental conditions assumed as optimum at a given activity level [15]. The metabolic rate (M) mainly corresponds to the heat production of the human body. Using PMV, the Percentage of People Dissatisfied (PPD) can be predicted. PPD indicates the number of people dissatisfied with the thermal environment and it is determined as follows:

$$PPD_{TC} = 100 - 95 \times exp(-0.03353 * PMV^4 - 0.2179 \times PMV^2) \qquad (2)$$

Depending on the ranges of PPD and PMV, three kinds of comfort zones can be identifies[16], as shown in Table 1. Category A corresponds to a high-performance building, while category B and category C corresponds respectively to a medium and moderate one. In this study, we refer to category B.

Parameters		Recommended Limits		
	Category	Summer		Winter
PMV	A	-0.2<PMV<0.2		
	B	-0.5<PMV<0.5		
	C	-0.7<PMV<0.7		
PPD, [%]	A	< 6		
	B	< 10		
	C	< 15		

Table 1. Thermal comfort according to PPD and PMV.

Visual Comfort –   Visual comfort is related to the quantity and quality of light within a given space at a given time. Light is a fundamental element in our capacity to see. Both too little and too much light can cause visual discomfort. Visual comfort mainly depends on photo-metric quantities such as the illuminance (i.e., the amount of luminous flux per unit area), luminance and, consequently, glare condition and contrast. Further aspects, such as views of outside space or light quality, are also considered. As concerns, the visual comfort, the European standard EN 12464-1[17] defines lighting requirements for indoor work areas in terms of quantity and quality of illumination. It covers offices, places of public assembly, restaurants/hotels, theaters/cinemas. For the aim of this paper, we refer to the standard for generic offices that recommends 500 lx on the work-plane.

Domain	Room/Space	NR Value	Recommended Noise Level (dB)
Industrial	Control Rooms	40 - 50	45 - 55 dB(A)
	Offices	35 - 40	40 - 45 dB(A)
Commercial and Leisure	Offices	35 - 40	40 - 45 dB(A)
	Private Offices	30 - 35	35 - 40 dB(A)

Table 2. Recommended noise levels.

Acoustic Comfort – Providing acoustic comfort consists of minimizing noise and maintaining satisfaction among residents. Acoustic comfort helps to improve concentration. Hence, buildings are designed to specific noise standards based on their use. For example, the noise level in a library may differ from noise specifications in a public hall. Noise Ratings (NR) [18] is one of the standards used to evaluate acoustic comfort. Each space or room has a recommended NR value, which is based upon the intended requirements, needs and function of the indoor environment. Table 2 shows an excerpt of the recommended NR levels for some application areas and the related mapping with the most common unit for measuring sound pressure level, the dB level.

2.2. Reinforcement Learning

Reinforcement Learning (RL) is a branch of machine learning inspired by behaviorist psychology, where human and animal behavior is studied from a reward and punishment perspective. According to this perspective, an RL approach concerned with how an autonomous agent, that senses and acts in a dynamic environment, learns to choose the most effective actions to achieve its goals by maximizing its overall rewards[12]. The agent learns the optimal control policy by trails during the interactions with the environment. This agent-environment interaction process is most commonly formulated as a Markov Decision Process (MDP), consisting of State, Action, and Rewards. One of the most popular learning algorithms is Q-learning [19]. Q-learning's objective is to find an optimal policy in the sense that the expected value of the total reward overall successive steps is the maximum achievable.

The core of the Q-learning algorithm is the Bellman Equation (see eq.3), according to which the optimal Q-value for a given state-action pair is equal to the sum of the maximum reward the agent can get from an action in the current state and the maximum discounted reward it can obtain from any possible state-action pair that follows.

$$Q(s_t, a_t) \leftarrow Q(s_t, a_t) + \alpha \times \left[R + \gamma \times \max_{a_{t+1}} Q(s_{t+1}, a_{t+1}) - Q(s_t, a_t) \right] \qquad (3)$$

where R is the reward received when an agent moves from the state s_t to the state s_{t+1}. The learning rate α determines to what extent newly acquired information overrides old information. The discount factor γ determines the current value of future rewards.

3. IEQ-based Q-Learning approach for organizing indoor space

In this section, we present the proposed RL approach for the optimization of indoor space configurations. In particular, such an optimized configuration is obtained using the Q-learning method driven by IEQ standards.

3.1. Problem Formulation

A practical furniture layout scheme has to meet functional requirements and respects some primary layout constraints (e.g., doors and windows cannot be impeded and furniture has to be reachable). In this paper, we consider interior layout design as the problem to find the best furniture layout that provides the most comfortable configuration for an area with office end-use in terms of thermal, acoustics and visual comfort. Moreover, we assume that a set F of furniture for equipping the room is selected by the occupant.

We express the problem of IEQ-based indoor layout configuration as a Markov Decision Process defined by the following tuple:

$$MDP =< S, S_0, S_f, F, Act(.), P, R > \tag{4}$$

In particular, S is a finite set of states of the world represented by different furniture locations. $S_0 \in S$ is an initial state of the world. It is represented by an initial placement of furniture on the room plane. $S_f \in S$ is a final state characterized by an optimal IEQ. F is the set of user furniture to be positioned in a room. $Act(f)$ is a finite set of possible movements for each furniture $f \in F$. In this paper, $Act(f) = \{moveUp, moveDown, moveRight, moveLeft\}$. P is a probability distribution function. When a movement of an item is executed, the world makes a transition from its current state S to a succeeding state S'. The probability of this transition is labeled as $P(S'|S,a)$. $R(S'|S,a)$ is the immediate reward obtained by moving a piece of furniture from a current position to a new position.

Let assume n different pieces of furniture $(f_1, f_2, ..., f_n)$. Our aim is to place these objects onto some locations $(l_1, l_2, ...l_n)$. We use RL to find a mapping $(f_1, f_2, ..., f_n) \rightarrow (l_1, l_2, ...l_n)$ that maximizes a reward function R subject to some constraints. Mainly, there is a unique location for placing each object f_i, but each location l_j can be assigned to any furniture piece. The proposed RL approach's objective is to find an optimal distribution of furniture on several locations such that it maximizes the rewards (in terms of thermal, acoustics and visual comfort) generated from each configuration.

For the sake of simplicity, we considered an indoor space as a rectangular room with predefined dimensions. A reference system is associated with the room plant, whose origin corresponds to the top left corner. Each piece of furniture has a location defined by a couple of coordinates (x, y) that corresponds to the top left corner of the furniture and size s represented by a couple $(width, depth)$. In a room, several kinds of objects may be located. Each step, the RL agent places a furniture f_j at a new location $l_j = (x_j, y_j)$. It is worth noting that some types of furniture can influence the sound pressure level, temperature, and environment's illuminance. Hence, after the agent action, the reward is evaluated considering the effect of the furniture displacement on comfort conditions.

In this paper, we address the indoor layout problem as the problem to configure a home office with office furniture (i.e., desk, printer, heater) to create a comfortable indoor space for home-working. In particular, the reward function is evaluated according to the desk's location, considered as the reference point. Indeed, we assume that the worker spends much of his/her working time in that position.

3.2. Reward Function

The reward function is related to the Indoor Environment Quality index, which is considered, in the most prominent literature, as an indicator of the level of comfort. IEQ index refers to the building's indoor environment quality considering the occupants' satisfaction level presented on a 0 -100% scale (as a percentage of users satisfied). It is composed of four indoor comfort sub-indexes (also presented on a 0 - 100% scale): thermal comfort TC_{index}, acoustic comfort AC_{index}, visual comfort VC_{index} and indoor air quality IAQ_{index} index. Some weights are associated with these indexes for the evaluation of the whole IEQ index. In this paper, we refer to the IEQ model proposed in [20]. Moreover, in this preliminary study, we do not consider the effect of the indoor air quality index on the IEQ. Hence, we formulated the reward function of the proposed RL approach as a weighted average of the comfort indexes as mentioned earlier:

$$R_{IEQ} = w_1 * TC_{index} + w_2 * ACc_{index} + w_3 * VC_{index} \tag{5}$$

TC_{index} refers to the percentage of people accepting the thermal environment:

$$TC_{index} = 100 - PPD_{TC} \tag{6}$$

where PPD is the Predicted Percentage Dissatisfied calculated according to Eq.2. AC_{index} relates to the building's ability to provide an environment with minimal unwanted noise. We assume that this noise source may be external or internal to the indoor space we are considering. Internal noise, for example, can be generated by an HVAC (Heating, Ventilation and Air Conditioning) system. Conversely, external noise can be propagated, for example, by a home appliance placed in a space adjacent to the room under study. According to the IEQ model [20], the AC_{index} is given by:

$$AC_{index} = 100 - PD_{Acc} \tag{7}$$

where PD_{Acc} is the predicted percentage of sound level dissatisfied occupants with the change in noise level from *Recommended* to *Actual* value on a 0-100% scale:

$$PD_{Acc} = 2 * (Actual_{SoundPressureLevel} - Reccomended_{SoundPressureLevel}) \tag{8}$$

Moreover, to evaluate the AC_{index}, the noise reduction due to the source's distance to the receiver has to be considered. For a continuing sound source in a room, the sound level is the sum of direct and reverberant sound [21]. The sound pressure level for a receiver is given by:

$$L_{Receiver} = L_{Source} + 10log(Q/(4\pi r^2) + 4/R) \tag{9}$$

where $L_{Receiver}$ is the sound pressure level (dB) received in a given point of the room at r distance from the source. L_{Source} is the sound power level from the source of the noise. Q is the directivity factor that measures the directional characteristic of a sound's source. It assumes predefined values for the specific location of the sound source. Finally, the room constant R expresses the acoustic property of a room according to:

$$R = S\bar{\alpha}_{abs}/(1 - \bar{\alpha}_{abs}) \tag{10}$$

where S is the total surface of the room (including walls, floor and ceiling) and $\bar{\alpha}_{abs}$ is the mean absorption coefficient of the room.

$$\bar{\alpha}_{abs} = \frac{\sum_{i=1}^{n} S_i \alpha_i}{\sum_{i=1}^{n} S_i} \tag{11}$$

where S_i is an individual surface in the room (m^2) made with a specific material that is characterized by an absorption coefficient α_i [3]. Moreover, we assume that several sources of noise can affect the environment. Thus, we have to consider each source's contribution to the receiver given by eq eq.12 with k is the number of sources and L_j is the sound pressure level of the jth source.

$$L_{TotReceiver} = 10 * log \sum_{j=1}^{k} 10^{0.1*L_j} \qquad (12)$$

Finally, the visual comfort index VC_{index} is based on the amount of light falling on the working plane as follows:

$$VC_{index} = 100 - PD_L \qquad (13)$$

where PD_L is the percentage of persons dissatisfied with minimum daylight or the probable percentage of people switching on artificial lighting on a scale of 0 -100%. The predicted percentage of dissatisfied occupants is calculated as a function of daylight luminance E_{min} with the following equation:

$$PD_L = \frac{(-0.0175 + 1.0361)}{1 + exp[4.0835 \times (log(E_{min}) - 1.8223)} \times 100 \qquad (14)$$

On the other hand, according to the interior design principles, we also considered the following layout rules for indoor space dedicated to office end-use:

1. make the desk near the window. Such rule allows having more natural light on the desk, but at the same time, it satisfies IEQ constraints about visual comfort that, as previously said, is also improved by views of outside space;
2. make the furniture far from the door for not impeding its regular use.

Two reward functions R_{window} and R_{door} associate a quantitative measure to the compliance with rule 1 and 2, respectively. Hence, the total reward is given by:

$$R_{Tot} = R_{IEQ} + R_{window} + R_{door} \qquad (15)$$

3.3. IEQ-based Q-Learning Algorithm

The pseudo-code of the IEQ-based Q-Learning algorithm is shown in Algorithm 1. Starting from the MDP model and the characteristics of the room to be configured, Algorithm 1 provides a mapping of the furniture $(f_1, f_2, ..., f_n)$ to some room locations $(l_1, l_2, ...l_n)$ that maximizes the environmental comfort of the occupant. Firstly, Algorithm 1 initializes a matrix representing the room under study. For each furniture f_i a set of available actions $Act(f_i)$ is available. Then, Algorithm 1 randomly selects the initial positions of the furniture that have been located in the room and initializes the Q-learning matrix. An initial reward is calculating for the initial state S_0 according to Eqs. 5 - 15. For each episode of the learning process, starting from the initial state S_0, the q-learning agent chooses an action from the list of the available actions that allows moving furniture in the selected action direction. The exploration strategy used in this paper is the ε -greedy policy. Hence, a new reward is evaluated for the new observed state S' according to Eqs. 5 - 15, and the Q-learning matrix is updated accordingly. An optimal final state is a state

---

[3]Common absorption coefficients can be found at https://www.acoustic.ua/st/web_absorption_data_eng.pdf

---

Algorithm 1: IEQ-based Q-Learning for interior design

Data: MDP model, Room Plant
Result: Opt Location Vector $(l_{1_f}, l_{2_f}, ..., l_{n_f})$
$< S, S_0, S_f, F, Act(.), P, R > \leftarrow MDP$;
$RoomMatrix \leftarrow matrix(Room.width, Room.depth)$;
$F \leftarrow (f_1, f_2, ..., f_n)$;
for $f_i \in F$ **do**
    | $Act(f_i) \leftarrow (moveUp(l_i), moveDown(l_i), moveLeft(l_i), moveRight(l_i))$;
    | $AvailableAction.append(Act(f_i))$
$S_0 \leftarrow (l_{1_0}, l_{2_0}, ..., l_{n_0})$;
$Q \leftarrow \varnothing$;
Compute Initial IEQ-Reward for S_0 (Eqs. 5 - 15);
$CurrentState \leftarrow S_0$;
// Learning Process
for *each episode t* **do**
    | *Choose act from Available Action based on* ε *- greedy policy*;
    | *Make Action act*;
    | *Observe the new state S'*;
    | $R_{Tot}(t) \leftarrow sum(R_{IEQ}, R_{window}, R_{door})$;
    | $Q(S, act)(t) = (1 - \alpha) \times Q(S, act)(t-1) + \alpha(R_{Tot}(t) + \gamma \max_{act'}(Q[S', act']))$
    | $CurrentState \leftarrow S'$
$S_f \leftarrow$ *Find S where Q(S, _) is Max*;
$(l_{1_f}, l_{2_f}, ..., l_{n_f}) \leftarrow S_f$;

---

having a max q-learning value. Hence, the mapping of the furniture on the locations of the room that maximizes the environmental comfort for that room and respects the layout design constraints is given by $(f_1, f_2, ..., f_n) \leftarrow (l_{1_f}, l_{2_f}, ..., l_{n_f})$

4. A pilot study on a simulated home office

To perform a preliminary evaluation of the proposed approach, we considered a simple case study built on a simulated room. A desk and a portable heater must be put for realizing a comfortable home office in such a room. The size of the simulated room is $4 \times 3 \times 3$ m. The room plant is represented as a grid where the dimension of each cell is 0.5×0.5 m. The room has 3 windows (yellow rectangles W) and a door (blue rectangle D) located as it is shown in Fig.1. The room has $12m^2$ of parquet on the floor with absorption coefficient $\alpha_{abs} = 0.2$, $12m^2$ of roof with absorption coefficient $\alpha_{abs} = 0.1$, $9m^2$ of large panes made of 4mm glass with absorption coefficient $\alpha_{abs} = 0.3$ and $33m^2$ of walls with absorption coefficient $\alpha_{abs} = 0.1$. The portable heater produces a noise of $50dB$, and, for this study, a setpoint temperature of $23°C$ has been supposed. In the adjacent space outdoor of the room near the door, there is a domestic appliance (i.e. a washing machine) that produces a noise of $50dB$. The occupant's metabolic rate for typical office task is 1.2 met and we assume a medium clothes thermal insulation $I_{cl} = 1.0clo$ for winter season and an air velocity $v_a = 0.1m/s$. The relative humidity is $p_v = 50\%$. Acquisition of daylight interior illuminance values is obtained by the lighting simulation software Dialux Evo [22]. In particular, the room's central space is characterized by illuminance

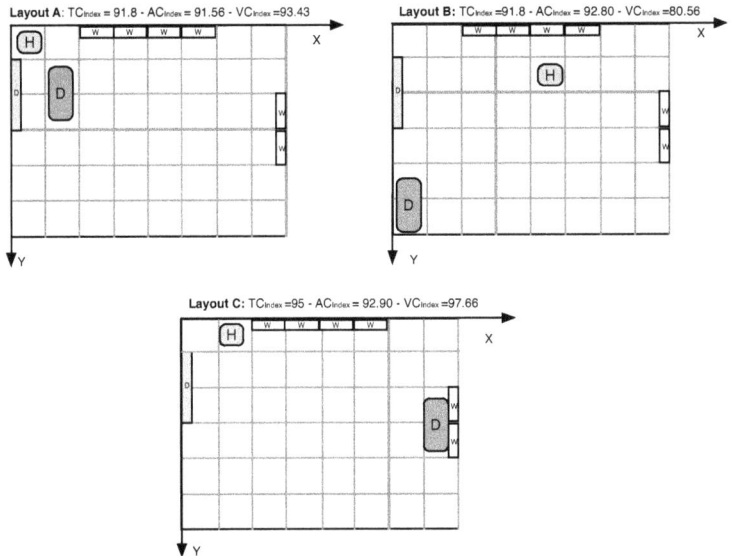

Figure 1. Layout configurations with different comfort conditions.

values around $300lx$, the area close to the windows shows about $500lx$. The area near the bottom left corner of the room has the lowest illuminance level (i.e., $150lx$). According to the IEQ standards, for the case under study, we consider as acceptable values of comfort index the following: $TC_{index} > 90$, $VC_{index} > 97$ and $AC_{index} > 95$.

In Fig.1, we show some three layout configurations at three different steps of the algorithm. *Layout A* shows an initial configuration of the room with the portable heater on the top left corner and the desk near the door. This furniture layout provides to an occupant that works at the desk the following index of comfort: $TC_{index} = 91.8$, $AC_{index} = 91.56$ and $VC_{index} = 93.43$. *Layout B* shows an intermediate step of the algorithm where the desk is on the bottom left corner of the room and the portable heater is near the windows of the opposite wall. In this case, we can see that $TC_{index} = 91.8$, $AC_{index} = 92.80$ and $VC_{index} = 80.56$. At it is expected, the low value of the $VC_{index} = 80.56$ is determined by the low illuminance level at the desk location. Finally, *Layout C* shows an optimal result obtained by the algorithm. In this case, the IEQ indexes show better values compared with the previous ones. In this state, the desk's location is near the window on the vertical wall, and the location of the portable heater is near the top left corner. We can see that $TC_{index} = 95$, $AC_{index} = 92.90$ and $VC_{index} = 97.66$. Since the standard for acoustics recommends $45dB$, the desk's distance from the portable heater and the door is not enough to reduce the combined effect of the two sources of noise at the recommended value. Thus, we cannot have a notable improvement in acoustic comfort for this room with the kind of furniture understudy. For this reason, the AC_{index} does not reach an adequate value. Conversely, a significant improvement is shown in the visual and thermal comfort determined by the windows' proximity, showing high illuminance values and better operative temperature. As we can see, the optimal configuration also respects the layout design constraints, as previously mentioned.

5. Conclusions

The paper presents an approach for reorganizing interior spaces for home-working. The problem is addressed using RL techniques to obtain optimized layout configuration in terms of Thermal, Visual and Acoustic comfort. Although the work is still preliminary, the results of some tests are encouraging. We are programming to conduct an extensive set of tests on different real cases and develop some software functionality that will allow the process to be automatized entirely without any external calculation and to cope with different kinds of environments.

References

[1] D. Mumovic, M. Santamouris, A Handbook of Sustainable Building Design and Engineering:" An Integrated Approach to Energy, Health and Operational Performance", Routledge, 2013.

[2] M. Frontczak, P. Wargocki, Literature survey on how different factors influence human comfort in indoor environments, Building and environment 46 (4) (2011) 922–937.

[3] A. Standard, 55-92. thermal environmental comfort conditions for human occupancy. ashrae (1992).

[4] B. Yang, L. Li, C. Song, Z. Jiang, Y. Ling, Automatic interior layout with user-specified furniture, Computers & Graphics 94 (2021) 124–131.

[5] Z. Liu, S. Tang, W. Xu, S. Bu, J. Han, K. Zhou, Automatic 3d indoor scene updating with rgbd cameras, in: Computer Graphics Forum, Vol. 33, Wiley Online Library, 2014, pp. 269–278.

[6] L. F. Yu, S. K. Yeung, C. K. Tang, D. Terzopoulos, T. F. Chan, S. J. Osher, Make it home: automatic optimization of furniture arrangement, ACM Transactions on Graphics (TOG)-Proceedings of ACM SIGGRAPH 2011, v. 30,(4), July 2011, article no. 86 30 (4).

[7] A. Coronato, G. Paragliola, A structured approach for the designing of safe aal applications, Expert Systems with Applications 85 (2017) 1–13.

[8] A. Coronato, G. De Pietro, Situation awareness in applications of ambient assisted living for cognitive impaired people, Mobile Networks and Applications 18 (3) (2013) 444–453.

[9] A. Coronato, M. Naeem, G. De Pietro, G. Paragliola, Reinforcement learning for intelligent healthcare applications: A survey, Artificial Intelligence in Medicine 109 (2020) 101964.

[10] M. Naeem, G. Paragliola, A. Coronato, A reinforcement learning and deep learning based intelligent system for the support of impaired patients in home treatment, Expert Systems with Applications 168 (2021) 114285.

[11] A. Coronato, G. De Pietro, Tools for the rapid prototyping of provably correct ambient intelligence applications, IEEE Transactions on Software Engineering 38 (4) (2011) 975–991.

[12] M. Naeem, S. T. H. Rizvi, A. Coronato, A gentle introduction to reinforcement learning and its application in different fields, IEEE Access (2020).

[13] J.-H. Choi, K. Lee, Investigation of the feasibility of poe methodology for a modern commercial office building, Building and Environment 143 (2018) 591–604.

[14] P. Li, T. M. Froese, G. Brager, Post-occupancy evaluation: State-of-the-art analysis and state-of-the-practice review, Building and Environment.

[15] P. O. Fanger, et al., Thermal comfort. analysis and applications in environmental engineering., Thermal comfort. Analysis and applications in environmental engineering.

[16] D. Markov, Practical evaluation of the thermal comfort parameters, Annual International Course: Ventilation and Indoor climate, Avangard, Sofia (2002) 158–170.

[17] E. UNI, 12464-1: 2011, Light and lighting. Lighting of work places. Part 1.

[18] L. L. Beranek, W. E. Blazier, J. J. Figwer, Preferred noise criterion (pnc) curves and their application to rooms, The Journal of the Acoustical Society of America 50 (5A) (1971) 1223–1228.

[19] C. J. Watkins, P. Dayan, Q-learning, Machine learning 8 (3-4) (1992) 279–292.

[20] M. Piasecki, K. Kostyrko, S. Pykacz, Indoor environmental quality assessment: Part 1: Choice of the indoor environmental quality sub-component models, Journal of Building Physics 41 (3) (2017) 264–289.

[21] H. B. Lewis, L. Bell, Industrial noise control, fundamentals and applications, new york: M (1994).

[22] DIAL, Dialux evo 8.2, professional lighting design software (2018). URL Available on line at https://www.dial.de/en/dialux-desktop/

Intelligent Environments 2021
E. Bashir and M. Luštrek (Eds.)
© 2021 The authors and IOS Press.
This article is published online with Open Access by IOS Press and distributed under the terms
of the Creative Commons Attribution Non-Commercial License 4.0 (CC BY-NC 4.0).
doi:10.3233/AISE210096

Inverse Reinforcement Learning Through Max-Margin Algorithm

Syed Ihtesham Hussain Shah [a,b,1], Antonio Coronato [b]

[a] *Dept. of ICT and Engineering, University of Parthenope, Italy*
[b] *Institute for high performance computing and networking (ICAR), National Research Council, Italy*

Abstract. Reinforcement Learning (RL) methods provide a solution for decision-making problems under uncertainty. An agent finds a suitable policy through a reward function by interacting with a dynamic environment. However, for complex and large problems it is very difficult to specify and tune the reward function. Inverse Reinforcement Learning (IRL) may mitigate this problem by learning the reward function through expert demonstrations. This work exploits an IRL method named Max-Margin Algorithm (MMA) to learn the reward function for a robotic navigation problem. The learned reward function reveals the demonstrated policy (expert policy) better than all other policies. Results show that this method has better convergence and learned reward functions through the adopted method represents expert behavior more efficiently.

Keywords. Reinforcement Learning, Inverse Reinforcement Learning, Markov Decision Process, Max-Margin Algorithm

1. Introduction

Machine Learning (ML) is an application of Artificial Intelligence (AI) that provides the ability to automatically learn and improve from experience without being explicitly programmed. ML focuses on the development of computer programs that can access data and use it to learn for themselves.

RL is one of the ML paradigms where agents solve problems by acquiring experiences via interactions with the environment [1]. The result of agent's interaction is a policy that can provide solutions to complex tasks without having specific knowledge about the underlying problem [2] in several fields (e.g., healthcare [3]). Due to the generalization capability of RL, it can be applied to more complex scenarios [4]. However, the specification of a reward function in advance is a problem in RL that may cause design difficulties.

Most of the time, researchers are supposed to specify the reward function manually to infer optimal decision [5,6]. However, the reward signals may be extremely scattered under this settings. Most rewards are almost zero. This technique may lead to the difficulty

---

[1]Corresponding Author: Syed Ihtesham Hussain Shah, Institute for high performance computing and networking (ICAR), National Research Council (CNR), Via Pietro Castellino, 111 Napoli – 80131, Italy. E-mail: ihtesham.shah@icar.cnr.it

of recognition which actions are useful in obtaining the ultimate feedback [7].

IRL can solve this problem where the derivation of the reward function is performed from expert demonstrations and behavior [8]. In the last 20 years, IRL has gained the attention of many researchers in fields of psychology, AI, ML and control theory. In IRL the goal is to learn the underlying rewards function for the expert demonstrations or trajectories.

In this paper, we adopt a technique of maximum margin IRL [9] for finding a best-fitting reward function for expert trajectories. The reward function can be considered as a linear combination of features function and weight vector. Features functions are predefined basis functions and the weight vector is computed by maximizing the margin between both the expert feature expectations and the estimated feature expectations. The solution is acceptable when this margin goes below a predefined threshold value. The result shows that learned reward functions under this scheme are true underlying reward functions for expert trajectories.

The rest of the paper is organized as follows: Related work in emerging and existing fields is discussed in section-2 that presents an overview of the state of the art. Section-3 comprises a brief review of the background and problem formulation, where basics about the Markov Decision Process (MDP) and IRL linear programming are presented. Section-4 introduces our proposed model. This section describes the problem of learning the reward function not explicitly, but through observing an expert demonstration. Experimental setup and results are discussed in section-5. We summarize the paper in section-6 by giving a conclusion and future directions.

2. Related work

RL has experienced growth in attention and interest due to promising results in intelligent environments [10–12] and the areas like: playing AlphaGo [13], controlling systems in robotics [14–16], medical [17], atari [18] and competitive video . A method of investigating challenges posed by reporting procedures, reproducibility and proper experimental techniques through Deep Reinforcement Learning (DRL) is discussed in [19]. Generally, Imitation Learning (IL) is categorized into three types: adversarial imitation learning (AIL) behaviour cloning (BC) and (IRL). Behaviour Cloning (BC) [20] directly maps states to an actions and learns policy through supervised learning. BC can avoid interacting with the environment. However, BC introduces a compounding error without considerable improvement during training.

In IRL [9, 21] the goal is to learn the reward function based on the expert demonstrations. It models the intention and preference of the demonstrator. Maximum Likelihood IRL (MLIRL) [22] estimate the gradient of the likelihood function. It defines that likelihood of the data set can be represented by the product of likelihood of the state-action pairs.

However, as compared with IRL, BC [23] learns policy directly by minimizing the Jensen-Shannon divergence between learned policy and expert policy [24]. Therefore different techniques is utilized to recover both uncertainty and rewards information. In a continuous state and action spaces Gaussian process [25] is utilized. Deep GP model [26] mounds multiple hidden GP layers. It has the ability of learning complicated reward structures with very limited demonstrations. Leveraged Gaussian processes [27] can learn from both negative and positive demonstrations. Apprenticeship Learning (AL) is to learn a reward function that illuminates the demonstrated policy [28] a margin better than alternative policies.

BC and RL [29, 30] are two major methods used now a days in robotics, autonomous car driving and healthcare sector. Dynamic Treatment Regime (DTR) [31–34] oversimplify personalized medicine and treatment is frequently tailored to a patient's dynamic-state. When the Electronic Health Record (EHR) are optimal and plentiful, BC can effectively recover the doctor's policies.

3. Preliminaries

3.1. Markov Decision Process.

MDP is used for decision making (e.g. [35]) and is based on Markov property, which does not consider past information when taking actions and depends on only current state. MDP is a form of tuple (S, A, T, R, γ) [36]. Where $S = (s_1, s_2, s_3, ...s_n)$ represent set of states that exist in the given environment. The agent can move to any of these states based upon the chosen action and transition function. $A = (a_1, a_2, ...a_n)$ is a set of actions that the agent can take in a state at every time step. $T(s_t, a_t, s(t+1))$ is a transition function which tells the probability of ending up in a state $s(t+1)$ if you take action a in state s in time step t. $\gamma \in [0, 1]$: is a discount factor that takes the value between zero and one. A value of zero gives more weight to immediate rewards and a value close to one gives more weight to long term rewards.

3.2. Reinforcement Learning

The rationale of RL is to interact with the environment without having any prior knowledge and to learn how to achieve a goal as stated in Figure 1. The component that interacts with the environment is called **Agent**. An agent takes an action a_t in a state s_t at time step t and the environment returns the next state and reward. The aim of the RL agent is to learn an optimal **Policy**, which tells which action to take in a state in order to maximize a long term reward. The aggregated reward is estimated as given in Eq. (1).

$$V^{\pi}(s) = E_{\pi}\{R_t | s_t = s\}$$

$$= \sum_{a \in A(s)} \pi(s, a) \sum_{s_{t+1} \in S} T^a_{ss_{t+1}} \{r(s, a) + \gamma V^{\pi}(s_{t+1})\} \tag{1}$$

Where V^{π} is the value function to estimate the accumulated reward when you start in state s_t and follow the current policy π. R_t is the reward function which is calculated as given below.

$$R_t = r_{t+1} + \gamma r_{t+2} + \gamma^2 r_{t+3} + ... \sum_{k=0}^{\infty} \gamma^k r_{t+k+1} \tag{2}$$

Similarly, the Q value function can be written as:

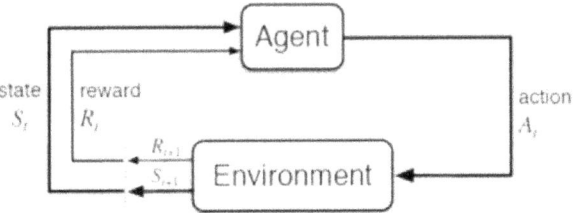

Figure 1. The Reinforcement Learning Problem

$$Q^{\pi}(s,a) = \sum_{\pi} \{R_t | s_t = s, a_t = a\} \tag{3}$$

Q function is the expected aggregated return when you take action a in state s under the current policy π for all state-action pairs afterwards. The goal is to find an optimal policy π^*. The state value function (V) or the state-action value function (Q) for optimal policy π^* using Bellman optimality equations are given below:

$$V^*(s) = \max_a \sum_{s_{t+1} \in S} T^a_{ss_{t+1}} (R^a_{ss_{t+1}} + \gamma V^*(s_{t+1})) \tag{4}$$

$$Q^*(s,a) = \sum_{s_{t+1} \in S} T^a_{ss_{t+1}} (R^a_{ss_{t+1}} + \gamma \max_{a_{t+1}} V^*(s_{t+1}, a_{t+1})) \tag{5}$$

Eqs. (4) and (5) explains the target of the optimal policy i.e. is to learn the policy that provides the maximum reward return and it is also considered as the fundamental strategy for several IRL methods.

3.3. Inverse Reinforcement Learning (IRL)

IRL is the problem of learning the preferences of an expert (agent) by observing its behaviour and avoiding the manual specification of a reward function. IRL flips the RL problem and attempts to extract the reward function from the observed behaviour of an agent. We have some expert trajectories that are supposed to generate the optimal policy. Therefore, the goal here is to find the reward function that is being implicitly optimised by the optimal policy π^*.

IRL is expressed as an optimization problem and Linear Programming (LP) [37] can be adopted to solve it by considering three types of state spaces. These state spaces are discussed next.

3.3.1. LP for finite state space

First, consider the optimal policy π^* for a MDP with a finite state space S and the policy transition matrix $T^{-\pi}_{n*n}$. For each state only one action provides an optimal solution from the set of K possible actions and all other $k-1$ actions are non-policy actions hence non policy transition matrix are $T^{-\pi} = (T^1, \ldots T^{k-1})$. An action is considered optimal only when the reward function satisfies the following equation.

$$(T^{-\pi} - T^{\pi})(I - \gamma T^{\pi})^{-1} R0 \tag{6}$$

3.3.2. LP for infinite state space

In infinite state spaces, transition probabilities can not be expressed in terms of matrix. The expected value based on sampled subset $s_0 \subset S$ is estimated and it is guaranteed that actions through the estimated policy always generate a lower expected value than the expert's policy actions.

$$\forall_{s_0 \in S}, \forall_{a \in A} \colon E_{\hat{s}T(\hat{s}|s,\pi(s))}\{V^{\pi}(\hat{s})\} \geq E_{\hat{s}\,T(\hat{s}|s,a)}\{V^{\pi}(\hat{s})\} \tag{7}$$

3.3.3. LP with sampled trajectories

It deals with the scenario where the knowledge about an exact policy is not given. An optimal policy based on some unknown reward function generates these trajectories. However, the goal is to calculate the empirical value of a trajectory rather than the expected value of the policy. We assume that there is a vector of features ϕ over states and by using the current estimated reward function $\hat{R}$ the corresponding value estimate $\hat{V}_i(s)$ for each value function can be calculated as:

$$\hat{V}_i(S) = \sum_{s_j \in S} \gamma^i \phi_i(s_j) \tag{8}$$

It is also considered that the optimal expert policy π^* always generates an higher empirical value $\hat{V}(\pi^*)$ than any other policy π^i.

$$\hat{V}(\pi^*) \geq \hat{V}(\pi^i) \tag{9}$$

4. System Model

Given the expert trajectories τ^E, the goal is to find the reward function that represents the best explanation of the expert behavior.

$$\begin{aligned}\tau^E &= [(s_1^1, a_1^1, s_2^1, a_2^1, .., s_d^1, a_d^1), (s_1^2, a_1^2, s_2^2, a_2^2, .., s_d^2, a_d^2), ...] \\ &= [\tau^1, \tau^2, \tau^3 ...]\end{aligned} \tag{10}$$

The rewards function can be considered as a linear combination of **k** known, fixed and bounded basis functions ϕ.

$$R(s,a) = w_1 \phi_1(s,a) + w_2 \phi_2(s,a) + \ldots\ldots + w_k \phi_k(s,a) \tag{11}$$

$$R(s,a) = \sum_i \{w_i \phi_i(s,a)\} \tag{12}$$

Where feature function $\phi_i : S \rightarrow \mathbf{R}$ and weights $w_i \in \mathbf{R}$. The policy π is defined as mapping states to an action and the value of a policy V is the sum of all the discounted rewards that is possible to collect by following that policy as given in Eq. (8). Expectation of a value function V^{π} is defined as:

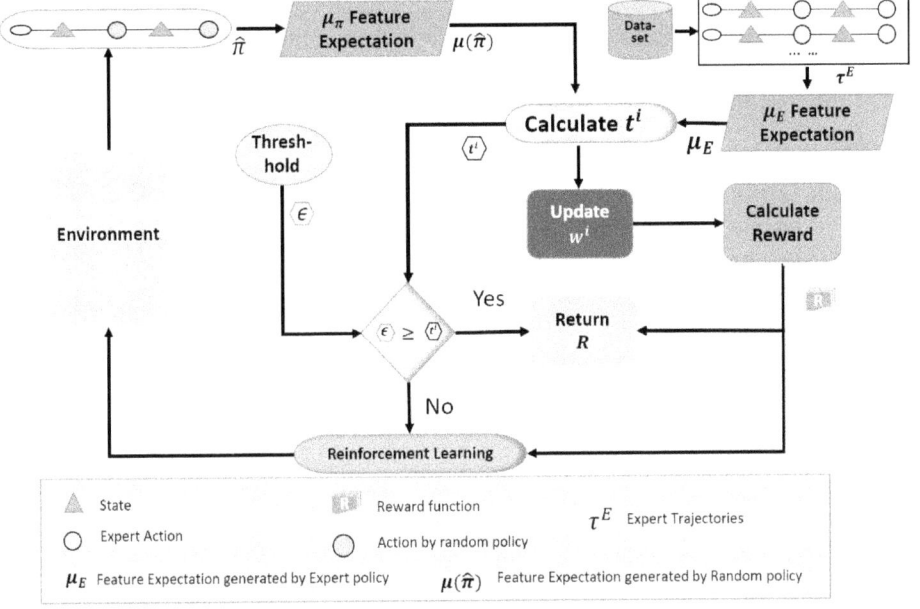

Figure 2. Proposed Layout for Max-Margin Algorithm

$$E[V^{\pi}(s_0)] = E\left[\sum_{t=1}^{\infty} {}^t R(s_t) | \pi\right] \tag{13}$$

$$= E\left[\sum_{t=1}^{\infty} {}^t w.\phi(s_t) | \pi\right] \tag{14}$$

$$= w.E\left[\sum_{t=1}^{\infty} {}^t \phi(s_t) | \pi\right] \tag{15}$$

$$= w.\mu_{\pi} \tag{16}$$

Algorithm-1 is used to find a weight vector w that minimizes the difference between the expert feature expectations μ_E and the estimated feature expectations $\mu_{\hat{\pi}}$. Let consider an MDP without reward MDP/R. Given the expert trajectory τ^E, we calculate expert's feature expectations μ_E. We also estimate the feature expectation $\mu(\hat{\pi})$ specifically by exploring a random policy $\hat{\pi}$. To find a policy $\hat{\pi}$ such that $||\mu_E - \mu\hat{\pi}||_2 \leq \varepsilon$ [38]:

$$E\left[\sum_{t=1}^{\infty} {}^t R(s_t) | \pi_E\right] - E\left[\sum_{t=1}^{\infty} {}^t R(s_t) | \hat{\pi}\right] \tag{17}$$

Algorithm 1 Max-margin Algorithm

1: **Given:** Expert trajectories τ^E generated by behavior policies π^E, discount factor γ, termination criteria ε,

2: **Initialize:** Feature matrix ϕ, number of iteration, $n = \infty$, Expert feature expectation $\mu_E = E[\sum_{t=1}^{\infty} {}^t R(s_t)|\pi_E]$

3: Randomly pick some policy $\hat{\pi}^0$.

4: **while** $(t \geq \varepsilon)$ **do**

5:      Compute t^i by using Eq. (20) and let the w^i be the value to attain this maximum.

6:      Compute Reward function $R(w^{(i)})^T \phi$.

7:      Using RL Algo. find policy $\hat{\pi}^{(i)}$ for this Reward

8:      Estimate Feature expectation $\mu(\hat{\pi}) = E[\sum_{t=1}^{\infty} \gamma^t \phi(s_t)|\hat{\pi}]$

9:      Set $i = +1$.

10: **end while**

11: **Return:** $R = (w^{(i)})^T \phi$

$$= |w^t \mu_E - w^t \mu(\hat{\pi})| \tag{18}$$

$$\leq ||w^t||_2 \, ||\mu(\hat{\pi}) - \mu_E||_2$$
$$\leq \varepsilon \tag{19}$$

Where $w : (||w||_1 \leq 1)$.

Maximum values of w that minimize the distance between the expert feature expectations and the estimated feature expectations are the true values for w [39].

$$t^i = \max_{w:||w||_2 \leq 1} \; \min_{j \in \{0..(i-1)\}} w^T (||\mu(\hat{\pi}) - \mu_E||_2) \tag{20}$$

The value of w generates the reward function through dot multiplication with the basis function ϕ.

The optimal policy for this reward function is computed through a RL algorithm. The estimated feature expectations for this kind of policy is calculated and compared with the expert feature expectations. We want to explore a policy which minimizes the euclidean distance stated in Eq. (20). A reward function for such a policy will be the true under laying reward function for the expert policy.

If the algorithm terminates with $t^{(n+1)} \leq \varepsilon$ then it means that there is at-least one policy from the set whose performance is as good as the expert's policy minus ε. For a solution μ_E is separated from μ^i by a margin of at most ε.

$$w^T \mu^i \geq w^T \mu_E - \varepsilon, \; \forall w \text{ with } ||w||_2 \leq 1 \tag{21}$$

$$\min ||\mu_E - \mu^i||_2 \leq \varepsilon \tag{22}$$

In this section we have introduced an approach of max-margin algorithm. Layout of proposed model is presented in Figure 2. At the initial stage, some policies are randomly

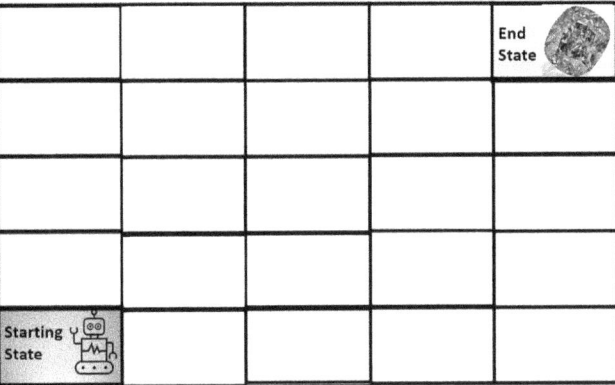

Figure 3. A sample gridworld problem

picked and feature expectation is calculated. The value of the threshold t^i is then computed and compared with a predefined value of ε. In case the value of t^i is not satisfactory, then the weight vector w^i is updated. The weight vector defines the values of the underlying reward function using dot multiplication with feature vector ϕ.

The optimal policy for the current reward function is estimated by utilizing the RL algorithm. The feature expectations for this policy are calculated and found the new value of t^i as in the previous step. Updating weight vector and learning optimal policy for the current reward function through RL is carried out repeatedly until it is found a policy whose feature expectations μ^i are nearly similar to the expert feature expectations μ_E and satisfies the equation given bellow:

$$t^i \leq \varepsilon \tag{23}$$

This policy achieves performance very close to those of the expert's policy. Therefore, the reward function that is used to estimate such an optimal policy through RL is considered the true underlying reward function for the expert trajectories.

5. Experiment

In this section, we report the experiment that we have conducted to evaluate the adopted model. First we have tested the algorithm on a gridword example followed by quantitative and qualitative studies.

A gridworld is a 2-D decision-making example for robotic navigation as stated in [40]. We have considered a 5x5 gridworld for our experiment as shown in Figure 3. The robot has to reach its destination, which is represented by the diamond. Except from the edge states, an agent can take four possible actions at each state including *up*, *down*, *left* and *right* with 30% chance of moving randomly. The grid is divided into non-overlapping regions and for each cell in the region there is a feature $\phi_i(s)$. We have generated different expert trajectories for these non-overlapping regions and hence calculated the expert feature expectation μ_E. On the other hand, some initial policies are also generated randomly and computed the estimated feature expectations.

The value of γ is 0.09 and the threshold is equal to 0.01. The adopted approach of MMA

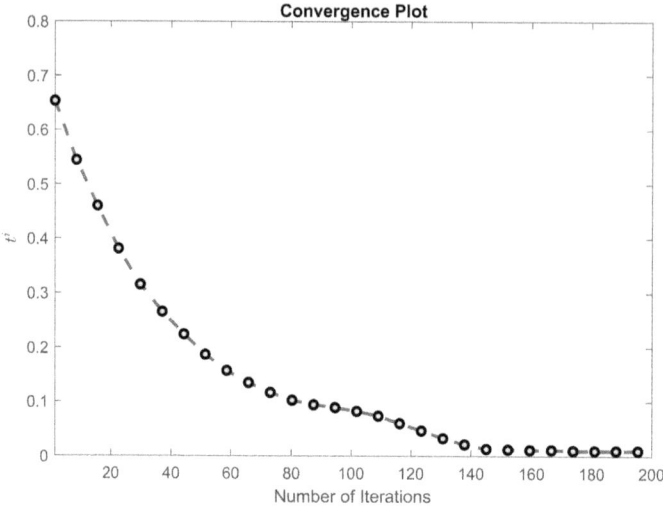

Figure 4. Difference between expert feature expectations and estimated feature expectations along with No. of iterations

is tested to recover the underlying reward function for the given expert trajectories. The difference between the expert feature expectations and the estimated feature expectations is calculated according to Eq. (20). The values of t along with the number of iterations are plotted in Figure 4. It can be seen that the algorithm convergences after a few iterations. The results for initial rewards and recovered rewards are shown in Figure 5 with different colours at each state.

The groundtruth reward represents the initial reward distribution for the gridworld. It can be seen that all states except the end state have zero reward initially. The top right corner (yellow box) represents the end state with the highest reward.

On the other hand, after having run successfully the adopted approach, the value of the

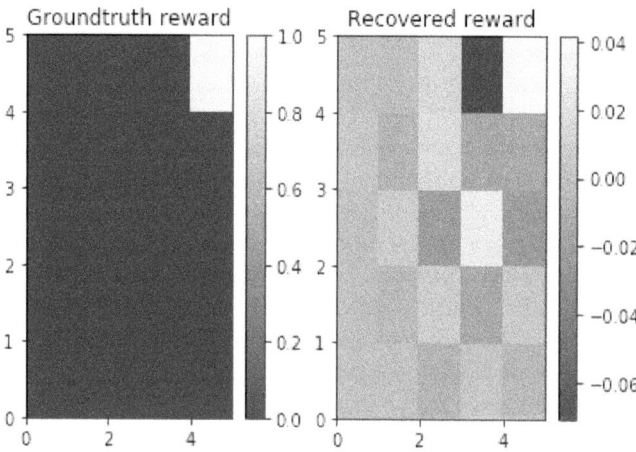

Figure 5. Groundtruth rewards represents the initial reward and after successful implementation of max-margin algorithm under-laying rewards are recovered

recovered rewards at each state is represented beside the "Groundtruth reward" and the scale of colors indicates the behavior and preference for the expert at each state.

6. Conclusion

In this paper, we have adopted a max-margin technique that is based on IRL. It focuses on determining the true underlying reward function for some given demonstrations. The MMA approach assumed that the reward function is shaped as a linear function of known features. This recovered reward function helps RL and MDP problems to mimic the expert behavior in an efficient way. Results achived for robotic navigation in the gridworld problem show that the MMA provides fast convergence.

List of Acronyms

AI Artificial Intelligence
MDP Markov Decision Process
RL Reinforcement Learning
DRL Deep Reinforcement Learning
DTR Dynamic Treatment Regime
IRL Inverse Reinforcement Learning
IL Imitation Learning
AL Apprenticeship Learning
ML Machine Learning
BC Behaviour Cloning
MLIRL Maximum Likelihood IRL
MMA Max-Margin Algorithm
LP Linear Programming

Acknowledgment

This work is partly supported by AMICO project which has received funding from the National Programs (PON) of the Italian Ministry of Education, Universities and Research (MIUR): code ARS0100900 (Decreen.1989, 26 July 2018)

References

[1] M. Naeem, S. T. H. Rizvi, and A. Coronato, "A gentle introduction to reinforcement learning and its application in different fields," *IEEE Access*, vol. 8, pp. 209 320–209 344, 2020.

[2] L. P. Kaelbling, M. L. Littman, and A. W. Moore, "Reinforcement learning: A survey," *Journal of artificial intelligence research*, vol. 4, pp. 237–285, 1996.

[3] A. Coronato, M. Naeem, G. De Pietro, and G. Paragliola, "Reinforcement learning for intelligent healthcare applications: A survey," *Artificial Intelligence in Medicine*, vol. 109, p. 101964, 2020.

[4] R. S. Sutton and A. G. Barto, *Reinforcement learning: An introduction.* MIT press, 2018.

[5]   M. K. Bothe, L. Dickens, K. Reichel, A. Tellmann, B. Ellger, M. Westphal, and A. A. Faisal, "The use of reinforcement learning algorithms to meet the challenges of an artificial pancreas," *Expert review of medical devices*, vol. 10, no. 5, pp. 661–673, 2013.

[6]   S. A. Murphy, "Optimal dynamic treatment regimes," *Journal of the Royal Statistical Society: Series B (Statistical Methodology)*, vol. 65, no. 2, pp. 331–355, 2003.

[7]   A. Raghu, M. Komorowski, L. A. Celi, P. Szolovits, and M. Ghassemi, "Continuous state-space models for optimal sepsis treatment-a deep reinforcement learning approach," *arXiv preprint arXiv:1705.08422*, 2017.

[8]   S. Zhifei and E. M. Joo, "A survey of inverse reinforcement learning techniques," vol. 5, no. 3, pp. 293–311, 2012.

[9]   P. Abbeel and A. Y. Ng, "Apprenticeship learning via inverse reinforcement learning," in *Proceedings of the twenty-first international conference on Machine learning*, 2004, p. 1.

[10] P. Ribino and C. Lodato, "A norm compliance approach for open and goal-directed intelligent systems," *Complexity*, vol. 2019, 2019.

[11] C. Di Napoli, P. Ribino, and L. Serino, "Customisable assistive plans as dynamic composition of services with normed-qos," *Journal of Ambient Intelligence and Humanized Computing*, pp. 1–26, 2021.

[12] P. Ribino and C. Lodato, "A distributed fuzzy system for dangerous events real-time alerting," *Journal of Ambient Intelligence and Humanized Computing*, vol. 10, no. 11, pp. 4263–4282, 2019.

[13] D. Silver, A. Huang, C. J. Maddison, A. Guez, L. Sifre, G. Van Den Driessche, J. Schrittwieser, I. Antonoglou, V. Panneershelvam, M. Lanctot *et al.*, "Mastering the game of go with deep neural networks and tree search," *nature*, vol. 529, no. 7587, pp. 484–489, 2016.

[14] N. Heess, G. Wayne, D. Silver, T. Lillicrap, Y. Tassa, and T. Erez, "Learning continuous control policies by stochastic value gradients," *arXiv preprint arXiv:1510.09142*, 2015.

[15] P. Ribino, M. Bonomolo, C. Lodato, and G. Vitale, "A humanoid social robot based approach for indoor environment quality monitoring and well-being improvement," *International Journal of Social Robotics*, pp. 1–20, 2020.

[16] M. Bonomolo, P. Ribino, and G. Vitale, "Explainable post-occupancy evaluation using a humanoid robot," *Applied Sciences*, vol. 10, no. 21, p. 7906, 2020.

[17] C. Lodato and P. Ribino, "A novel vision-enhancing technology for low-vision impairments," *Journal of medical systems*, vol. 42, no. 12, pp. 1–13, 2018.

[18] V. Mnih, K. Kavukcuoglu, D. Silver, A. Graves, I. Antonoglou, D. Wierstra, and M. Riedmiller, "Playing atari with deep reinforcement learning," *arXiv preprint arXiv:1312.5602*, 2013.

[19] P. Henderson, R. Islam, P. Bachman, J. Pineau, D. Precup, and D. Meger, "Deep reinforcement learning that matters," in *Proceedings of the AAAI Conference on Artificial Intelligence*, vol. 32, no. 1, 2018.

[20] D. A. Pomerleau, "Efficient training of artificial neural networks for autonomous navigation," *Neural computation*, vol. 3, no. 1, pp. 88–97, 1991.

[21] Y. Gao, J. Peters, A. Tsourdos, S. Zhifei, and E. M. Joo, "A survey of inverse reinforcement learning techniques," *International Journal of Intelligent Computing and Cybernetics*, 2012.

[22] H. Ratia, L. Montesano, and R. Martinez-Cantin, "On the Performance of Maximum Likelihood Inverse Reinforcement Learning," in *arXiv preprint arXiv*, 2012, p. 1202.1558.

[23] J. Ho and S. Ermon, "Generative adversarial imitation learning," in *Advances in neural information processing systems*, 2016, pp. 4565–4573.

[24] I. Goodfellow, J. Pouget-Abadie, M. Mirza, B. Xu, D. Warde-Farley, S. Ozair, A. Courville, and Y. Bengio, "Generative adversarial nets," in *Advances in neural information processing systems*, 2014, pp. 2672–2680.

[25] Z.-J. Jin, H. Qian, and M.-L. Zhu, "Gaussian processes in inverse reinforcement learning," in *2010 International Conference on Machine Learning and Cybernetics*, vol. 1.   IEEE, 2010, pp. 225–230.

[26] M. Jin, A. Damianou, P. Abbeel, and C. Spanos, "Inverse reinforcement learning via deep gaussian process," *arXiv preprint arXiv:1512.08065*, 2015.

[27] K. Lee, S. Choi, and S. Oh, "Inverse reinforcement learning with leveraged gaussian processes," in *2016 IEEE/RSJ International Conference on Intelligent Robots and Systems (IROS)*.   IEEE, 2016, pp. 3907–3912.

[28] J. A. Bagnell, N. Ratliff, and M. Zinkevich, "Maximum margin planning," in *Proceedings of the International Conference on Machine Learning (ICML)*.   Citeseer, 2006.

[29] L. Wang, W. Zhang, X. He, and H. Zha, "Supervised reinforcement learning with recurrent neural network for dynamic treatment recommendation," in *Proceedings of the 24th ACM SIGKDD International*

Conference on Knowledge Discovery & Data Mining, 2018, pp. 2447–2456.

[30] Y. Zhang, R. Chen, J. Tang, W. F. Stewart, and J. Sun, "Leap: Learning to prescribe effective and safe treatment combinations for multimorbidity," in *proceedings of the 23rd ACM SIGKDD international conference on knowledge Discovery and data Mining*, 2017, pp. 1315–1324.

[31] J. M. Robins, "Causal inference from complex longitudinal data," in *Latent variable modeling and applications to causality*. Springer, 1997, pp. 69–117.

[32] J. Robins, "A new approach to causal inference in mortality studies with a sustained exposure period—application to control of the healthy worker survivor effect," *Mathematical modelling*, vol. 7, no. 9-12, pp. 1393–1512, 1986.

[33] J. M. Robins, "Correcting for non-compliance in randomized trials using structural nested mean models," *Communications in Statistics-Theory and methods*, vol. 23, no. 8, pp. 2379–2412, 1994.

[34] S. M. Coghlan Jr, M. J. Connerton, N. H. Ringler, D. J. Stewart, and J. V. Mead, "Survival and growth responses of juvenile salmonines stocked in eastern lake ontario tributaries," *Transactions of the American Fisheries Society*, vol. 136, no. 1, pp. 56–71, 2007.

[35] A. Coronato and A. Cuzzocrea, "An innovative risk assessment methodology for medical information systems," *IEEE Transactions on Knowledge and Data Engineering*, pp. 1–1, 2020.

[36] M. L. Puterman, *Markov decision processes: discrete stochastic dynamic programming*. John Wiley & Sons, 2014.

[37] A. Y. Ng, S. J. Russell *et al.*, "Algorithms for inverse reinforcement learning." in *Icml*, vol. 1, 2000, p. 2.

[38] N. D. Ratliff, J. A. Bagnell, and M. A. Zinkevich, "Maximum margin planning," in *Proceedings of the 23rd international conference on Machine learning*, 2006, pp. 729–736.

[39] D. Choi, T.-H. An, K. Ahn, and J. Choi, "Future trajectory prediction via rnn and maximum margin inverse reinforcement learning," in *2018 17th IEEE International Conference on Machine Learning and Applications (ICMLA)*. IEEE, 2018, pp. 125–130.

[40] P. Crook and G. Hayes, "Learning in a state of confusion: Perceptual aliasing in grid world navigation," *Towards Intelligent Mobile Robots*, vol. 4, 2003.

Intelligent Environments 2021
E. Bashir and M. Luštrek (Eds.)
© 2021 The authors and IOS Press.
This article is published online with Open Access by IOS Press and distributed under the terms
of the Creative Commons Attribution Non-Commercial License 4.0 (CC BY-NC 4.0).
doi:10.3233/AISE210097

An Overview of Inverse Reinforcement Learning Techniques

Syed Ihtesham Hussain Shah [a,b,1], Giuseppe De Pietro [b]

[a] *Dept. of ICT and Engineering, University of Parthenope, Italy*
[b] *Institute for high performance computing and networking (ICAR), National Research Council, Italy*

Abstract. In decision-making problems reward function plays an important role in finding the best policy. Reinforcement Learning (RL) provides a solution for decision-making problems under uncertainty in an Intelligent Environment (IE). However, it is difficult to specify the reward function for RL agents in large and complex problems. To counter these problems an extension of RL problem named Inverse Reinforcement Learning (IRL) is introduced, where reward function is learned from expert demonstrations. IRL is appealing for its potential use to build autonomous agents, capable of modeling others, deprived of compromising in performance of the task. This approach of learning by demonstrations relies on the framework of Markov Decision Process (MDP). This article elaborates original IRL algorithms along with their close variants to mitigate challenges. The purpose of this paper is to highlight an overview and theoretical background of IRL in the field of Machine Learning (ML) and Artificial Intelligence (AI). We presented a brief comparison between different variants of IRL in this article.

Keywords. Inverse Reinforcement Learning, Markov Decision Process, Intelligent Environment

Introduction

ML is an application of AI that focuses on learning and improving itself from experience and without being explicitly programmed. ML emphasizes on developing algorithms that can access data and use it for self-learning [1,2,3] in an intelligent environments [4,5,6]. We are dealing with a certain number of sensors, which enable the IE [7] to be aware of the user's current action and goal. Human activities are observable through different sensors [8] and observations can be assumed to teach another environmental device or system to perform the task in a better way [9]. A typical way of teaching a system in a decision-making problem requires direct coding. However, another paradigm called Learning from Demonstrations has emerged to teach devices via demonstrations [10].

RL techniques provide a solution for decision-making problems [11]. RL agent interact with the dynamic environment, gains experience and improve itself [12]. The result of an agent's interaction is a policy that can provide solutions to complex tasks without having

---

[1]Corresponding Author: Syed Ihtesham Hussain Shah, Institute for high performance computing and networking (ICAR), National Research Council (CNR), Via Pietro Castellino, 111 Napoli – 80131, Italy. E-mail: ihtesham.shah@icar.cnr.it

specific information about the underlying problem [13]. Unlike supervised learning, RL does not need output labels that are sometimes not available or maybe expensive to find in real-time application. RL has better generalization abilities and can easily be applied to more complex scenarios [14] e.g. minimize medication errors [15], [16] and risk management [17].

However, the problem associated with RL is the reward function that has to be specified in advance. For complex and large problems, it is very difficult to specify and exhaustive to tune the reward function. To counter design difficulties of RL, IRL is introduced. IRL is an extension of RL problem where reward function is learned through expert demonstrations. In this paper, we present fundamental and advanced techniques of IRL [18] for finding the best-fitting reward function for expert trajectories.

The rest of the paper is organized as follows: Section-1 comprises a brief review of the background and problem formulation, where basics about the MDP is presented. Original IRL algorithm is discussed in section-2. This section describes the problem of learning the reward function not explicitly, but through observing an expert demonstration. Section-3 introduces an overview of different variants and extensions of IRL in emerging and existing fields. We summarize the paper in section-4 by giving a conclusion.

1. Theoretical Background

1.1. Markov Decision Process

Generally, RL algorithms satisfy Markov Decision Process (MDP) that are based on Markov property. It does not consider past information while taking actions in current state. MDP is a form of tuple (S, A, T, R, γ) [19] explained below:

- $S = s_1, s_2, s_3, \ldots s_n$ represent set of all possible states in the given environment. Based upon the chosen action and transition function the agent can move to any of these states.
- $A = a_1, a_2, a_3, \ldots a_n$ is a set of all possible actions that an agent can take in a state at every time step.
- $T(s_t, a_t, s(t+1))$ is a transition function which tells the probability of having into the state $s(t+1)$ by taking action a in current state s in time step t.
- $R(s_t, a_t, s(t+1))$ is a reward function which tells about the cost of taking action a in state s in time step t.
- $\gamma \in [0, 1]$: is a discount factor that takes the value between zero and one. A value close to zero gives more weight to current rewards while a value close to one gives more weight to long term rewards.

1.2. Value Function

A policy π is a function of mapping state s to an action a. It can be stochastic $\pi : S \rightarrow Prob(A)$ or deterministic $\pi : S \rightarrow A$ [20]. For a policy π, V^π is the value function which estimates the accumulated reward value in state s_t and follow the policy π. The value of a policy V^π for a given state s is given as:

$$V^{\pi}(s) = E_{\pi}\{R_t|s_t = s\} = \sum_{a \in A(s)} \pi(s,a) \sum_{s_{t+1} \in S} T^a_{ss_{t+1}}\{r(s,a) + \gamma V^{\pi}(s_{t+1})\} \quad (1)$$

Where V^{π} is the value function to estimate the accumulated reward and R_t is the reward function which is calculated as given below.

$$R_t = r_{t+1} + \gamma r_{t+2} + \gamma^2 r_{t+3} + \ldots \sum_{k=0}^{\infty} \gamma^k r_{t+k+1} \quad (2)$$

Q function is the expected aggregated return when you take action a in state s.

$$Q^{\pi}(s,a) = \sum_{\pi}\{R_t|s_t = s, a_t = a\} \quad (3)$$

The goal of solving MDP is to find optimal policy π^*. The state-action value function (Q) or state value function (V) for optimal policy π^* can be calculated by using Bellman equations as given below:

$$V^*(s) = \max_a \sum_{s_{t+1} \in S} T^a_{ss_{t+1}}(R^a_{ss_{t+1}} + \gamma V^*(s_{t+1})) \quad (4)$$

$$Q^*(s,a) = \sum_{s_{t+1} \in S} T^a_{ss_{t+1}}(R^a_{ss_{t+1}} + \gamma \max_{a_{t+1}} V^*(s_{t+1}, a_{t+1})) \quad (5)$$

Eqs. (4) and (5) explains the objective of learning the policy that provides the maximum reward return. Other machine learning techniques have successfully been applied to many real-world problems i.e. skill transfer to robots [8] and autonomous navigation [7] etc. However, these techniques are usually suffering from fewer training samples and poor generalization. The conventional terminologies of MDP and RL are usually accepting fix and predetermined reward functions. However, it is difficult to specify appropriate reward functions particularly for complex and large problems [10]. Therefore, the researcher gave a solution in form of IRL to tackle these limitations. IRL methods flip the RL problem and rather than finding optimal policy it tries to find underlying rewards for some given policies.

2. Inverse Reinforcement Learning (IRL)

Generally, IRL supposes that the expert is acting according to an underlying policy π_E. In some cases, the policy may not be known, and learners observe sequences of the expert's state-action pairs called trajectories. It follows some composition of a model that helps in learning the unknown rewards function for these trajectories. Conventional models include, a linearly weighted combination of reward features, non-linear (probability distribution over reward functions) or a neural network representation [21].
Typical framework of IRL is shown in Figure-1. Expert trajectories are supposed to generate optimal policy. Rather than the policy, the reward function is the briefest, robust,

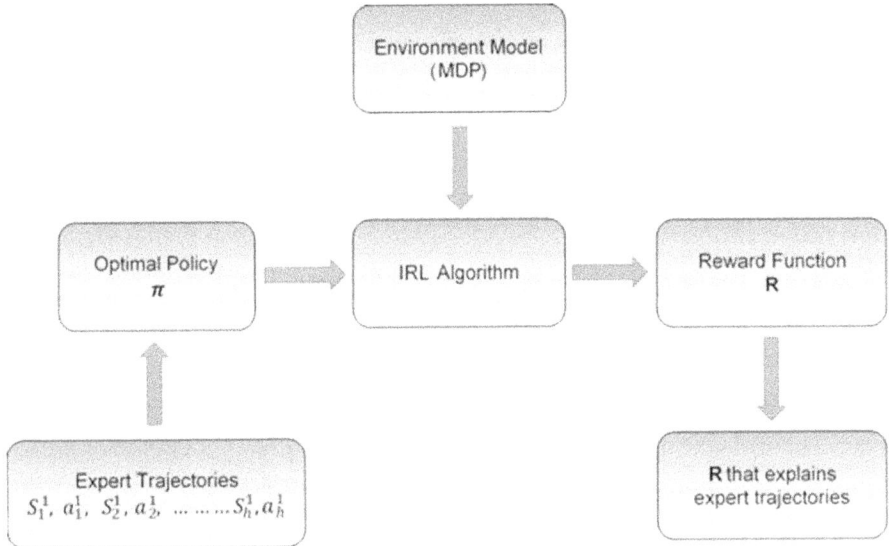

Figure 1. Inverse Reinforcement Learning Problem

and convenient definition of the task. It computes how bad or good certain actions are. Therefore, the goal here is to find the reward function that is being implicitly optimized by the optimal policy π^*. Once we have the true reward function, the problem is concentrated on finding the right policy, and standard reinforcement learning methods can be applied to solve the problem.

IRL as an optimization problem and proposed Linear Programming (LP) approach to solve the problem by considering three types of state spaces [22].

2.1. LP for finite state space

First, consider the set of optimal policies π for a MDP with a finite state space S and policy transition matrix $T_{n*n}^{-\pi}$. For each state only one action provide an optimal solution from set of K possible actions and all other $k-1$ actions are non-policy actions hence non policy transition matrix are $T^{-\pi} = T^1, \ldots T^{k-1}$. Action is considered optimal only when the reward function satisfy the following equation:

$$(T^{-\pi} - T^{\pi})(I - \gamma T^{\pi})^{-1} R0 \tag{6}$$

Where $T^{-\pi}$ is non policy transition matrix and T^{π} represents policy transition matrix. Many solutions satisfy Eq. (6). let say assigning zero reward value to all the states is always a solution. To remove this problem two methods are proposed.

Costly single-step deviation. It maximizes the distance between the optimal policy Q value and the second-best value among all the others.

$$\textbf{maximize} : \Sigma_{s \varepsilon S} \, Q^{\pi}(s, \pi(s)) - \max_{a \varepsilon A | \pi(s)} Q^{\pi}(s, a) \tag{7}$$

Eq. (7) successfully removes several degenerate reward functions but the reward function that is learned by this kind of heuristics might be different from the real one. To tackle this issue, a second method is introduced which assumes that reward function with **many small rewards** are more natural and should be preferred. Maximization of many terms leads to a minimization of the reward vector's norm i.e. $l_1 - norm: -\lambda \|\hat{R}\|_1$. A suitable value for parameter λ can be searched automatically. In addition, an upper bound value for reward function is also imposed.

$$\forall (s \varepsilon S) \quad |\hat{R}| \leq R_{max} \tag{8}$$

However, this approach cannot be applied to very large state spaces.

2.2. LP for Infinite State Space

For this kind of state-space researcher have used the function approximation to find a linear combination of d known, fixed and bounded basis function Φ_i [23].

$$\hat{R}(s) = \alpha_1 \phi_1(s) + \alpha_2 \phi_2(s) \ldots \ldots + \alpha_d \phi_d \tag{9}$$

Where $\alpha \to [0,1]$ is a constant. However, this linear combination comes with a huge disadvantage. We are not sure that the estimated reward function is the one with true value. Eq. (9) only expresses a smaller subset of all the possible reward functions. On the other hand, V^π can be expressed in term of basis function as follows:

$$V^\pi = \alpha_1 V_1^\pi + \ldots \ldots + \alpha_d V_d^\pi \tag{10}$$

Transition probabilities can't be expressed in the term matrix anymore because of infinite state space. Instead, the expected value based on sampled subset $s_0 \subset S$ is estimated and it is enforced that non-optimal actions always lead to lower expected value than optimal actions.

$$\forall_{s_0 \varepsilon S}, \forall_{a \varepsilon A} : E_{\acute{s} \, T(\acute{s}|s,\pi(s))} \{V^\pi(\acute{s})\} \geq E_{\acute{s} \, T(\acute{s}|s,a)} \{V^\pi(\acute{s})\} \tag{11}$$

A penalization factor p is introduced that define how much constraints can be penalized [18].

$$\max \sum_{s \in S_0} \min_{a \in A | \pi(s)} p \left[E_{\acute{s} \, T(\acute{s}|s,\pi(s))} (V^\pi(\acute{s})) - (E_{\acute{s} \, T(\acute{s}|s,a)}) (V^\pi(\acute{s})) \right] \tag{12}$$

This formulation maximizes the smallest difference between the expected value generated by policy action and all the other expected values it could find.

2.3. LP with Sampled Trajectories

This third algorithm is more closer to the real-world problem. It deals with the scenario where the knowledge about the exact policy is not given. we only observe trajectories that are combinations of states and actions. Optimal policy based on some unknown reward function generates these trajectories.
However, the goal is to calculate the empirical value of a trajectory rather than the ex-

---

Algorithm 1 Generic Algorithm for IRL

1: **Given:** Expert demonstrations, discount factor γ, termination criteria ε,
2: **Initialize:** Feature matrix ϕ, number of iteration n.
3: **while** (Termination criteria fulfil) **do**
4:     **for** $\forall\,(s,a)$ **do**
5:         Solve optimal value function V^* for MDP.
6:         Use V^* to define policy $\hat{\pi}$.
7:         Choose parameter α_i to make $\hat{\pi}$ more similar to demonstration;
8:     **end for**
9: **end while**
10: **Return:** $R(\hat{s},a)$

---

pected value of the policy. By using current estimated reward function $\hat{R}$ corresponding value estimate $\hat{V}_i(\xi)$ for each value function can be calculated as given below:

$$\hat{V}_i(\xi) = \sum_{s_j \in \xi} \gamma^j \phi_i\,(s_j) \tag{13}$$

γ is a discount factor. Empirical value for overall trajectory is :

$$\hat{V}(\xi) = \sum_{i=1}^{d} \alpha_i\,\hat{V}_i(\xi), \quad |\alpha_i| \leq 1, \; i = \{1,....,d\} \tag{14}$$

There are might be many expert trajectories generated at different initial positions. Now the goal is to find such a value of i that yields a higher empirical reward for optimal trajectory.

$$\hat{V}(\xi_{\pi^*}) \geq \hat{V}(\xi_{\pi^i}) \tag{15}$$

Eq. (15) represent that optimal expert trajectory always generates higher empirical value $\hat{V}(\xi_{\pi^*})$ than any other policy. Linear programming is used to find the best fitting parameter α_i. There are two important assumption made here [23]:

1. For any given policy trajectories can be generated.
2. Given any reward function, a policy which is optimal for this reward function can be generated.

To find the true empirical value, limit the maximum value of factor α_i and a penalization factor p (as stated in the previous section)is also introduced.

$$maximize\,[\sum_{i=1}^{m} \mathrm{p}\,\{\hat{V}(\xi_{\pi^*}) - \hat{V}(\xi_{\pi^i})\}] \tag{16}$$

A generic algorithm for IRL is given in Algorithm-1. The key idea is to use observation from the expert policy and find information about the underlying MDP. Estimation of the optimal policy is performed through the execution of the algorithm iteratively.

3. Foundation of IRL Variants

Many changes have been made in the fundamental IRL algorithm depending upon the nature of the application. We will try to give a brief introduction to some of them in this section.

3.1. Maximum Margin Planning

Let consider an MDP without reward MDP/R. In this context we have the same component as regular MDP except for the reward model. Given some expert's feature expectation μ_E and feature mapping ϕ, the goal is to find a policy that explains expert behavior perfectly on an unknown reward function. Reward function might be a linear combination of features.

$$R(s,a) = w_1\phi_1(s,a) + w_2\phi_2(s,a) + \ldots\ldots + w_k\phi_k(s,a) \tag{17}$$

$$R(s,a) = \sum_i \{w_i\phi_i(s,a)\} \tag{18}$$

Where feature function $\phi_i : S \rightarrow R$ and weights $w_i \in R$. To find a policy $\hat{\pi}$ such that $||\mu\hat{\pi} - \mu_E||_2 \leq \varepsilon$ we define:

$$E[\sum_{t=1}^{\infty}{}^t R(s_t)|\pi_E] - E[\sum_{t=1}^{\infty}{}^t R(s_t)|\hat{\pi}] = ||W^t|| \, ||\mu_E - \mu(\hat{\pi})||_2 \tag{19}$$

where $W : (||w||_1 \leq 1)$. Maximum margin algorithm are used to find such a policy $\hat{\pi}$ that minimize the difference between expert feature expectation μ_E and other random chosen policy expectation $\mu\hat{\pi}$.

3.2. Maximum Entropy IRL

Maximum entropy uses the probability approach to resolve the ill-posed issue associated with original IRL. Maximum entropy [24] is based on the principle that makes it free from the arbitrary assumption about the missing information of the system. The trajectories of experts also weighted with the estimated rewards and preference given to those policies with high rewards [25].

$$P_x(O_x|\theta) = [1/Z_i]exp(\alpha_x E(O_x,\theta)) \tag{20}$$

In Eq. (20), θ represents the parameters of the reward function and R is a linear combination of features as discussed in the previous section. Through the maximum likelihood approach of observed trajectory, we can get the optimum value of θ [26].

$$\theta^* = \arg\max_\theta L(\theta) \tag{21}$$

$$\theta^* = \arg\max_{\theta} \sum_o \log P_x(O_x|\theta) \tag{22}$$

Gradient-based optimization method can be used to obtain the optimum values for deterministic MDP. The gradient is presented by the difference between expected feature values by the learner and expected empirical feature values. Early IRL based methods are countered with the problem of label bias mentioned in [27]. It only considers actions in run time while trajectories are compared later-on by taking the actions instead of comparing before. Consequently, the best policy having the highest reward may not be the one with the highest probability [28]. Maximum entropy encounters this problem by focusing on distribution over trajectories rather than actions. This algorithm applied in many applications i.e. to predict the driver behavior, route recommendation for taxi driver [29].

3.3. Non-linear Programming

Max-Margin Planning (MMP) still assumes a linear form of reward function but another approach LEArning and seaRCH (LEARCH) [30] have introduced for nonlinear behavior of reward function. This algorithm is tested in autonomous navigation where an agent (vehicle) was operated in complex unstructured terrain [31]. A parallel approach was also introduced and used in a visual navigation system where costs to detected objects and a suitable path to the drivers in the current situation have been assigned automatically [32]. Self-Imitation Learning (IL) [33] proposes that without any feedback from an external expert, the policy can be learned iteratively after the agent makes mistakes and decisions.

3.4. Maximum Likelihood IRL

Maximum Likelihood IRL (MLIRL) [34] problem utilizing an estimate of the gradient of the likelihood function. The algorithm defines that likelihood of the expert data-set $\mathscr{L}_\theta(O)$ can be represented by the product of the likelihood of the state-action pairs.

$$\mathscr{L}_\theta(O) = \prod_{i=1}^{k} l_\theta(s_i, a_i) \tag{23}$$

Where O represents experts demonstration. The reward function is estimated by Maximizing the log-likelihood function.

$$R_\theta^* = \arg\max_{R} \log \mathscr{L}_\theta(O) \tag{24}$$

Eq. (24) assumes that actions are more likely to be selected having higher Q^* values.

3.5. Gradient-based IRL

The neural gradient [29] approach was used to refine the reward function instead of directly modifying the policy. Policies are derived from the RL algorithm for the current reward function in MMP. Therefore, policy space depends on the reward function, and any change in the parameter space θ^* affects the policy space. So a neural network is used to solve θ^* by gradient approach. By using the gradient descent method, IRL stretched to several target settings [35]. In [27] an idea was introduced where any deviation from the expert's trajectory was corrected.

3.6. Monte-Carlo Markov Chain (MCMC)

To approximate the posterior $P(R|O_x)$ a Monte-Carlo Markov Chain (MCMC) algorithm proposed in [36]. This algorithm generates a sample set of reward function distributions $\{r_1, r_2, \ldots . . r_N\}$ according to the targets distribution.

$$P(R|O_x) \approx [1/N] \sum_{i=1}^{N} \delta(R, r_i) \tag{25}$$

These r_i samples are related to a trajectory of a Markov chain. It is considered that its invariant distributions are the counterpart of the target distribution [36]. For a large dimension problem, the drawback of this algorithm is that it requires a large number of sample rewards distributions to guarantee that the estimation is precisely represented by the sample set [37].

4. Conclusion

The reward function is an essential parameter for RL to estimate the best policy. In many applications, it is difficult to specify the true reward function. IRL provide a solution to this problem and has been an attractive field for a researcher for the last decades. It focuses on determining the true underlying reward function for given demonstrations in IE. In this paper, we have presented a conventional model of IRL while some modification and advancement of existing IRL techniques are also discussed here. The experiences of IRL have altered a lot from its first introduction. Improvements empower IRL to be implemented in more complex and practical applications.

List of Acronyms

AI Artificial Intelligence
MDP Markov Decision Process
RL Reinforcement Learning
IRL Inverse Reinforcement Learning
IL Imitation Learning
ML Machine Learning
MLIRL Maximum Likelihood IRL
MMP Max-Margin Planning
LP Linear Programming
IE Intelligent Environment
LEARCH LEArning and seaRCH

References

[1] P. Ribino, M. Bonomolo, C. Lodato, and G. Vitale, "A humanoid social robot based approach for indoor environment quality monitoring and well-being improvement," *International Journal of Social Robotics*, pp. 1–20, 2020.

[2] M. Bonomolo, P. Ribino, and G. Vitale, "Explainable post-occupancy evaluation using a humanoid robot," *Applied Sciences*, vol. 10, no. 21, p. 7906, 2020.

[3] C. Lodato and P. Ribino, "A novel vision-enhancing technology for low-vision impairments," *Journal of medical systems*, vol. 42, no. 12, pp. 1–13, 2018.

[4] P. Ribino and C. Lodato, "A norm compliance approach for open and goal-directed intelligent systems," *Complexity*, vol. 2019, 2019.

[5] C. Di Napoli, P. Ribino, and L. Serino, "Customisable assistive plans as dynamic composition of services with normed-qos," *Journal of Ambient Intelligence and Humanized Computing*, pp. 1–26, 2021.

[6] P. Ribino and C. Lodato, "A distributed fuzzy system for dangerous events real-time alerting," *Journal of Ambient Intelligence and Humanized Computing*, vol. 10, no. 11, pp. 4263–4282, 2019.

[7] A. Coronato and G. De Pietro, "Tools for the rapid prototyping of provably correct ambient intelligence applications," *IEEE Transactions on Software Engineering*, vol. 38, no. 4, pp. 975–991, 2011.

[8] M. Al-Faris, J. Chiverton, D. Ndzi, and A. I. Ahmed, "A review on computer vision-based methods for human action recognition," *Journal of Imaging*, vol. 6, no. 6, p. 46, 2020.

[9] A. Coronato and G. De Pietro, "Formal design of ambient intelligence applications," *Computer*, vol. 43, no. 12, pp. 60–68, 2010.

[10] A. G. Billard, S. Calinon, and R. Dillmann, "Learning from humans," *Springer handbook of robotics*, pp. 1995–2014, 2016.

[11] A. Coronato, M. Naeem, G. De Pietro, and G. Paragliola, "Reinforcement learning for intelligent healthcare applications: A survey," *Artificial Intelligence in Medicine*, vol. 109, p. 101964, 2020. [Online]. Available: https://www.sciencedirect.com/science/article/pii/S093336572031229X

[12] M. Naeem, S. T. H. Rizvi, and A. Coronato, "A gentle introduction to reinforcement learning and its application in different fields," *IEEE Access*, 2020.

[13] L. Pack Kaelbling, M. L. Littman, A. W. Moore, and S. Hall, "Reinforcement Learning: A Survey," *Journal of Artiicial Intelligence Research*, vol. 4, pp. 237–285, 1996.

[14] R. S. Sutton and A. G. Barto, "Reinforcement Learning: An Introduction, Second Edition," 2018, vol. 258, no. 6685, pp. 675–676.

[15] M. Naeem, G. Paragliola, and A. Coronato, "A reinforcement learning and deep learning based intelligent system for the support of impaired patients in home treatment," *Expert Systems with Applications*, p. 114285, 2020.

[16] M. Naeem, G. Paragiola, A. Coronato, and G. De Pietro, "A cnn based monitoring system to minimize medication errors during treatment process at home," in *Proceedings of the 3rd International Conference on Applications of Intelligent Systems*, 2020, pp. 1–5.

[17] G. Paragliola and M. Naeem, "Risk management for nuclear medical department using reinforcement learning algorithms," *Journal of Reliable Intelligent Environments*, vol. 5, no. 2, pp. 105–113, 2019.

[18] P. Abbeel and A. Y. Ng, "Apprenticeship learning via inverse reinforcement learning," in *Proceedings of the twenty-first international conference on Machine learning*, 2004, p. 1.

[19] M. L. Puterman, *Markov decision processes: discrete stochastic dynamic programming*. John Wiley & Sons, 2014.

[20] S. Zhifei and E. M. Joo, "A review of inverse reinforcement learning theory and recent advances," in *2012 IEEE congress on evolutionary computation*. IEEE, 2012, pp. 1–8.

[21] A. Coronato and A. Cuzzocrea, "An innovative risk assessment methodology for medical information systems," *IEEE Transactions on Knowledge and Data Engineering*, pp. 1–1, 2020.

[22] A. Y. Ng, S. J. Russell *et al.*, "Algorithms for inverse reinforcement learning." in *Icml*, vol. 1, 2000, p. 2.

[23] S. Zhifei and E. M. Joo, "A survey of inverse reinforcement learning techniques," *International Journal of Intelligent Computing and Cybernetics*, 2012.

[24] E. T. Jaynes, "Information theory and statistical mechanics. II," *Physical Review*, vol. 108, no. 2, pp. 171–190, oct 1957.

[25] Z. Shao and M. J. Er, "A review of inverse reinforcement learning theory and recent advances," *2012 IEEE Congress on Evolutionary Computation, CEC 2012*, pp. 10–15, 2012.

[26] B. D. Ziebart, A. Maas, J. A. Bagnell, and A. K. Dey, "Maximum entropy inverse reinforcement learning," in *Proceedings of the National Conference on Artificial Intelligence*, vol. 3, 2008, pp. 1433–1438.

[27]   J. Lafferty, A. Mccallum, and F. C. N. Pereira, "Departmental Papers (CIS) Conditional Random Fields : Probabilistic Models for Segmenting and Labeling Sequence Data," in *Machine Learning*, vol. 2001, no. Icml, 2001, pp. 282–289.

[28]   B. D. Ziebart, A. L. Maas, A. K. Dey, and J. A. Bagnell, "Navigate like a cabbie: Probabilistic reasoning from observed context-aware behavior," in *UbiComp 2008 - Proceedings of the 10th International Conference on Ubiquitous Computing*, 2008, pp. 322–331.

[29]   F. S. Melo and M. Lopes, "Learning from demonstration using MDP induced metrics," in *Machine Learning and Knowledge Discovery in Databases*, vol. 6322 LNAI, no. PART 2, 2010, pp. 385–401.

[30]   N. D. Ratliff, D. Silver, and J. A. Bagnell, "Learning to search: Functional gradient techniques for imitation learning," *Autonomous Robots*, vol. 27, no. 1, pp. 25–53, jul 2009.

[31]   D. Silver, J. A. Bagnell, and A. Stentz, "Learning from demonstration for autonomous navigation in complex unstructured terrain," *International Journal of Robotics Research*, vol. 29, no. 12, pp. 1565–1592, 2010.

[32]   C. Pradalier, R. Siegwart, and G. Hirzinger, "Springer Tracts in Advanced Robotics: Preface," Berlin, Heidelberg, 2011.

[33]   J. Oh, Y. Guo, S. Singh, and H. Lee, "Self-imitation learning," in *International Conference on Machine Learning*.   PMLR, 2018, pp. 3878–3887.

[34]   H. Ratia, L. Montesano, and R. Martinez-Cantin, "On the Performance of Maximum Likelihood Inverse Reinforcement Learning," in *arXiv preprint arXiv*, 2012, p. 1202.1558.

[35]   A. Boularias, J. Kober, and J. Peters, "Relative entropy Inverse Reinforcement Learning," in *Journal of Machine Learning Research*, vol. 15, 2011, pp. 182–189.

[36]   C. Andrieu, N. De Freitas, A. Doucet, and M. I. Jordan, "An introduction to MCMC for machine learning," *Machine Learning*, vol. 50, no. 1-2, pp. 5–43, jan 2003.

[37]   D. Ramachandran and E. Amir, "Bayesian inverse reinforcement learning," *IJCAI International Joint Conference on Artificial Intelligence*, pp. 2586–2591, 2007.

Intelligent Environments 2021
E. Bashir and M. Luštrek (Eds.)
doi:10.3233/AISE210098

Prediction of Breast Cancer Using AI-Based Methods

Sanam AAMIR [a], Aqsa RAHIM [a], Sajid BASHIR [b] and Muddasar NAEEM [c]

[a] *National University of Science And Technology*
[b] *National University of Technology*
[c] *ICAR-CNR*

Abstract. Breast cancer has made its mark as the primary cause of female deaths and disability worldwide, making it a significant health problem. However, early diagnosis of breast cancer can lead to its effective treatment. The relevant diagnostic features available in the patient's medical data may be used in an effective way to diagnose, categorize and classify breast cancer. Considering the importance of early detection of breast cancer in its effective treatment, it is important to accurately diagnose and classify breast cancer using diagnostic features present in available data. Automated techniques based on machine learning are an effective way to classify data for diagnosis. Various machine learning based automated techniques have been proposed by researches for early prediction/ diagnosis of breast cancer. However, due to the inherent criticalities and the risks coupled with wrong diagnosis, there is a dire need that the accuracy of the predicted diagnosis must be improved. In this paper, we have introduced a novel supervised machine learning based approach that embodies Random Forest, Gradient Boosting, Support Vector Machine, Artificial Neural Network and Multilayer Perception methods. Experimental results show that the proposed framework has achieved an accuracy of 99.12%. Results obtained after the process of feature selection indicate that both preprocessing methods and feature selection increase the success of the classification system.

Keywords. Breast Cancer, WDBC, RFE, k-NN, Support Vector Machines, Random Forest, Decision Tree, ANN, MLP, Machine Learning

1. Introduction

In the modern world, individuals are more vulnerable to cancer than they have ever been. Cancer is a fatal disease that is present worldwide. Approximately 9.6 million people died because of cancer in 2018. One out of six deaths are caused by cancer globally [36]. Nearly 70% of all cancer-related deaths occur in middle-income and low-income countries. Other reasons contributing to cancer-related deaths are low fruit and vegetable consumption, body mass index, lack of physical activity and alcohol use. A research estimated the death of approx. 40,920 women in 2018 due to breast cancer alone [36]. According to statistics of the World Health Organization (WHO), 2.09 million women are diagnosed with breast cancer every year [1]. As is the case with any type of cancer, early diagnosis is the only cure to breast cancer as well. Researchers and scientists have conducted a variety of experiments for early detection of breast cancer, so that it can be effectively cured and risks to patients' life may be significantly reduced.

Cancer is a term that is used for a large group of diseases that affect various parts of the human body. Cancer is mainly characterized by spread of abnormal cells rapidly. This spread of abnormal cells is so rapid that cells go beyond their normal limits and invade adjacent parts of the body and spread to other organs which causes deterioration finally leading to death. This process is known as metastasis [1]. The main cause of cancer-related deaths is metastasis. Other terms used for cancer are malignant tumors and neo-plasms [1]. Breast cancer in women has an extremely high mortality rate. In Breast Cancer, rapidly dividing cells form breast masses in breast cancer. Such masses are named as tumors that are malignant (cancerous) of benign (non-cancerous) Malignant Tumors penetrate healthy tissues in the body and cause damage. Cancer spreads when malignant cells infect healthy cells, and these malignant cells spread very quickly [16]. Therefore, it is very important that not only cancer is diagnosed, but also that it is diagnosed in its early stage so that it can be cured. Evolution in the field of technology and medicine has led to availability of a large amount of data that is stored and provided to researchers that make innovative use of several techniques for detection of the disease.

The main challenge is to detect if the tumor is benign or malignant. Various models have been developed so for detection of breast cancer. The models have used risk factors, blood analysis data and features extracted from x-ray images to detect cancer. Features are extracted from Breast Cancer mammograms (x-ray images), that include many at-tributes i.e. Clump Thickness, Marginal Adhesion, Uniformity of Cell Size/ Shape, Sin-gle Epithelial Cell Size, Bland Chromatin, Normal Nucleoli, Bare Nuclei and Mitosis. They are normally used for the diagnosis and detection of breast cancer. Other researches have used risk factors i.e. age, number of previous biopsies, race, number of first-degree relatives affected with breast cancer, not only to predict breast cancer at the first instance but to predict its recurrence as well. Researchers have also used blood analysis data, in-cluding BMI, age, HOMA, Glucose, Leptin, Resistin, MCP.1 and Adiponectin to diag-nose Breast Cancer using Machine Learning techniques. Other information that has been utilized for aiding diagnosis of Breast Cancer include race/ethnicity, pregnancy history, breastfeeding history, being overweight, exposure of chest or face to radiation before the age of 30, exposure to chemicals, low vitamin D levels, menstrual history and lack of exercise.

Recently, the development in computing technology and the introduction of new machine learning algorithms e.g. reinforcement learning [23], neural network [21] the goal of Ar-tificial Intelligence (AI) has become a step closer. AI has important application in diverse fields including:healthcare [12], [22], robotics and autonomous control, vision enhancing method for low vision impairments [19], natural language processing, dynamic norma-tive environments [31], risk management [26], intelligent environments [10], games and self-organized system [11], ambient assisted living techniques to improve the quality of life of elderly [13], Social humanoid robot [9] can help to monitor indoor environmental quality [29] and distributed fuzzy system able to infer in real-time critical situations [30].

In this paper, we propose a novel approach for the diagnostic prediction of Breast Cancer by careful feature selection and data handling. The diagnostic characteristics from the WDBC dataset [5] are used. Our aim is to predict the tumor as malignant or benign with a reduced set of features and improved accuracy.

The remaining part of the paper is organized as follows: section 2 discusses literature, section 3 discusses the proposed work of this study, section 4 explains the methodology and experimentation, section 5 gives detailed information about the experimental results and analyzes the results of the experiment and provides comparative analysis with previous studies, section 6 discusses results of application of the framework on other datasets and section 7 concludes the work.

2. Literature Review

Cancer spreads when malign cells infect healthy cells, and these malign cells spread very quickly [1]. Therefore, it is very important that not only cancer is diagnosed, but also that it is diagnosed in its early stage so that it can be cured. Evolution in the field of technology and medicine has led to availability of a large amount of data that is stored and provided to researchers that contributed to innovative use of several techniques for detection of the symptoms of the disease. A challenge that radiologists usually face is that after the tumor is discovered, how to distinguish if it is a malignant or benign tumor [16].

In recent years, modern research and machine learning techniques have been taken into consideration for the treatment of Breast Cancer. Breast Cancer is not only detected earlier but also predictions can be made about whether a person will be able to survive it and about how likely is that the cancer cells will start recurring [25]. Careful analysis of different models and features can help in the development of a model that achieves high accuracy. SVM and ANN are two of the most widely used classification algorithms for solving the breast cancer prediction problem [24].

The most common datasets used for prediction of Breast Cancer are the Breast Tissue Dataset (BTD) [3], the Coimbra Breast Cancer (CBC) dataset [2], the SEER Breast Cancer (SEERBCD) dataset [4], the Wisconsin (Original) Breast Cancer (WOBC) dataset [6], the Wisconsin (Diagnostic) Breast Cancer (WDBC) dataset and the Wisconsin (Prognostic) Breast Cancer (WPBC) dataset [7]. BTD [3] contains the impedance measurement of freshly excised breast tissue that were obtained at various different frequencies. It comprises of a total of 6 classes. The CBC [2] dataset contains anthropometric data and parameters that are gathered in routine blood analysis. There are a total of 10 predictors. The predictors are quantitative and the presence or absence of breast cancer is indicated by a binary dependent variable. The SEERBCD was obtained in November 2017 under the SEER program of NCI.

This program provides information on population-based cancer statistics. It contains the data of 4024 patients that includes their age, race, marital status, tumor size, estrogen status, progesterone status, information about regional nodes and some other factors. The WOBC dataset was obtained from clinical cases reported by Dr. Wolberg. The databases reflect chronological grouping of data with 8 groups having number of cases recorded between January 1989 and November 1991. The WPBC contains information about 30 features that are computed from digitized images. Each record is a representation of the follow-up data for a single breast cancer case. It consists of a total of 34 attributes. The WDBC dataset is similar to WPBC.

Authors in [17] predicted a patient survival chance, by feeding various prognostic variables involving time factor, to the neural network. The predict results of the neural network were compared to that of a regression model, using data of 1373 patients to determine the performance of the models. For predictions of malignant probabilities for

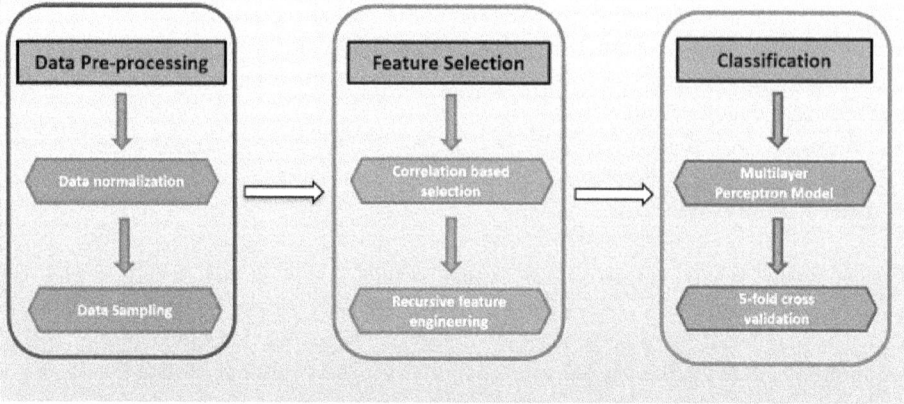

Figure 1. Flowchart of BCAD Framework

non-recurring cases and the recurring time period of diseases, the work of [28] designed a linear diagnostics model. This model was tested on the dataset for 569 patient and involved a cross-validation approach, giving the accuracy of up to 97.5%.

Quinlan in [20] also developed a model for medical diagnostics and predictions by incorporating Minimum Description length (MDL) penalty to the C4.5 decision tree algorithm, gaining accuracy of 94.74%. Delen in [27] then compared the Decision Tree model such as C4.5 with Neural Networks and Linear Regression models, using large datasets of around 200000 patient records. They concluded that a Decision tree algorithm such as C4.5 can often outperform the other two, on large datasets, by achieving accuracy of around 93.6% or more.

Feature selection and classification effect accuracy up to a great extent. Feature selection improves the overall classification accuracy and processing time since if proper features are selected, then it will not only benefit accuracy but also improve overall training time. Most of the above studies focused mainly on preparing the data using feature selections, feature extraction, various forms of data representations, and tuning model parameter, instead of dealing with model structures and changing them to obtain better performances. Even in model tuning, processes such as differential evolution leads to greater risk of overfitting the model.

3. System Model

Figure 1 shows the flowchart of BCAD Framework including data preprocessing steps, feature selection using filter-based feature selection methods and classification using Multilayer perceptron model.

First, data pre-processing is performed, data is checked for any inconsistencies or missing values, followed by random sampling. Sampling generates a unique sampling distribution that is based on the actual data. Sampling is done so that the result can be a more accurate estimate of the features selected. The data is then normalized.

Feature selection is then performed by correlation-based selection and recursive feature elimination methods. First, a correlation analysis of the features is done. The features that are highly co-related are set aside and one of them is selected in such a manner that

Table 1. COMPARISON OF MACHINE LEARNING ALGORITHMS ON WISCONSIN BREAST CAN-
CER DATASET

Paper	Features	Classifier	Validation	Accuracy
[8]	30	Gradient Boosting	10-fold	98.88%
[34]	20	Random Forest	10-fold	98.77%
[14]	–	K-means Clustering	–	92.00%
[32]	30	SMO	10-fold	97.71%
Our approach	11	MLP	5-fold	99.12%

if three of the features are highly correlated, then one of them is selected. The reason for this is that highly correlated features have the same impact on the result. Since there contribution to the result will be same, using all the features will not be a sensible approach, as the impact on the result would be the same. This results in a reduced feature set of 16 features. Recursive feature elimination (RFE) is then applied to the data in order to obtain the optimal features. This results in the resulting set of 11 features that are then sent as input to the classification models.

4. Results

Table 1 represents various machine learning models proposed by different authors and their achievements. In the approach proposed in this paper, for the purpose of feature selection, I have first eliminated the features that were highly correlated in such a manner that only one of the features that are highly correlated are selected i.e. if three of the features are highly correlated, one of them is selected. This results in a reduced feature set of 16 features. Further, I have applied recursive feature selection method to further reduce the number of features. This results in the selection of 11 best features for the purpose of classification that are sent to the classification models. The best accuracy is achieved by random forest.

The proposed approach was aimed at Feature space selection in order to test the influence of the feature space. For this purpose, a hybrid of correlation-based feature selection and Recursive feature elimination (RFE) were used to reduce the feature space to 11 features. This is particularly important to cater the problem of overfitting in Machine Learning.

Table 2 represents the comparison of our proposed framework with various machine learning based approaches proposed by various different authors and their achievements on the WPBC and WOBC datasets.

Table 2 shows the results of our approach on WOBC and WPBC datasets. We have applied our framework on both the datasets and recorded the results. First, standardization is done to ensure good normalization of features followed by feature selection. The analysis to identify the strongest predictors is on filter-based feature selection methods: correlation analysis followed by recursive feature elimination. It is evident that our proposed EDFBC framework outperforms state-of-the-art approaches. Optimal feature selection enhances the overall classification process, increasing the accuracy and reducing overall training time.

Table 2. COMPARISON OF BCAD FRAMEWORK ON WOBC AND WPBC DATASETS

Paper	Features	Classifier	Validation	Accuracy
[33]	10	J48 and MLP	10-fold	97.28%
[15]	–	Random Forest	10-fold	96.70%
Our approach	8	MLP	5-fold	98.20%
[35]	–	Fuzzy C-means clustering	4-fold	97.13%
[18]	19	Linear Regression	10-fold	84.34%
Our approach	16	MLP	5-fold	98.33%

5. Conclusion and Future Work

We have present BCAD framework for classification of breast cancer using machine learning models with focus on careful feature selection. Using a hybrid of correlation-based feature selection and recursive feature elimination, useful features are extracted from the WDBC dataset and used for classification using the Multilayer Perceptron Models. Experiments have shown that our framework outperforms state-of-the-art methods. We used the Wisconsin Diagnostic Breast Cancer (WDBC) Dataset that is used as a benchmark to improve the early diagnosis of breast cancer and eliminate challenges faced by radiologists in determining if tumor is malignant or benign. We compare the performance of proposed models including SVM, Random Forest, Gradient Boosting, Artificial Neural Network and Multilayer Perceptron Model. The best performing algorithm was the Multilayer Perceptron Model with an accuracy of 99.12%.

In future, it is recommended that large standardized public datasets must be constructed and then combined with the application of different feature selection and classification techniques to provide promising tools for the detection of breast cancer in its early stages.

References

[1] (2018, 10.01.2018). cancer. available: $http : //www.who.int/en/news - room/fact - sheets/detail/cancer$.

[2] Coimbra breast cancer dataset. available at: $https : //archive.ics.uci.edu/ml/datasets/breast + cancer + coimbra$.

[3] $https : //archive.ics.uci.edu/ml/datasets/breast + tissue$.

[4] Seer breast cancer dataset. available at: $https : //ieee - dataport.org/open - access/seer - breast - cancer - data$.

[5] Wisconsin diagnostic breast cancer dataset. available at: $http : //archive.ics.uci.edu/ml/datasets/breast + cancer + wisconsin + (diagnostic)$.

[6] Wisconsin original breast cancer dataset. available at: $https : //archive.ics.uci.edu/ml/datasets/breast + cancer + wisconsin + \%28original\%29$.

[7] Wisconsin prognostic breast cancer dataset. available at: $https : //archive.ics.uci.edu/ml/datasets/breast + cancer + wisconsin + \%28prognostic\%29$.

[8] Samyam Aryal and Bikalpa Paudel. Supervised classification using gradient boosting machine: Wisconsin breast cancer dataset. 2020.

[9] Marina Bonomolo, Patrizia Ribino, and Gianpaolo Vitale. Explainable post-occupancy evaluation using a humanoid robot. *Applied Sciences*, 10(21):7906, 2020.

[10] A. Coronato and G. De Pietro. Tools for the rapid prototyping of provably correct ambient intelligence applications. *IEEE Transactions on Software Engineering*, 38(4):975–991, 2012.

[11] A. Coronato and G. D. Pietro. Formal design of ambient intelligence applications. *Computer*, 43(12):60–68, 2010.

[12] Antonio Coronato, Muddasar Naeem, Giuseppe De Pietro, and Giovanni Paragliola. Reinforcement learning for intelligent healthcare applications: A survey. *Artificial Intelligence in Medicine*, 109:101964, 2020.

[13] Claudia Di Napoli, Patrizia Ribino, and Luca Serino. Customisable assistive plans as dynamic composition of services with normed-qos. *Journal of Ambient Intelligence and Humanized Computing*, pages 1–26, 2021.

[14] Ashutosh Kumar Dubey, Umesh Gupta, and Sonal Jain. Analysis of k-means clustering approach on the breast cancer wisconsin dataset. *International journal of computer assisted radiology and surgery*, 11(11):2033–2047, 2016.

[15] P Hamsagayathri and P Sampath. Performance analysis of breast cancer classification using decision tree classifiers. *Int J Curr Pharm Res*, 9(2):19–25, 2017.

[16] Murat Karabatak. A new classifier for breast cancer detection based on naïve bayesian. *Measurement*, 72:32–36, 2015.

[17] Yılmaz Kaya and Murat Uyar. A hybrid decision support system based on rough set and extreme learning machine for diagnosis of hepatitis disease. *Applied Soft Computing*, 13(8):3429–3438, 2013.

[18] Rafaqat Alam Khan, Nasir Ahmad, and Nasru Minallah. Classification and regression analysis of the prognostic breast cancer using generation optimizing algorithms. *International Journal of Computer Applications*, 68(25):42–47, 2013.

[19] Carmelo Lodato and Patrizia Ribino. A novel vision-enhancing technology for low-vision impairments. *Journal of medical systems*, 42(12):1–13, 2018.

[20] Olvi L Mangasarian, W Nick Street, and William H Wolberg. Breast cancer diagnosis and prognosis via linear programming. *Operations Research*, 43(4):570–577, 1995.

[21] Muddasar Naeem, Giovanni Paragiola, Antonio Coronato, and Giuseppe De Pietro. A cnn based monitoring system to minimize medication errors during treatment process at home. In *Proceedings of the 3rd International Conference on Applications of Intelligent Systems*, pages 1–5, 2020.

[22] Muddasar Naeem, Giovanni Paragliola, and Antonio Coronato. A reinforcement learning and deep learning based intelligent system for the support of impaired patients in home treatment. *Expert Systems with Applications*, page 114285, 2020.

[23] Muddasar Naeem, S Tahir H Rizvi, and Antonio Coronato. A gentle introduction to reinforcement learning and its application in different fields. *IEEE Access*, 2020.

[24] Abdullah-Al Nahid and Yinan Kong. Involvement of machine learning for breast cancer image classification: a survey. *Computational and mathematical methods in medicine*, 2017, 2017.

[25] World Health Organization et al. *WHO position paper on mammography screening*. World Health Organization, 2014.

[26] Giovanni Paragliola and Muddasar Naeem. Risk management for nuclear medical department using reinforcement learning algorithms. *Journal of Reliable Intelligent Environments*, 5(2):105–113, 2019.

[27] J Ross Quinlan. Improved use of continuous attributes in c4. 5. *Journal of artificial intelligence research*, 4:77–90, 1996.

[28] Peter M Ravdin and Gary M Clark. A practical application of neural network analysis for predicting outcome of individual breast cancer patients. *Breast cancer research and treatment*, 22(3):285–293, 1992.

[29] Patrizia Ribino, Marina Bonomolo, Carmelo Lodato, and Gianpaolo Vitale. A humanoid social robot based approach for indoor environment quality monitoring and well-being improvement. *International Journal of Social Robotics*, pages 1–20, 2020.

[30] Patrizia Ribino and Carmelo Lodato. A distributed fuzzy system for dangerous events real-time alerting. *Journal of Ambient Intelligence and Humanized Computing*, 10(11):4263–4282, 2019.

[31] Patrizia Ribino and Carmelo Lodato. A norm compliance approach for open and goal-directed intelligent systems. *Complexity*, 2019, 2019.

[32] Gouda I Salama, M Abdelhalim, and Magdy Abd-elghany Zeid. Breast cancer diagnosis on three different datasets using multi-classifiers. *Breast Cancer (WDBC)*, 32(569):2, 2012.

[33] Gouda I Salama, M Abdelhalim, and Magdy Abd-elghany Zeid. Breast cancer diagnosis on three different datasets using multi-classifiers. *Breast Cancer (WDBC)*, 32(569):2, 2012.

[34] Ahmet SAYGILI. Classification and diagnostic prediction of breast cancers via different classifiers. *International Scientific and Vocational Studies Journal*, 2(2):48–56, 2018.

[35] PB Tintu and R Paulin. Detect breast cancer using fuzzy c means techniques in wisconsin prognostic breast cancer (wpbc) data sets. *International Journal of Computer Applications Technology and*

Research, 2(5):614–617, 2013.

[36] Haowen You and George Rumbe. Comparative study of classification techniques on breast cancer fna biopsy data. 2010.

1st International Workshop on Artificial Intelligence and Machine Learning for Emerging Topics (ALLEGET'21)

Intelligent Environments 2021
E. Bashir and M. Luštrek (Eds.)
© *2021 The authors and IOS Press.*
This article is published online with Open Access by IOS Press and distributed under the terms
of the Creative Commons Attribution Non-Commercial License 4.0 (CC BY-NC 4.0).
doi:10.3233/AISE210100

Preface to the Proceedings of 1st International Workshop on Artificial Intelligence and Machine Learning for Emerging Topics (ALLEGET'21)

Raquel MARTÍNEZ-ESPAÑA, Andrés BUENO-CRESPO, Fernando TERROSO-SAÉNZ[1] and Andrés MUÑOZ

Polytechnic School, Catholic University of Murcia (UCAM), Murcia, Spain

It is our pleasure to welcome you to the 1st International Workshop on Artificial Intelligence and Machine Learning for Emerging Topics (ALLEGET'21), co-located in the 17th International Conference on Intelligent Environments (IE'21). A major goal of this workshop is to bring academic scientists, engineers and industry researchers together to exchange and share their experiences and research results about the use of intelligent systems to overcome the issues related to relevant emergent topic in our society, such as precision agriculture, the use of intelligent techniques and models to provide solutions that actually profit from open and crowdsourced location data in many different perspectives and intelligent systems applied to social media.

This volume presents the papers that have been accepted in the inaugural edition of this workshop. It consists of four high quality papers, each one providing a different view, from hardware-related aspects to machine-learning approaches, of how intelligent systems can be now applied to several scenarios in the need of emerging smart technologies.

We would like to thank all authors who submitted papers, the IE organization staff, the members of the technical program committee and especially our reviewers. They have worked very hard in reviewing papers and making valuable suggestions for the authors to improve their work.

As a result of all these efforts, the 1st edition of ALLEGET was very successful. Given the rapidity with which science is advancing in all of the areas covered by this workshop, we expect that these future ALLEGET edition will be as stimulating as this first one.

---

[1] Corresponding author: fterroso@ucam.edu

Acknowledgments

This workshop is partially funded by the Spanish Ministry of Science, Innovation and Universities under projects WaterOT (RTC-2017-6389-5) and GlobalOT (RTC2019-007159-5) and by the the Fundación Séneca del Centro de Coordinación de la Investigación de la Región de Murcia under Project 20813/PI/18.

Intelligent Environments 2021
E. Bashir and M. Luštrek (Eds.)

225

© 2021 The authors and IOS Press.
doi:10.3233/AISE210101

Tourism Recommender System Based on Cognitive Similarity Between Cross-Cultural Users

Luong Vuong Nguyen, Tri-Hai Nguyen and Jason J. Jung [1]

Department of Computer Engineering, Chung-Ang University, Seoul 156-756, Republic
of Korea

Abstract. Nowadays, the speedy increasing information in tourism services since a massive amount of data is constructed by tourists experiences. The recommendation systems are widely applied to tourism services and focus on determining personalized user preferences to handle this extensive information. Exploiting the different cultural effects rarely consider in recent studies despite this factor influences recommendation based on user preferences. Furthermore, existing research only evaluates the relevance of cultural differences to their recommendation, rather than using the cross-cultural factors to recommendations systems. This paper proposes the collaborative filtering recommendation system based on similar tourist places where users from different cross-cultural can share their spatial experiences. To do that, we first collect user feedback about similar tourist places from many nationalities (consider as the cultures). We then exploit this feedback to define similar cross-cultural users (neighbors) based on a cognitive similarity. Finally, the system generates personalized recommendations based on user experiences and their neighbors. The initial dataset collected from TripAdvisor, consisting of four types such as hotels, restaurants, shopping malls, and attractions, is provided to the feedback collection function in our experiment. We were using the classical method, user-based Pearson correlation, as a baseline to demonstrate the performance of our proposed method. The result shows that the proposed system outperforms the baseline in terms of MAE and RMSE metrics.

Keywords. recommendation system, cross-cultural, crowdsourcing, cognitive similarity

1. Introduction

Today, information about travel places and relevant resources, such as accommodations, restaurants, shopping malls, is usually searched by tourists who plan a trip. However, this information increases very fast every day, leading to a very complicated and time-consuming evaluation. Many systems are created to handle and make suitable recommendations depending on tourist preferences to solve this problem. In recent researches, the personalized techniques applied to collaborative filtering recommendation systems (RS) are commonly used in the tourism domain [1,3,4]. In proposing schemes, collabo-

---

[1]Corresponding Author: Professor, Chung-Ang University, Seoul 156-756, Republic of Korea; E-mail: j2jung@gmail.com.

rative filtering (CF) approaches are the most used. They make use of user-provided details related to assessments (or ratings). This can cause the issue of sparsity when user assessments are inadequate. Traditional CF approaches based on nearest neighbor algorithms, on the other hand, show significant issues with efficiency and scalability. In the last few years, several recommendation strategies have been suggested to improve the efficiency of the recommendations [2,5] and address other typical disadvantages of the recommended systems.

In this study, we propose a novel approach that determines the cognitive similarity between cross-cultural users to improve the performance of CF effectively. To do it, we first collect feedback about the tourist places from users who have visited or know these places. This feedback content includes what they think about the tourist places (e.g., they like it, and they think it similars others). Based on this feedback, the second step is to construct the cognition pattern of each user that is the priority of them with the extracted factor from tourist places. These extracted factors from tourist places are the area ranking, the user rating, and the number of reviews. The cognitive similarity between the users (the similar cognition pattern) is estimated using cosine similarity measurement. Finally, we make the recommendations to active users according to the k-nearest neighbor algorithm. The initial dataset for the crowdsourcing platform is collected from TripAdvisor consists of four types of tourist places: hotels, restaurants, shopping malls, and attractions. All feedback from users about tourist places stored in the database and provide for the experiments. The main contributions of this paper are summarized as follows.

1. We deploy the online crowdsourcing platform, where users from cross-cultural can share and give their feedback about tourist places that they have visited or known.
2. We do extensive experiments that are modeling the cognitive similarity-based CF recommendation system of cross-cultural users for tourism services

The remainder of this paper is structured as follows. Section 2 introduces the related work on the CF recommendation systems and recent studies in improving the user-based method. Section 3 presents the tourism recommendation system based on the cognitive similarity between cross-cultural users. The experimental results are details describe in Section 4. In Section 5, we provide some conclusions and directions for future work.

2. Related work

RS techniques, such as content-based, collaborative filtering, demographic filtering, have been applied to tourism services to provide personalized recommendations to tourists. Among these techniques, the CF approach was widely applied in the tourism area. In general, CF can be classified into two classes model-based and neighborhood-based methods. Model-based methods make recommendations by learning a predictive model according to user-item ratings. There are several recent studies in model-based, such as latent semantic analysis (LSA), Bayesian clustering, support vector machines (SVM), latent Dirichlet allocation (LDA), and singular value decomposition (SVD) [7,8,9,10]. Neighborhood-based methods make recommendations by measuring the similarities among users or among items (know as user-based and item-based, respectively). In recent research, the effectiveness of CF stability-enhancing was mentioned in [13] that focused on clustering users method. In particular, they presented a novel bio-inspired cluster-ing based on swarm intelligence and fuzzy clustering models. Besides, in [14], ontology and dimensionality reduction techniques were used to handle both popular RS problems: sparsity and scalability. This research aims to increase the accuracy in the recommending process by using the ontology, combine with Singular Value Decomposition (SVD), the dimensionality reduction technique, to improve the scalability of recommending method. In [15], the cognitive similarity between users was considered to define the similar users that improve the performance of collaborative filtering in the movie recommendation system. This issue was also mentioned and exploited in [11] that proposed the crowd-sourcing platform to collecting feedback from users who have experiences in the movies domain. In [12], the cross-cultural contextualization was exploited for the recommen-dation system. This study proposed a new definition of cross-cultural contextualization, then based on that, they compute the cross-cultural factor affecting users. To do that, they presented a model based on the combination of matrix factorization and clustering techniques.

This work focuses on improving a user-based recommendation system that proposes a better method to define the most similar items for active users. We define the k-nearest neighbor dynamically based on cognitive similarity among users in each cluster of users and the whole users in the system because the cognitive similarity between users dy-namically changes according to their history activities. In this way, the proposed method improves the accuracy in define similar users, and therefore, the accuracy in the recom-mendation process is enhancing. The evaluation demonstrates that the proposed method achieves the outperformance in comparison with the traditional method, user-based user-based Pearson correlation.

3. Recommendations based on Cognitive Similarity between Cross-Cultural Users

3.1. Similarity Measurement

In our dataset, there are four types of tourist places, and therefore we decided to use the overlap extracted factors from these tourist places, which are (R) user ratings, (K) area rankings, and (N) the number of reviews. To calculate the similarity between tourist place p_i and p_j, we following these steps *(i)* each tourist place p represented by these overlap extracted factors, *(ii)* determine the similarity of each pair of extracted factors in the first step by applying the cosine similarity formulation, *(iii)* the average similarity score of all extracted factors is assigned to a similarity score between tourist place p_i and p_j. The formulation for measuring the similarity score $(\Im)$ between tourist place p_i and p_j is described by

$$\Im(p_i, p_j) \equiv \langle R_{ij}, K_{ij}, N_{ij} \rangle, \tag{1}$$

where R_{ij}, K_{ij}, and N_{ij} represent the similarity measurement of these factors user ratings, area rankings, and the number of reviews between tourist place p_i and p_j, respectively. In this study, we measure the similarity between extracted factors by using the soft co-sine metric defined as the cosine of the angle between two non-zero vectors of an inner product space. For example, considering the users rating factor, the similarity of these extracted factors between tourist places p_i and p_j measured by soft cosine metric, which described as

$$R_{ij} = \frac{\sum_1^n R_i R_j}{\sqrt{\sum_1^n (R_i)^2} \sqrt{\sum_1^n (R_j)^2}}.$$

(2)

We repeat Eq. 2 with the remaining factors that are area ranking (K_{ij}) and the number of reviews (N_{ij}). Finally, we obtain the ($\mathfrak{I}$) between these tourist places according to Eq. 1. According to these steps above, the formulation Eq. 1 is repeated to measuring similarity between p_i and each tourist place in the whole of our dataset. We then obtain a set $\{\mathfrak{I}(p_i, p_{i+j}) \mid j \in [1, .., n-1]\}$, where n is number of the tourism places in the database.

3.2. Cognitive Similarity-based Cross-Cultural Users Clustering

We collect feedback from cross-cultural users that they think about the tourist places (e.g., similar, interest) based on the crowdsourcing platform using the initial dataset. The process to collect feedback is as follows.

1. The system randomly shows four tourist places (consist of hotels, restaurants, shopping malls, and attractions). The user selects the tourist place that he/she has visited or known (implicit that he/she was interested in this place).
2. The platform will suggest four tourist places based on the type of tourist places that he/she selected and the list $N - places$ constructed as described in Section 3.1.
3. He/she selects one of four tourist places that they think similar and submit, the system store this information and assigned it to the history of user activities (pair of similar places). If he/she does not select any tourist place, the system assigns the tourist place that he/she chose in the first step to the history of user activities (interest tourist place).

Following this process, we have datasets of each user in the system. The essential property in these datasets is the priority model measured during selecting similar tourist places' process from each user (called the user cognition pattern). This cognition pattern is updated according to the activities of users in the platform. We then cluster the users based on these similar cognition patterns of the cross-cultural users. In this study, the similarity between cognition patterns, which we called cognitive similarity between cross-cultural users, is determined.

Definition 1 - Cognitive Similarity between Cross-Cultural Users.

Given user u and v, the cognitive similarity $\mathfrak{I}$ between these users is the similarity between their cognition pattern that represents by the similarity of their priorities in selecting similar tourist places. By using the cosine similarity, we have the following formula

$$\mathfrak{I}(u, v) = \frac{\sum_i^N \rho_u^i \rho_v^i}{\sqrt{\sum_i^N (\rho_u^i)^2} \sqrt{\sum_i^N (\rho_v^i)^2}},$$

(3)

where ρ represent the priorities in selecting similar tourist places of user u and v.

For instance, given i is the number of tourist places in the datasets $I(u)$ of user u, the cognition pattern ρ_u^i of user u is defined as a set $\{R, K, N\}$ which each item in a set have value is the average similarity score of each extracted factor from each pair of similar tourist places in $I(u)$. When user u has new activities (new data inserted to $I(u)$), the cognition pattern of user u automatically re-calculated and updated to the database. A set $\{R, K, N\}$ has three value scores re-ordered from highest to smallest and is an ordered vector, which is described as

$$\rho_u^i = \{R_u^i; K_u^i; N_u^i\}. \tag{4}$$

In this work, we aim to make the user clustering automatically, and dynamic re-calculate for updating to the database since any new activities of the user will make the cognition pattern changing. We deploy the calculation function to determine the cognition pattern of each user according to any new activities that happened in the whole platform. It means on the servers side, and the platform has to manage the database structure to store and handle the long complicate process.

4. Experiment

4.1. Dataset

To construct the initial datasets provided for the crowdsourcing platform, we decide to crawl tourist places information from TripAdvisor, which has the practical information for our purpose. We have collected approximately eighteen thousand tourist places from 84 cities of 24 countries distributed on five continents. The detailed statistic of the initial tourist places dataset is shown in table 1.

We keep collecting data from more than 150 active users on our platform and enrich our database. Our purpose is to deploy the system scalability to as many users as possible for collecting more valuable feedback from many different users from different countries (different cultures). The collected feedback from cross-cultural users have the format $(U_i, p_a, p_b, \rho_{U_i}^{a,b}, \gamma)$ inside U_i is the id of the user. p_a and p_b are id of tourist places, respectively. $\rho_{U_i}^{a,b}$ is a vector representing the cognition pattern of user U_i in selecting a pair similar tourist places (p_a, p_b).

Table 1. Statistics of the tourist places dataset

#	Title	Amount
1	Number of countries / cities / tourist places	24 / 84 / 18,286
2	Number of reviews	239,446
3	Number of ratings	166,428
4	Number of users	153
5	Number of feedback from users	2210

4.2. Evaluation

To construct the evaluation report, we experimented on the dataset described above. The steps to deploy the experiment is described as follows. Firstly, the dataset is separated into five mutually exclusive folds because we decide to carry out the evaluation, which uses the five-fold cross-validation. We split the dataset into two parts which 80% for the training-set and the rest 20% for the testing-set each fold. In the second step, the tourist places prediction in the testing-set according to the corresponding tourist places in the training-set for each fold. Finally, the performance is evaluated based on comparing the predictions and the authenticity available in the test-set for each fold. The final result is an average of all predicted in all folds.

The classic method, User-based Pearson Similarity (UBPS), is identified as the basis for the comparative analysis in this section. Among the different criteria used to evaluate the accuracy of CF approaches, the mean absolute error (MAE) and the root mean square error (RMSE) is selected as the evaluation metric for the comparative analysis report. The formulations of MAE and RMSE are described by

$$\textbf{MAE} = \frac{1}{n} \sum_{i=1}^{n} |y_i - y_i^p|, \tag{5}$$

$$\textbf{RMSE} = \sqrt{\frac{1}{n} \sum_{i=1}^{n} (y_i - y_i^p)^2}, \tag{6}$$

where n is the number of tourist places, y_i is the real values and y_i^p is the predicted values in the test set. The results of MAE and RMSE range from 0 to infinity. Infinity is the maximum error according to the scale of the measured values. The baseline has the best performance in approximately 50 neighbors of the active user. Therefore, we designate the experiments with the neighborhood sizes to be $\{5, 10, 20, 30, 50\}$. The comparison between the proposed method and the baseline is shown in Figure 1.

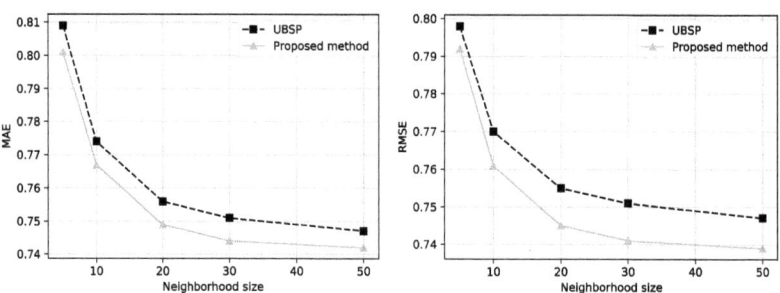

Figure 1. Performance comparison between the proposed method and the baseline in term of MAE and RMSE

5. Conclusion and Future work

This paper presented a novel approach to improving user-based CF that focuses on user clustering based on the cognitive similarity between cross-cultural users. We deployed a crowdsourcing platform to gather feedback from cross-cultural users for this purpose. The platform has the simplest and easiest steps in the feedback collection process from users. The experiments performed on the data set verified that the proposed approach meets the expectations that improve performance over the baseline UBPS, which has only global similarity.

In the future study, the valuation methods used to form a user cluster shall be examined based on data obtained from cross-cultural users. For example, the method will consider the cognition pattern of users and consider the demographic or other personal information that is allowed to collect from users. Moreover, a more advanced and sophisticated approach to splitting the list into the preferred user cluster can be introduced.

Acknowledgments   This work was supported by the National Research Foundation of Korea (NRF) grant funded by the Korea government (NRF-2019K1A3A1A80113259).

References

[1]   Chen R, Hua Q, Chang YS, Wang B, Zhang L, Kong X. A Survey of Collaborative Filtering-based Recommender Systems: From Traditional Methods to Hybrid Methods based on Social Networks. IEEE Access. 2018 Oct 24;6:64301-20.

[2]   Haruna K, Akmar Ismail M, Suhendroyono S, Damiasih D, Pierewan AC, Chiroma H, Herawan T. Context-Aware Recommender System: A Review of Recent Developmental Process and Future Research Direction. Applied Sciences. 2017; 7(12):1211.

[3]   Wang X. Personalized Recommendation Framework Design for Online Tourism: Know You Better Than Yourself. Industrial Management & Data Systems. 2020 Oct; 120(11):2067-2079.

[4]   Nguyen LV, Jung JJ, Hwang M. OurPlaces: Cross-Cultural Crowdsourcing Platform for Location Recommendation Services. ISPRS International Journal of Geo-Information. 2020 Nov; 9(12):711.

[5]   Nguyen LV, Nguyen TH, Jung JJ. Content-Based Collaborative Filtering using Word Embedding: A Case Study on Movie Recommendation. Proceedings of the International Conference on Research in Adaptive and Convergent Systems (RACS '20); 2020 Oct 13-16; Gwangju, South Korea. ACM; 2020. p. 96—100.

[6]   Herlocker JL, Konstan JA, Borchers A, Riedl J. An Algorithmic Framework for Performing Collaborative Filtering. Proceedings of SIGIR Forum; 2017 Aug 2; New York, USA. ACM; 2017. p. 227–234.

[7]   Hofmann T. Collaborative Filtering via Gaussian Probabilistic Latent Semantic Analysis. Proceedings of the 26th Annual International ACM SIGIR Conference on Research and Development in Information Retrieval; 2003 Aug 1; Toronto, Canada. ACM p.259–266

[8]   Breese JS, Heckerman D, Kadie C. Empirical Analysis of Predictive Algorithms for Collaborative Filtering. arXiv preprint arXiv:1301.7363. 2013 Jan 30.

[9]   Blei DM, Ng AY, Jordan MI. Latent Dirichlet Allocation. Journal of Machine Learning Research. 2003 Jan 3:993-1022.

[10]   Adomavicius G, Kwon Y. New Recommendation Techniques for Multicriteria Rating Systems. IEEE Intelligent Systems. 2007 Jun; 22(3):48-55.

[11]   Nguyen LV, Jung JJ. Crowdsourcing Platform for Collecting Cognitive Feedbacks from Users: A Case Study on Movie Recommender System. In: Reliability and Statistical Computing. Springer; 2020. p.139-150.

[12]   Hong M, An S, Akerkar R, Camacho D, Jung JJ. Cross-cultural Contextualisation for Recommender Systems. Journal of Ambient Intelligence and Humanized Computing. 2019 Sep;9:1-2.

[13]   Logesh R, Subramaniyaswamy V, Malathi D, Sivaramakrishnan N, Vijayakumar V. Enhancing Recommendation Stability of Collaborative Filtering Recommender System through Bio-inspired Clustering Ensemble Method. Neural Computing and Applications. 2020 Apr;32(7):2141-64.

[14]   Nilashi M, Ibrahim O, Bagherifard K. A Recommender System based on Collaborative Filtering using Ontology and Dimensionality Reduction Techniques. Expert Systems with Applications. 2018 Feb 1;92:507-20.

[15]   Nguyen LV, Hong MS, Jung JJ, Sohn BS. Cognitive Similarity-based Collaborative Filtering Recommendation System. Applied Sciences. 2020 Jan;10(12):4183.

Intelligent Environments 2021
E. Bashir and M. Luštrek (Eds.)
© *2021 The authors and IOS Press.*
This article is published online with Open Access by IOS Press and distributed under the terms
of the Creative Commons Attribution Non-Commercial License 4.0 (CC BY-NC 4.0).
doi:10.3233/AISE210102

Low Power LoRa Transmission in Low Earth Orbiting Satellites

Rifath Shaarook[1] and Dr. Srimathy Kesan[2]

[1,2] *Space Kidz India*

Abstract. In the recent past, a modulation technique called "LoRa" has become popular in the communication industry due to its long range capabilities with low power consumption, a few satellites have carried LoRa and worked successfully inspite of being skeptical about Doppler shift, packet loss, etc. This paper discusses the preliminary results of LoRa communications established from a recently launched cubesat. Satish Dhawan Satellite (SD Sat) is a 3U Cubesat built by Space kidz India, weighing 1.9 kg, it carried a radiation counter and magnetometer as its payload. It carried a LoRa transmitter which transmits beacon at 100 mW. It is also transmitting FSK RTTY and CW RTTY. The satellite was launched into sun-synchronous orbit at 510 x 498 km. Multiple ground stations around the world have confirmed the signal reception.

Keywords. Cubesat, LoRa, Satellites, LEO, IoT

1. Introduction

Space Kidz India is an organization from India working towards promoting Space Education and awareness among the students across the globe. Space Kidz India has previously worked on 12 high-altitude balloon missions to test various subsystems for cubesats at near space environment. It has also successfully completed two suborbital space flights, SKISAT and KALAMSAT which tested the subsystems apart from launching an orbital satellite "KalamSAT V2" launched through ISRO on PSLV C44 in 2019. Satish Dhawan SAT (SDSat) is the second satellite built by students at Space Kidz India, named after one if the founding fathers of India's space agency Indian Space Research Organization, as part of his centenary birth celebration. The aim of the mission was to test the various aspects of the newly designed satellite which was indigenously fabricated end to end. One of the experiment was to test how the low power transmission in LoRa mode would perform. Other experiments includes measuring the ionizing radiation experienced by the satellite from various sources like cosmic radiation, solar flares, etc. and testing how the magnetometer present in the satellite would perform. The team intends to use the magnetometer data for developing an Attitude Control System for cubesats in the future. The satellite's engineering model and flight model are shown in the Figure 1. Both the models are identical and the engineering model is used for all sorts of testing like vibration testing, thermovaccum testing and shock testing while the flight model will the satellite which actually gets to fly onboard the rocket and reaches the orbit. You can also see that the solar panels are attached outside for the power

[1,2] Rifath Shaarook, Dr. Srimathy Kesan are with Space Kidz India, Chennai, India, e-mail: cto@skilabz.in, e-mail: ceo@spacekidzindia.in

generation. Figure 2 shows the internals of the satellite where it is covered with Multi Layer Insulation material for thermal insulation to maintain an operatable temperature inside the satellite.

The satellite is a 3U cubesat measuring 340 mm x 100 mm x 100 mm and weighing 1.9 kg. [1] It was launched onboard ISRO's PSLV-C51 launch vehicle on 28th February 2021. Following the launch, the satellite was deployed from the cubesat deployer at 12.15 PM IST over Madagascar. The satellite was inserted into a 510 km x 498 km sun-synchronous orbit with an inclination of 97.5 degrees, making it possible for the satellite to cover the entire earth in the subsequent orbits. After about eight and half hours post the launch, first FSK signals were confirmed over Australia by SatNOGS Network [2] and one of the first LoRa signals were received over USA by an amateur radio operator

known as N6RFM by his call sign [3].

Figure 1. Engineering and Flight Model of the SDSat

Before the launch the satellite has to be integrated with the launch vehicle, Figure 2 shows the satellite getting integrated into the deployer which will be further attached with the rocket and Figure 3 shows the satellite getting deployed into orbit from the rocket.

Figure 2. SDSat getting inserted into the deployer

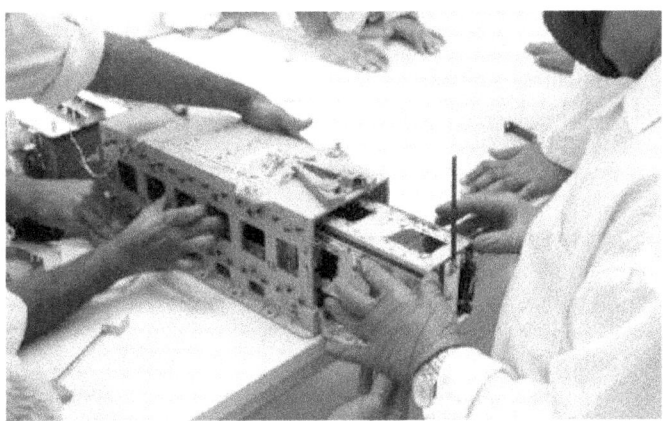

Figure 3. SDSat getting inserted into the deployer

The satellite uses a custom transceiver built using SX1278 from Semtech as it's heart for sending telemetry and receiving telecommands. SX1278 and it's family of chips can transmit in both FSK and LoRa, since LoRa was an experimental feature in this mission, another primary mode of transmission was required and FSK was selected as it has a long heritage of usage in space. The telemetry includes both health data like battery levels, solar panel voltage, etc and sensor data from the main payloads. The transmitter also transmits CW beacon as it is easier to locate the satellite from ground and perform Doppler calculations. The transceiver carries a single SX1278 chip, thus it operates in half duplex mode, transmission and reception both share the same frequency of 435.5 MHz. Transceiver parameters are given in Table 1.

Table 1. Transceiver Parameters

Tx Frequency	435.5 MHz
Rx Frequency	435.5 MHz
Antenna Type	Tape Measure Dipole
Antenna Gain	2.15 dbi
Radiation Pattern	Omnidirectional
EIRP	-7.85 dbW

2. LoRa Transmission

LoRa is a type of modulation technique developed and patented by Semtech Corporation based in USA and it has become quite popular in the recent days due to its long range transmission capabilities by using small amount of power compared to other types of modulations. [4] LoRa's spread spectrum modulation techniques are based on chirp spread spectrum technology. Since spacecrafts have limited power production capability in space, saving power and making de- vices efficient are one of the most important thing in designing spacecrafts and recent LoRa developments have attracted many teams to use LoRa in their satellites. [5] As LoRa is a favourable technique for IoT systems, many companies intend using space based LoRa gateways for enabling IoT networks, but due to the effect of Doppler shift produced by high orbital velocity in lower earth orbit and wide bandwidth of LoRa modulation has made many skeptical about the capabilities of using LoRa with low cost ground equipment and antenna systems. Parameters of the LoRa transmissions are given in the Table 2 and antenna in its deployed form is given in the Figure 4.

Table 2. LoRa Parameters

Carrier Frequency	435.5 MHz
Bandwidth	125 KHz
Spreading Factor	9
Coding Rate	7
Output Power	100 mW/20 dam
Preamble Length	8 Symbols
Sync Word	12
Gain	0

An Atmel ATMEGA-32u4 micro controller has been con- nected with SX1278 via SPI interface and it acts as a data handler and communication system manager while the main onboard computer is an Atmel ATMEGA-328.

Antenna is a half-wave dipole and it is made out of simple steel measuring tape. It was cut half the wavelength of the transmitting frequency and tuned using a VNA. Antenna and the transmitter are connected using RG-58 coaxial cable using SMA connectors on both the sides. Threadlocker was used while tightening the SMA to make sure it will not get loosened during the launch due to the vibration and shocks produced.

Figure 4. Antenna of the Satellite when it's deployed

The satellite first transmits in FSK RTTY at 100 mW power and it includes health data like battery power, solar panel power production, on-board computer reset counter, etc and it transmits sensor data which includes data from all the payloads. Then the satellite switches to CW mode where it transmits its name first in high power mode which is 100 mW and then in low power mode which is 20 mW. We did this to study how easily we can receive 20 mW beacons so that our future missions can operate beacons in that power alone to save power. The last transmission mode is LoRa where it sends a static message with its name and signifying that LoRa is active. There is also a LoRa repeater mode present in the satellite which we are testing right now and it will become available soon where it will be transmitting short messages sent to it by uplink telecommands. The whole process of transmitting in different modes take around 20 seconds and it will switch to receive mode for 15 seconds to receive telecommands sent from the ground. It acts according to the command it receives or else it will continue the same loop once the 15 seconds receive mode is completed.

The satellite is designed to comply with the rules of International Telecommunication Union. Hence, during the receiving period if a particular telecommand with a secret code is been sent, the satellite will go on a receiving mode and stop the transmission on the whole, until you send another telecommand with a secret code to put the satellite back into normal operations.

3. LoRa Reception

The satellite was deployed from the launch vehicle at 12:15 PM IST over Madagascar, the orbit it was inserted into was   a 510 km x 498 km sun-synchronous orbit. The satellite was turned on immediately after the deployment but transmission was programmed to be disabled due to safety reasons for 30 minutes. Hence, the satellite was turned on after 30 minutes and first beacons of the satellite were heard over Australia in SatNOGS Network and the signals were confirmed using second reception over Argentina. The first signals from LoRa were received by the Ham Radio operator Mr. Bob known by his call sign N6RFM in TinyGS network. TinyGS is an open- source distributed ground station network having multiple ground stations around the world. [6] During the initial period after the launch, it is hard to exactly locate the satellite, people who receive the satellite's signal uses Two-Line element (TLE) for tracking, it is an orbital parameter to identify the position of the satellite. After the launch, this TLE will not be accurate enough to track, but after a few initial signal receptions, the position of the satellite became accurate enough that many people around the Globe started receiving the signals from SDSat. A typical reception of LoRa beacon during nominal operation of the satellite is shown in the Figure 5. The beacon contains the data "SDSAT, LORA ACTIVE:". Once the repeater is activated, this will also contain the messages sent to the satellite.

On the hardware side setup of the reception, You need a good antenna tuned to receive the specific frequency of the satellite with a low noise amplifier. But many have received the signals using just a generic UHF dipole antenna without any LNA. Figure 6 (left) shows a 10 element Yagi Uda antenna with azimuth/elevation rotator and an LNA while figure 6(right) shows a generic UHF dipole antenna without any LNA mounted inside a window of a building. A study about the comparison of antenna setup and signal reception quality is underway.

You also need a receiver setup with antenna, a SX1278 module with any micrcontroller running generic LoRa recep- tion software with the parameters shown in Table 2, would be able to receive the signals. We are using SX1278 with ESP32 Microcontroller running TinyGS software for the reception and the setup of one such hardware with a simple antenna is shown in the Figure 7. It is located in the Robotics Lab of Middlesex University, Dubai where students are receiving our satellite signals. It shows a simple dipole antenna made up of Co-axial cable with a band pass filter and low noise amplifier connected to a ESP32 board known as TTGO which also contains a SX127x LoRa chip and an OLED display for easy operation. How to install the software and setup your own receiving station can be found at the website of TinyGS.

Figure 5. LoRa beacon received in TinyGS Network

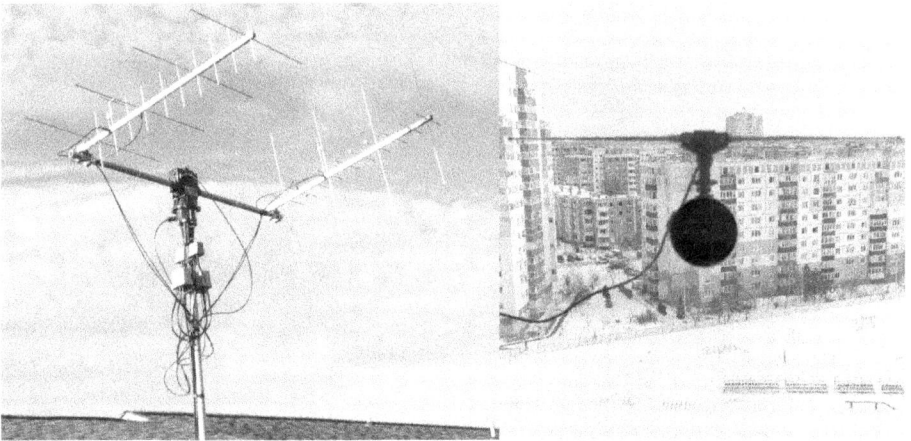

Figure 6. Yagi UDA antenna and simple dipole antenna

4. Tracking the Satellite

The satellite is in lower earth orbit, traveling at a velocity of 28,000 km/h, which means that the satellite will orbit the earth around 16 times a day and cross a given location two times a day. To receive the signals from satellite you should know when to look at the sky and get ready for receiving the satellite so that you can switch on your receiving hardware at that particular time so you can save a lot of electrical power and also be sure that the signals are from your satellite, to know this you can use lot of free and open source software like Orbitron, GPredict, etc. The TLE file of the given satellite can be downloaded from the Celestrak or similar TLE providing website and loaded up on the tracking software to find the exact location of the satellite in the orbit at any given moment. There are similar software available for smartphones also, like Satorbit which allows you to track the satellite using your mobile devices. Figure 8 shows the SDSat being tracked using the SatOrbit software in an Android phone.

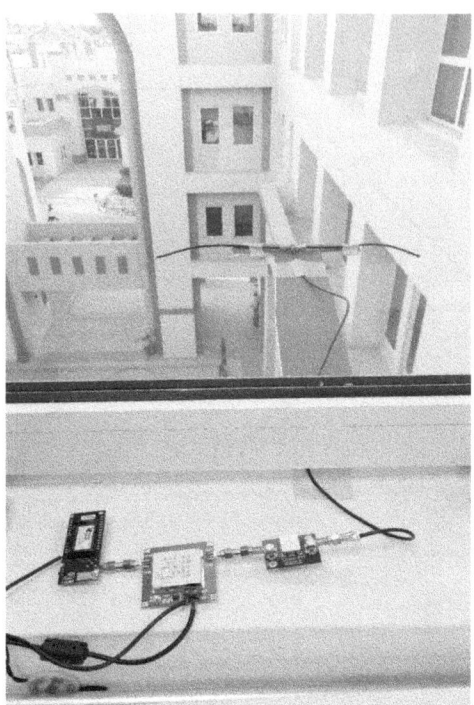

Figure 7. Ground Station setup at the Robotics Lab, Middlesex University, Dubai.

5.Downlink Signal Waterfall

The downlink signals of SDSat during nominal operations were received using a Software Defined Radio and various formats of the downlink are noted in the waterfall. Figure 9 shows the SDSat's signal received over US East Coast by amateur radio operator Scott Chapman, known as K4KDR by his callsign. The figure shows a series of data from the bottom and a 15-second interval where the satellite will go to receive mode and wait for any telecommands that user may uplink. Then it transmits it's health data in RTTY format which con- tains data like Satellite's onboard temperature, battery level, solar panel power level, reset counter, special messages from the previous uplinked commands if any. And then the payload data gets transmitted in which it contains data from sensors like Intertial Measurement Unit, Onboard Radiation Counter, etc. Both the RTTY are transmitted with a transmission power of 100 mW. Followed by this, two Morse code beacons are sent which contains the word "SDSAT". First it is transmitted with a power of 100 mW then with the power of 20 mW. Lastly the satellite transmits LoRa signal with 100 mW of transmission power. Due to the LoRa's wide bandwidth of 125 KHz, it's hard to monitor it on the SDR waterfall but it gets received by the LoRa hardware, again it shows the sensitivity and high possibility of reception in long range with LoRa technology.

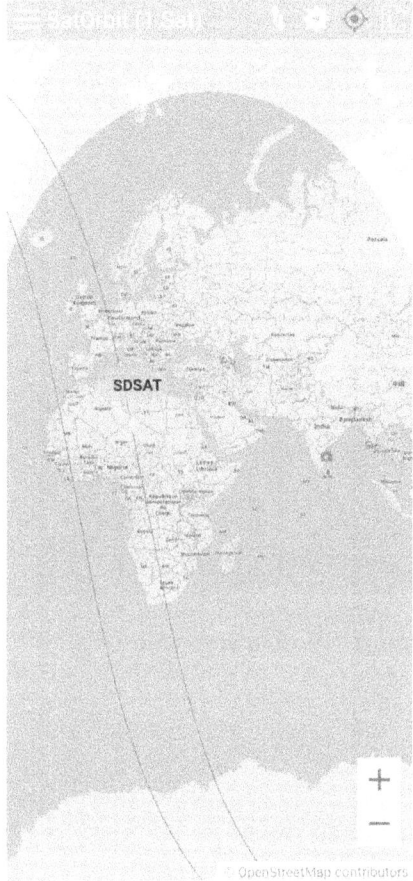

Figure 8. SDSat being tracked in the SatOrbit App

6.Conclusion

Initial observations have shown that it is possible to receive 100 mW LoRa signals from a satellite deployed in low earth orbit. We've also seen successful receptions and decoding of FSK RTTY at 100 mW and CW in both 100 mW and 20 mW of output power. People have come up with many antenna ideas from simple dipole to complex directional antenna with satellite trackers. More study about the antenna system and reception hardware is underway.

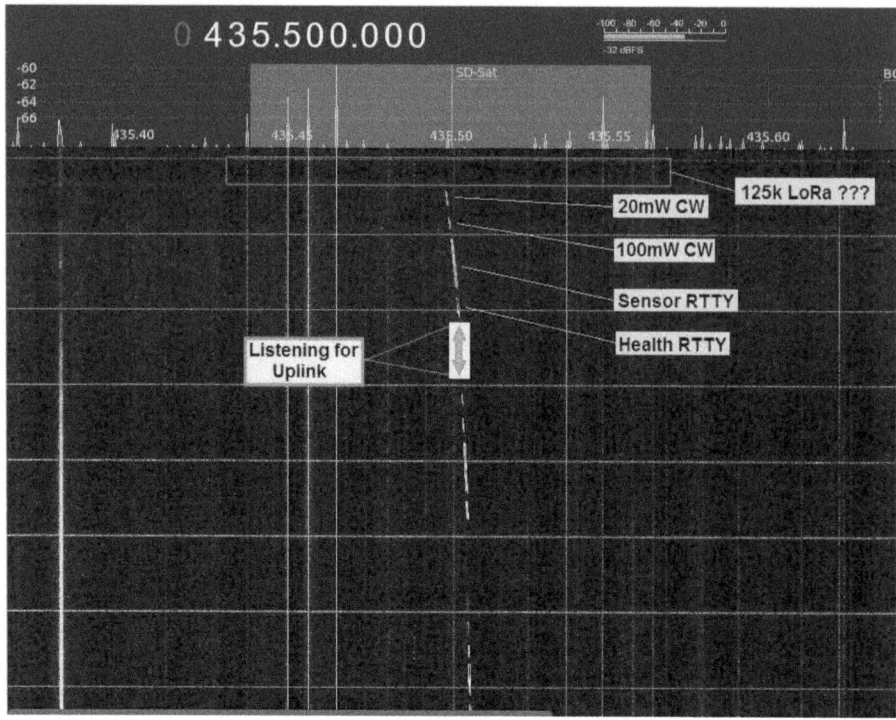

Figure 9. SDSat's signal waterfall received using SDR

Acknowledgement

Satish Dhawan SAT was supported by the ISRO and IN- SPACe. [7]

References

1. C. P. S. University, "Cubesat Design Specification." [Online]. Available:
    https://www.cubesat.org/cubesatinfo
2. SatNOGS, "Observation #3703386." [Online]. Available: https://network.
    satnogs.org/observations/3703386/
3. B. N6RFM, "SDSat First LoRa signals." [Online]. Available:
    https://twitter.com/n6rfm/status/1366063930829918210
4. Semtech, "What is LoRa?" [Online]. Available: https://www.semtech. com/lora/what-is-lora
5. T. Wu, D. Qu, and G. Zhang, "Research on LoRa Adaptability in the LEO Satellites Internet of Things,"
    IEE Xplore. [Online]. Available: https://ieeexplore.ieee.org/document/8766462
6. "Welcome to TinyGS, the Open Source Global Satellite Network." [Online]. Available:
    https://tinygs.com/
7. "Satish Dhawan SAT (SDSAT)." [Online]. Available: https://www.isro. gov.in/Spacecraft/satish-
    dhawan-sat-sdsat

Intelligent Environments 2021
E. Bashir and M. Luštrek (Eds.)
doi:10.3233/AISE210103

nLORE: A Linguistically Rich Deep-Learning System for Locative-Reference Extraction in Tweets

Nicolás José FERNÁNDEZ-MARTÍNEZ [a,1] and Carlos PERIÑÁN-PASCUAL [b]

[a] Universidad Católica San Antonio de Murcia (Spain)
[b] Universitat Politècnica de València (Spain)

Abstract. Location-based systems require rich geospatial data in emergency and crisis-related situations (e.g. earthquakes, floods, terrorist attacks, car accidents or pandemics) for the geolocation of not only a given incident but also the affected places and people in need of immediate help, which could potentially save lives and prevent further damage to urban or environmental areas. Given the sparsity of geotagged tweets, geospatial data must be obtained from the locative references mentioned in textual data such as tweets. In this context, we introduce nLORE (neural LOcative Reference Extractor), a deep-learning system that serves to detect locative references in English tweets by making use of the linguistic knowledge provided by LORE. nLORE, which captures fine-grained complex locative references of any type, outperforms not only LORE, but also well-known general-purpose or domain-specific off-the-shelf entity-recognizer systems, both qualitatively and quantitatively. However, LORE shows much better runtime efficiency, which is especially important in emergency-based and crisis-related scenarios that demand quick intervention to send first responders to affected areas and people. This highlights the often undervalued yet very important role of rule-based models in natural language processing for real-life and real-time scenarios.

Keywords. location extraction, geolocation, named entity recognition, deep learning

1. Introduction

The large volume of user-generated content on Twitter can be exploited in social sensing settings where crisis management and tracking become of utmost importance for disaster relief operations [1,2]. Having a rapid understanding about crisis-related and emergency events can help handle human and economic resources effectively through immediate and timely decisions and actions taken by aid organizations and competent authorities. As a result, emergency responders can coordinate effective aid and help allocate resources in the affected areas and/or to the affected people, potentially saving lives and preventing further damage to environmental or urban areas. Because of the lack of geotagged data, e.g. geotagged tweets represent only around 1% of tweets [3], and its recent sharing restrictions, it becomes necessary to turn to other geospatial evidence, such as that found in tweets in the form of locative references, which are usually much more frequent than geotagged data [4]. Twitter has in fact been

---

[1] Corresponding author. E-mail: njfernandez@ucam.edu

exploited in many geolocation systems that handle real-life scenarios, e.g. natural or human-made disaster detection and tracking in floods, earthquakes, storms, civil unrest, war, crime, etc. [1,2,5], health surveillance and disease tracking, including the current COVID-19 pandemic [6], marketing and advertising [7], or traffic incident detection, road traffic control and/or traffic congestion [8]. In this regard, the location dimension proves to be critical for raising situation awareness of crisis-related events and understanding their impact, i.e. gaining insight into where the incident happened, where people require assistance, and/or which areas were affected.

Departing from previous work on location detection from tweets [9,10], we introduce a deep-learning implementation of LORE (i.e. neural LORE, or nLORE), which, by means of a bi-directional Recurrent Neural Network (bi-RNN) with linguistic feature engineering, automatically learns the linguistic features and rules provided by LORE to extract locative references of any type from English tweets with even greater accuracy. Indeed, nLORE slightly outperforms LORE, which already outperformed other entity recognizers such as Stanford NER, spaCy, Stanza, OpenNLP, and Google Entity Recognizer, offering a great qualitative and quantitative advantage over these general-domain entity recognizers and qualitatively over tweet-specific location-detection models. However, LORE shows much better speed and efficiency, being the preferred model in real-life crisis-related and emergency-based contexts, where quick deployment and allocation of resources by competent authorities is essential to come to the rescue of affected people or areas.

The article is organized as follows. Section 2 presents the major works in locative reference detection from tweets. Section 3 explains the methodology followed in the training phase of nLORE. Section 4 describes the evaluation phase. Section 5 presents the results obtained by nLORE in comparison with LORE and discusses the results and implications derived from this study, including future lines of research. Finally, Section 5 presents some conclusive remarks.

2. Related Work

Many works propose rule-based systems that rely on a combination of linguistic rules (e.g. locative preposition + proper nouns) and/or lexical resources (e.g. gazetteers of place names or lists of locative cues) together with supportive natural language processing (NLP) tasks, such as the part-of-speech (POS) tagging of n-grams [3,11,12]. Other systems are based on probabilistic frameworks, using Conditional Random Fields (CRF) or neural networks such as Convolutional Neural Networks (CNN) or Recurrent Neural Networks (RNN) with or without additional linguistic feature engineering [7,13]. Others present hybrid systems, which combine rule-based methods with machine- or deep-learning techniques and/or other named entity recognition (NER) tools [8,14]. What all these works share is the limited semantic coverage of fine-grained location types, as well as the lack of a sound linguistic-based approach for their identification. In the remainder of this section, we describe the most prominent works in locative reference extraction from tweets.

A microtext location-detection model was proposed by [14] using regex-based rules, the Open Calais NER software, machine-learning techniques for abbreviation disambiguation, and the National Geospatial Intelligence Agency gazetteer for the identification of places in New Zealand and Australia at and within city level, such as geopolitical entities, buildings, and streets. It was tested on two evaluation datasets: one about the 2011 Christchurch dataset in New Zealand and another about the 2011 Texas wildfire in the US. In the first evaluation dataset, the model achieved an F1 score of 0.85 for streets, 0.86 for buildings, 0.96 for geopolitical entities, and 0.88 for place abbreviations, giving an average of 0.9. In the second dataset, it obtained an F1 score of 0.71.

A linguistic-based unsupervised location-detection model was presented by [11] based on linguistic techniques and rules such as noun-phrase extraction and n-gram matching techniques using regex-based rules and the GeoNames database. It targeted geopolitical entities, points of interest (POIs), addresses, and surrounding distance and direction markers, giving an F1 score of 0.79.

A rule-based location-detection model for English tweets was developed by [3] using the OpenStreetMaps database. They used NLP techniques such as the NLTK sentence tokenizer, entity-based matching with OpenStreetMaps using an n-gram-based module, their own corpus of building and street types, and the NLTK stopword list enriched with a list of names. They focused on geopolitical entities, buildings, and streets. The evaluation stage was carried out with separate corpora of tweets about different incidents (i.e. blackout, earthquake, and hurricane) in different geographic areas (i.e. Christchurch, Milan, New York, and Turkey). Precision numbers were impressive, ranging from 0.93 to 0.99, and F1 scores ranged from 0.90 to 0.97, except for the Turkey earthquake dataset, where an F1 score of 0.28 was achieved.

A location extractor called NeuroTPR was built by [13], who used a bi-directional RNN with Long Short Term Memory (LSTM) enriched with linguistic features. Apart from building a tweet corpus from the 2017 Hurricane Harvey dataset, they also used GeoCorpora [4]. The targeted location types were geopolitical entities, natural landforms, POIs, and a few traffic ways. In the evaluation stage, they compared their model against standard off-the-shelf NER tools such as Stanford NER, spaCy NER, a basic bi-directional LSTM-CRF model and another deep-learning model from the 2019 SemEval geoparsing competition. The evaluation phase was carried out using the exact matching of location references. The best NeuroTPR model, trained on 3000 Wikipedia articles and 599 tweets, achieved a precision score of 0.787, a recall score of 0.678, and an F1 score of 0.728 on the *Harvey2017* corpus.

3. Method

We implemented a hybrid approach to NER by means of a bi-directional RNN with LSTM as a hidden layer structure and a CRF layer on top exploiting linguistic feature engineering and semantic information contained in word embeddings. Our primary goal was to check the power of linguistic feature engineering in nLORE. Derived from this goal, our secondary goals were (i) to elucidate whether linguistic feature engineering could overcome the sparsity of data in probabilistic models and (ii) to compare the performance of nLORE against its rule-based counterpart LORE, which outperformed well-known general-domain NER systems [9], with the aim of resolving whether our rule-based model LORE is as powerful and accurate as nLORE without the computational cost, time, and resources required in probabilistic approaches.

3.1. Architecture of the Model

The underlying idea in the use of RNNs in NLP scenarios is that language must be treated as a temporal phenomenon, that is, a linguistic realization is perceived as a sequence of tokens that are combined one after another, where the prediction of a given word is dependent on earlier words. A bi-directional RNN consists of two simple RNNs stacked on top of each other, where not only the previous values are taken into account but also the following ones by means of backpropagation and forward-propagation. For the hidden layer structure, LSTM served to keep in memory distant contextual information, which is appropriate given the nature of language, since nearby contextual information is not sufficient to predict sequences of words. For the output layer structure, we used a CRF layer on top of the bi-directional RNN with LSTM for its great predictive power when contextual information is provided (Figure 1).

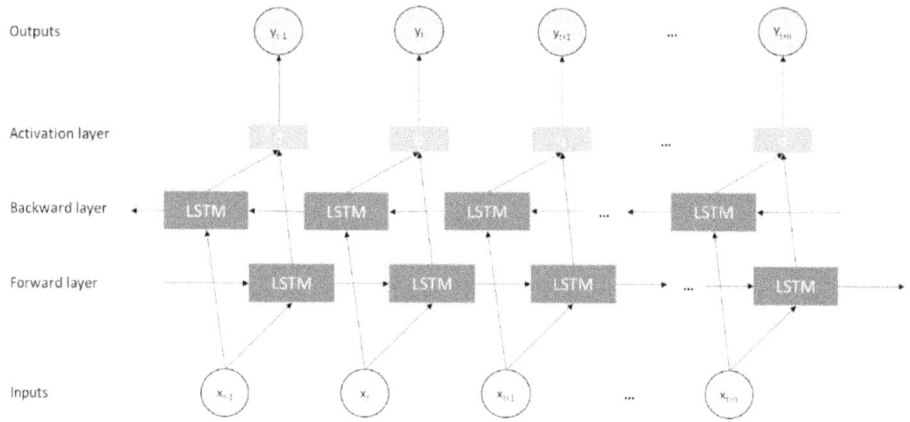

Figure 1. Neural architecture of the model.

3.2. Training Stage

Table 1 summarizes the optimal hyperparameters used for training nLORE. These values were chosen taking into account not only performance but also computational time and cost, striking a balance between both. Dropout values typically range from 0.2 to 0.5, representing a tiny fraction of output values which are set to zero during training; 0.5 provided the best performance. As for the hidden LSTM layer(s), we set only one hidden LSTM layer with a size of 200 neurons, to find a compromise between time and performance. The learning rate, whose value typically ranges from 0.001 to 1, affects the learning speed by determining how quick the weights shift in each training iteration. The lower the learning rate, the more accurate the estimation because of smaller increments in weights, but the more time it takes to train the network; 0.1 gave the best results. Mini-batch size selects a random subset of training data (i.e. a mini-batch) to be used in each iteration. Optimal performance in terms of speed and accuracy was observed with mini-batch sizes of 2 up to 32; 16 was the optimal number in terms of the cost-benefit ratio.

Table 1. Hyperparameters in the training phase of nLORE

Parameters	Values
Network type	Bidirectional RNN
Dropout	0.5
Hidden layer type, number and size	LSTM, 1, 200
Output layer type	CRF
Learning rate	0.1
Mini-batch size	16

We trained nLORE through eight models resulting from the combination of different corpus sizes (i.e. 1,000, 3,000, 5,000, and 7,000 tweets) and linguistic features (i.e. basic or extended features). The whole collection of tweets was compiled using a set of keywords related to a variety of crisis and emergency-related events: *earthquake, flood, car accident, bombing attack, shooting attack, terrorist attack,* and *incident.* After pre-processing the collection of tweets, which involved removing duplicates and nearly identical tweets, we obtained 8,063 microtexts, of which 7,000 were used for training and 1,063 were used for validation. The corpora were annotated taking into consideration the location types shown in Table 2.

Table 2. Location types in LORE and nLORE

Location type	Subtypes	Examples
Geopolitical entities	country, state, region, province, city, town, kingdom, villa…	*China, New York, Buenos Aires*
Natural landforms	mountain, mount, ridge, volcano, valley, lake, river, shore, beach, park, canyon…	*Mount Everest, Grand Canyon, Lake Michigan*
POIs	building, museum, school, station, stadium, garden, café, tavern, hospital, court, theater, residence, zoo, casino, square…	*Victoria Coach Station, Fox Valley Animal Referral Center, Hotel Park Villa*
Traffic ways	Highways, roads, addresses…	*I-90 SW, 11 Croydon Road, northbound J19*

The locative markers surrounding these location types were also annotated on the basis of the following taxonomy:

- Distance markers: kilometre(s), kilometer(s), km(s), metre(s), meter(s), m(s), mile(s), mi(s),yard(s), yd(s)…
- Directional markers: North, N, Northeast, NE, Northwest, NW, South, S…
- Movement markers: Northbound, NB, Southbound, SW…
- Temporal markers: hour(s), hr(s), h(s)…

In this regard, the complexity of the phrasal structure of a locative reference is illustrated in Figure 2, where the asterisk is used to mark optionality, and the double asterisk refers to the optional presence of locative markers either at the beginning or at the end of the locative reference.

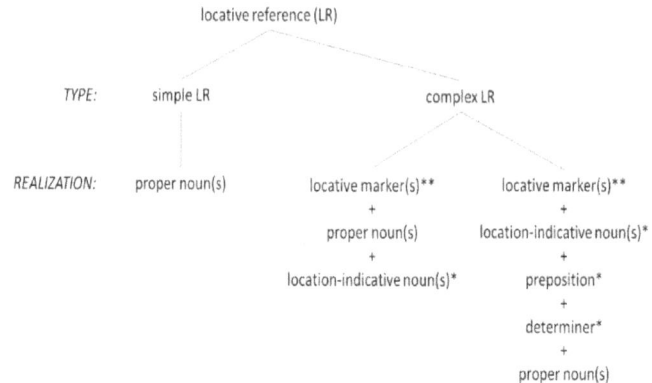

Figure 2. Phrasal structure of locative references

We adjusted the size of the training corpus (i.e. 1,000, 3,000, 5,000, and 7,000 tweets) in the training phase to determine whether linguistic feature engineering can overcome the problem of sparse data in a probabilistic framework. Moreover, considering that one of our purposes was to examine the usefulness of linguistic features, we differentiated between a basic model of features, focusing on those typically used in NER (i.e. token form and POS tag), and an extended model, including features such as token form, POS tag, presence of locative marker, and record in our place-name dataset obtained from GeoNames or in our location-indicative noun dataset obtained from WordNet (including words such as *street, restaurant, highway*, etc.)[2].

Other features employed were template features, context template features and word embeddings trained on a corpus of over 3.8m English tweets (Cheng et al., 2010)[3]. The template features for the basic model consisted of the preceding, current and following token form (i.e. t_{i-1}, t_i, and t_{i+1}) as well as their POS tags (i.e. p_{i-1}, p_i, and p_{i+1}) and the combination of both. For the extended model, template features also included:

- the preceding, current and following tokens belonging or not to the place-name dataset (i.e. pn_{i-1}, pn_i, and pn_{i+1}) in combination with their POS tag,
- the current and following token that may belong in the location-indicative noun dataset (i.e. li_i, li_{i+1}) in combination with their token form, and
- the preceding, current and following token that may be a locative marker (i.e. lm_{i-1}, lm_i, lm_{i+1}).

With respect to context template features, the context window considered the preceding token, the current token, and the following token, so that template features previously discussed could also be combined with the preceding and following tokens. Regarding

---

[2] We refer the reader to [9,10] for a much more detailed explanation on the linguistic-based criteria for the development of LORE, i.e. the semantic coverage of locative references and the compilation of lexical resources.

[3] Available on the following link: https://archive.org/details/twitter_cikm_2010

the trained word embeddings, we used the skip-gram model with Word2Vec vectors [16]. The intuition behind the role of word embeddings in our model was that it could help detect tokens that are similar to those which are place names in GeoNames, location-indicative words or locative markers, and also neighboring tokens that are part of or that predict a given locative reference, such as prepositions, proper nouns, etc. We specified a context window for word embeddings that included the current token, the preceding two tokens, and the following two tokens. All these features provided the best results in terms of computational resources and time in the training and evaluation phases. The resulting eight models were saved when they achieved the best result in the validation corpus. Iterations ranged from 20 to 25: the larger the corpus and the higher the number of feature types that were taken into account, the greater time it took to train each model.

4. Evaluation

The evaluation corpus was compiled on a different date to obtain different types of incidents with different locative references, thus ensuring that there was no overfitting. The compilation, pre-processing and annotation steps were the same, resulting in a corpus of 1,372 tweets.

We followed the evaluation measures most widely used in NER (i.e. precision, recall, and F1), as shown in Eqs. (1)-(3), where TP, FP and FN denote the number of true positives, false positives and false negatives, respectively.

$$Precision = \frac{TP}{TP+FP} \tag{1}$$

$$Recall = \frac{TP}{TP+FN} \tag{2}$$

$$F1 = 2 \cdot \frac{Precision \cdot Recall}{Precision+Recall} \tag{3}$$

As can be noted, F1 is the harmonic mean of precision and recall. All the measures return a value ranging from 0 to 1. The evaluation method was strict, thus considering full matches only.

5. Results and Discussion

Figure 3 shows the precision, recall, and F1 scores of nLORE in the eight models resulting from the combination of different corpus sizes and linguistic-based features.

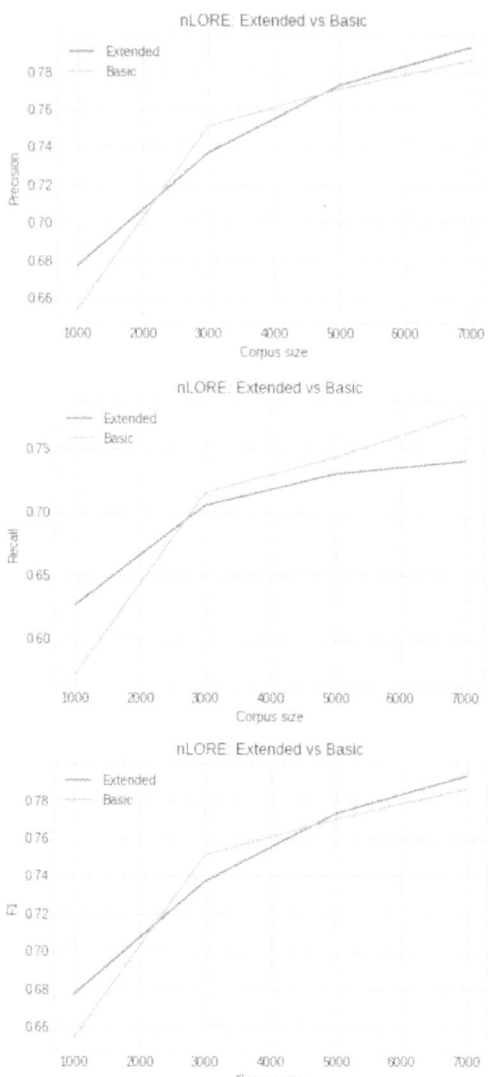

Figure 3. Precision, Recall, and F1 scores of nLORE, respectively.

It should be noted that precision was higher in the basic 1k and 3k models but increased exponentially in the extended model when the training dataset was larger, outperforming the basic 5k and 7k models. In the case of recall, the opposite is true. Whereas recall was higher for the extended 1k model, it was eventually outperformed by the basic 3k, 5k and 7k models. Therefore, the contribution of the extended linguistic features seems to be, at best, poorly significant and, at worst, only incidental. However, a trend can be noted with the 1k extended model. In fact, if we were to sketch the main reason in support of the use of extended linguistic features, this would be the better performance of the extended model over the basic model when corpus size is fairly small. In contrast, the difference in performance between basic and extended models tends to diminish as corpus size becomes increasingly larger and, even in this scenario, providing extra linguistic features might seem a bit counterproductive, since they worsen performance very slightly, as can be concluded from the recall scores, despite improving precision.

Our intuition behind the higher precision of the extended 5k and 7k models is that the extended linguistic features may help avoiding the extraction of wrong instances that cannot be properly discarded by the token form and POS tag features alone. However, adding extended linguistic features might have a negative impact –though very slight– on the identification of right locative references, as spotted in the recall scores. Incorporating these features may have incidentally made the process of locative-reference extraction stricter, which explains the fewer number of FPs and the greater number of FNs.

In the light of these results, we selected the 7k extended model because it offered the best F1 score. Therefore, we compared the 7k extended model against our rule-based model (Figure 4), where we can observe that nLORE slightly outperformed LORE in terms of precision (0.85 vs 0.73) and F1 (0.79 vs 0.76) but not recall (0.74 vs 0.79).

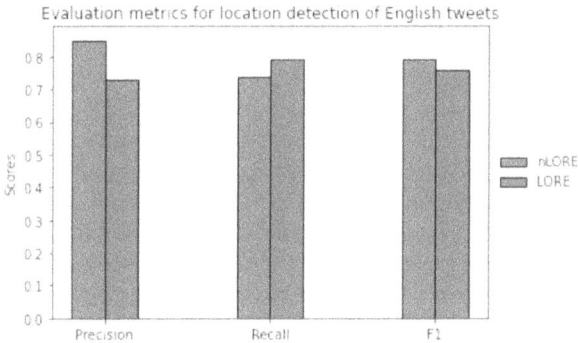

Figure 4. nLORE vs LORE.

The 7k extended nLORE model was able to extract all the locative references in Example 1 and Example 2, whereas LORE missed part of the locative references *NY 22, NY 138,* and *Northbound Exit Ramp.*

(1) Incident on <u>I684 NB</u> at <u>Exit 6A</u> – <u>NY 22</u> to <u>NY 138</u> – <u>Goldens Bridge</u> (<u>Northbound Exit Ramp</u>).
(2) earthquake <u>995 km north-east of Whangarei</u>

At times, nLORE failed in the extraction of locative references such as *Port Harcourt aiport* (Example 3), whereas LORE was able to detect it.

(3) FAAN shuts down <u>Port Harcourt airport</u> temporarily over bush fire incident

Sometimes, LORE misidentified person names as locative references. In Example 4, *Clancy* was extracted as a locative reference because the rule-based module could not filter it out. However, nLORE was able to prevent its extraction.

(4) I've read all of Clancy novels

In Example 5, LORE relies on rule-based patterns to detect *I-39B* as a traffic way, but nLORE missed this one, despite traffic ways being annotated as locative references.

(5) Police activity on I-39B

In short, nLORE incorporates in the system symbolic-based knowledge through linguistic feature engineering. This serves to intelligently infer and predict locative patterns in tweets by avoiding falling into errors caused by POS tags mislabeled by the Stanford POS tagger or mismatched locative items from the place-name dataset, which LORE could not prevent. The rule-based system tended to overmatch, negatively affecting precision but correlated with better recall, whereas nLORE was stricter, avoiding the extraction of wrong items but missing a few instances, hence negatively affecting recall but improving precision.

However, in terms of speed and efficiency, LORE greatly outperformed nLORE. On the one hand, LORE did not need a training phase, hence saving a great deal of time in its development. On the other hand, whereas LORE was able to process and extract all the locative references from the evaluation corpus in 12 seconds, nLORE lasted for almost 2 minutes. In other words, the rule-based model was almost 10x times quicker than its probabilistic counterpart. This supports the claim that rule-based models show better runtime efficiency than probabilistic models [17]. This is especially important in real-life, real-time emergency-based or crisis-related scenarios where time management becomes of utmost importance for emergency responders to locate the incident and act as quickly as possible. Moreover, LORE was already tested with a different evaluation corpus of English tweets [9], achieving similar evaluation numbers, which confirms its generalizability to other unseen collections of tweets. In addition to that, LORE already supports Spanish and French, having better performance than other NER systems with evaluation corpora in those languages [10]. Therefore, nLORE would be preferred over LORE in terms of performance, whereas LORE would be the preferred option when speed and efficiency are the most important concerns, which is indeed the case in emergency- or crisis-related settings. Even if rule-based approaches are seen as a dead-end technology in academia, they are nevertheless employed in companies and industries for real-life scenarios that demand efficiency, quick performance and scalability [17].

For future research, we could use larger training corpora with nLORE to check if performance can be enhanced and how speed is affected. We also leave for future research the use of different novel approaches in NLP, such as contextualized word embeddings and pre-trained language models, e.g. BERT, or the adaptation of nLORE to other languages such as Spanish or French, so that nLORE could become a multilingual NER system.

6. Conclusion

We have presented nLORE, a deep-learning system for fine-grained locative-reference extraction in tweets, feeding off the linguistic knowledge provided by LORE and trained on a relatively small corpus of English tweets, so that locative references can be inferred and extracted with slightly greater accuracy than LORE. In some respects, nLORE overcame some of the limitations presented by LORE, as evidenced by evaluation results and the examples above discussed. This research also demonstrates that linguistic feature engineering in probabilistic approaches may still provide a much-valued added benefit, especially when confronted with data sparsity, which could pave the way for more linguistic-oriented computational work in the field of NER. This approach goes in line with recent calls in linguistic and computational communities, requesting a greater interaction between linguistics and artificial intelligence [18]. At the same time, LORE still shows promising speed and efficiency and a performance almost as good as that of nLORE, which means that rule-based models are still a valuable and viable solution, despite the often underappreciated value of such models in the current NLP landscape. Given the importance of an efficient and quick response of emergency-based services and authorities in crisis-related scenarios, rule-based models such as LORE can help provide great value to systems oriented to give aid to affected people as quickly as possible, potentially saving lives or preventing further damage in affected areas.

Acknowledgements

Financial support for this research has been provided by the Spanish Ministry of Science, Innovation and Universities [grant number RTC 2017-6389-5: project WATERoT] and by the European Union's Horizon 2020 research and innovation program [grant number 101017861: project SMARTLAGOON].

References

[1]   Zhang C, Fan C, Yao W, Hu X, Mostafavi A. Social media for intelligent public information and warning in disasters: An interdisciplinary review. International Journal of Information Management. 2019; 49: 190-207.

[2]   Martínez-Rojas M, Pardo-Ferreira M del C, Rubio-Romero JC. Twitter as a tool for the management and analysis of emergency situations: A systematic literature review. International Journal of Information Management. 2018; 43: 196-208.

[3]   Middleton SE, Kordopatis-Zilos G, Papadopoulos S, Kompatsiaris Y. Location Extraction from Social Media. ACM Transactions on Information Systems. 2018; 36(4): 1-27.

[4]   Wallgrün JO, Karimzadeh M, MacEachren AM, Pezanowski S. GeoCorpora: building a corpus to test and train microblog geoparsers. International Journal of Geographical Information Science. 2018; 32(1): 1-29.

[5]   Imran M, Castillo C, Diaz F, Vieweg S. Processing Social Media Messages in Mass Emergency: A Survey. ACM Computing Surveys. 2014; 47: 1-38.

[6]   Singh L, Bansal S, Bode L, Budak C, Chi G, Kawintiranon K, et al. A first look at COVID-19 information and misinformation sharing on Twitter. 2020; Available from: http://arxiv.org/abs/2003.13907

[7]   Inkpen D, Liu J, Farzindar A, Kazemi F, Ghazi D. Location detection and disambiguation from twitter messages. Journal of Intelligent Information Systems. 2017; 49 (2): 237-253.

[8]   Das RD, Purves RS. Exploring the Potential of Twitter to Understand Traffic Events and Their Locations in Greater Mumbai, India. IEEE Transactions on Intelligent Transportation Systems. 2019; 1-10.

[9]   Fernández-Martínez NJ, Periñán-Pascual C. LORE: a model for the detection of fine-grained locative references in tweets. Onomázein. 2021.

[10]   Fernández-Martínez NJ, Periñán-Pascual C. Knowledge-based rules for the extraction of complex, fine-grained locative references from tweets. Revista Electronica de Linguistica Aplicada. 2020; 19(1): 136-63.

[11]   Malmasi S, Dras M. Location mention detection in tweets and microblogs. In: Hasida K, Purwarianti A, editors. Communications in Computer and Information Science. Singapore: Springer; 2016; 123-134.

[12]   Al-Olimat HS, Thirunarayan K, Shalin V, Sheth A. Location Name Extraction from Targeted Text Streams using Gazetteer-based Statistical Language Models. In: Proceedings of the 27th International Conference on Computational Linguistics. Association for Computational Linguistics; 2018; 1986-1997.

[13]   Wang J, Hu Y, Joseph K. NeuroTPR : A Neuro-net ToPonym Recognition Model for Extracting Locations from Social Media Messages. Transactions in GIS. 2020; 1-22.

[14]   Gelernter J, Balaji S. An algorithm for local geoparsing of microtext. GeoInformatica. 2013; 17(4): 635-67.

[15]   Cheng Z, Caverlee J, Lee K. You are where you tweet: A content-based approach to geo-locating Twitter users. In: Proceedings of the 19th ACM International Conference on Information and Knowledge Management. 2010; 759-768.

[16]   Mikolov T, Chen K, Corrado G, Dean J. Efficient Estimation of Word Representations in Vector Space. In: Proceedings of the International Conference on Learning Representations (ICLR 2013). 2013; 1-12.

[17]   Chiticariu L, Li Y, Reiss FR. Rule-based information extraction is dead! Long live rule-based information extraction systems! In: EMNLP 2013 - 2013 Conference on Empirical Methods in Natural Language Processing, Proceedings of the Conference. 2013; 827-832.

[18]   Linzen TAL. What can linguistics and deep learning contribute to each other? Response to Pater. Language. 2019; 95(1): 99-108.

Intelligent Environments 2021
E. Bashir and M. Luštrek (Eds.)
© 2021 The authors and IOS Press.
This article is published online with Open Access by IOS Press and distributed under the terms
of the Creative Commons Attribution Non-Commercial License 4.0 (CC BY-NC 4.0).
doi:10.3233/AISE210104

Solar Powered 4-Wheel Drive Autonomous Seed Sowing Robot for Rough Terrain

Fawad NASEER[a,1], Akhtar RASOOL[b], M. Zia QAMMAR[b] and Rafay SHAHOOD[c]

[a] *Head of Computer Science Department, Beaconhouse International College, Pakistan*
[b] *CS Department Government College University Faisalabad, Pakistan*
[c] *CS Department, Beaconhouse International College, Faisalabad, Pakistan*

Abstract. In the era of fourth industrial revolution, automation is a growing trend in almost every industry. The innovation in agriculture equipment is one of the major phase for civilized life, and the development of agricultural tools is a fundamental need towards the improvement of agriculture. Farmers use the same traditional methods and equipment for all ages, for example: seeds, spraying, weeding, etc., which have problems such as slow growth rate, irrigation, fertilization, crop monitoring of large areas. An autonomous, low-maintenance and portable robot can serve this purpose more accurately and efficiently with much better performance output. This article introduces the proposed model and design of a solar powered 4-wheel drive robot which works to plant seeds while cultivating the soil in a multi terrain surface. The robot avoids human effort ranging from the field path following to uniformly sowing of seeds at equal distance intervals using field area constraints prescribed by the farmer. This article presents the step by steps designing a robot and the parameters to be considered before creating a prototype. This robot increases the efficiency of seeding by exact measurement and distance, flattening of a surface, spraying of initial water, and also reduces random spreading of seeds problems.

Keywords. 4-Wheel Robot, Intelligent Farming, Arduino, Seed Sowing, Solar-powered, Agriculture, 4IR.

1. Introduction

World population is expected to grow about more than 8 billion by 2035, which will accelerate the global food consumption. To increase yields, the entire planting process needs to be systematically improved to maximize the growing capability to overcome this scenario. Agriculture is adapting automation in agricultural products and field cultivation methods. Subjective processing has become the most practical technology in agricultural services as it allows understanding, studying and responding to different circumstances to improve efficiency. Proximity sensor, humidity sensor and image sensor are different technologies which are mainly used as a judicious combination of information. An example of such high resolution data is soil analysis. Proximity detection requires the sensors to be in contact with the ground or very close. It helps characterize the soil below the surface at a specific location in a field [1]. Generally experienced farmers are familiar with traditional multi-row planting equipment.

---

[1] Fawad Naseer, Head of Computer Science Department, Beaconhouse International College, Recognized Teaching Centre of University of London, 54- West Canal Road, Faisalabad, Pakistan; E-mail: fawad.naseer@bic.edu.pk

Farmers have following restrictions in normal classical sowing methods [2]:
•   Inconsistent distribution of seeds in the corps field.
•   Due to Inconsistent distribution of seeds, which will result in gaps between plants.
•   Dead seeds due to placement of seeds at incorrect depth of soil.

(a)   (b)   (c)

Figure 1. Classical methods of sowing seeds.

Classical methods of sowing and spraying are available in almost all parts of the country, as shown in Fig 1, which results in low yields [3]. Further factors, such as insufficient energy availability and low mechanization rates, also affect crop productivity. Delays in sowing, harvesting, threshing, and improper nursery preparation also reduce crop productivity. Therefore, agricultural modernization cannot be avoided to meet population growth and food needs due to rapid industrialization [3].

Several different agricultural mechanisms have been established to speed up the seed distribution method in the corps field. The manual planter [4] requires farmers to operate and wheel rotation is used to operate the seed release mechanism. In contrast, electric seed growers use diesel engines or electric motors to drive internal mechanisms and are often mounted on tractors. The main drawback of these systems is their low energy efficiency as they require a lot of labor or fossil fuels to operate.

2. Related Works

Because there is no effective equipment to support farmers, new technology must be implemented to empower farmers with the capability to ensure and cope up the required demand of global need.

In [5], H. Pota R Eaton, J Katupitiya and S D Pathirana conclude that sowing bulks of seeds is necessary because the number of qualified sowers is greatly reduced. Planting distance and plant population are important factors in maximizing crop yield. By using the microcontroller, 8051 is used for communication between input and output devices. The main disadvantage of this model is that it has only one mechanism.

In [6], authors discussed the mechanisms of automatic tillage, sowing, fertilization and watering. Automatic sowing and fertilization is carried out using solenoids. The soil moisture sensor is used for automatic watering applications using Rasbeery pi and the Internet. The Arduino mega 2560 is used for the robot process and the Rasberry pi is used for communication with the robot. Internet systems are used by farmers to communicate with robots to run processes automatically. It reduces the burden on farmers.

In [7], the development of a GSM-based automatic drip irrigation system was the main focus. The system uses multiple sensors to get the state of the field and the irrigation schedule is based on this data. Microcontroller and App / DTMF are used for control and communication respectively. The author has developed a system for planning and scheduling irrigation processes based on real-time information on the spot. Several irrigation issues have been resolved, including the physical labor of farmers to manage irrigation, wasted water and wasted time.

The above research articles have helped us to understand the different aspects induced by research on agricultural robots. The robot designed in the literature search above has many issues with the robot's movement, sowing, power issues and other accuracy concerns. These questions are effectively addressed in this task. This work also sheds light on the future range of robots.

3. Theory of Seed Sowing

Sowing seeds is defined as the process of placing seeds under the soil so that it grows to become the plant. For comparison, planting is an acquaintance with the placing the propagules of plants in the soil for growing plants. which can be seedlings, roots, tubers, leaves or cuttings. Seeds may be sown directly or transplanted. The methods of sowing are enlisted and detailed as under:

- Drilling
- Dibbling
- Broadcasting
- Sowing with bamboo plough

As all mentioned above techniques to sow seeds in field, we come to know that conventional sowing has restrictions such as not being able to sow uniformly by hand and insufficient control of sowing depth, and labor demand is high, which results in sowing at uneven depths and have a negative effect on the total grown plants.

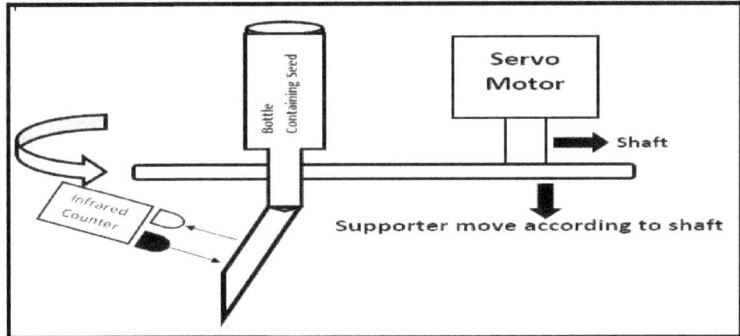

Figure 2. Seed dropping mechanism from seed container along with infrared based seed counter.

The technique which we have develop is that all seeds are collected in a vertical bottle which is mounted on a robot and filled by the farmers as required and dropped using a servo motor to flip the slider and drop seed one by one as shown in Fig. 2. There is an infrared sensor mounted on this mechanism which counts every seed and dropped precisely according to the need of a farmer. Our technique to sow a seed in a soil is the drilling mechanism, which is mounted on a robot, it drills the soil and then insert the seed by using the stated mechanism.

4. Design & Development of a Proposed Robot

There are a lot of fields where Artificial Intelligence is impacting human life with better performance results along with less human interaction. The proposed design of our robot is a large scale model to portray working capabilities with better performance including high accuracy of placing a seed under soil and placing seeds with accurate measured distance.

4.1. Construction

At initial stage robot mechanical design was developed using computer-aided design (CAD) and later we built it by using steel pipes, which is shown in Figure 3. The cylindrical seed boxes have been chosen to pour the seeds in bulk and for that 1.5-liter PET bottle have been used because they are light, cheap and easy to obtain. The payload consisting of seed boxes and electronic boxes is assigned a maximum weight of 23 kg.

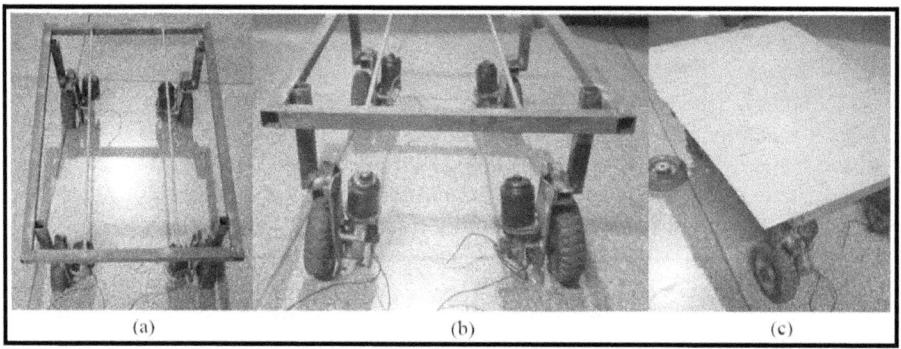

(a)   (b)   (c)

Figure 3. Steel rectangular pipe based robot chassis frame with four dc motors and tires.

We have designed the size of our robot frame as 20-inch x 30-inch x 13-inch to fit the payload frame. The frame is made of steel, which provides strong support for tough terrain, prevents the robot from rolling over, and has the power to effectively control its movements. Each tire is equipped with four DC motors, enabling the robot to move dynamically over rugged terrain and control it in a variety of circumstances at any cost. In the figure 3 shows a drive system consisting of four DC motors connected to all four buses. The diameter of the sprocket connected between the DC motor and bus is determined using a formula:

$$d = \frac{p}{\sin\left(180/_z\right)} \tag{1}$$

Where p is the pith of the sprocket, d is the diameter of the sprocket and z is the number of teeth on the sprocket. All the variables are decided on the concern of the required speed of the robot to be smooth and stable over different rough terrain. We can extract the dimension of the sprocket to be 93.76 mm by using the Eqn. (1) by using the number of teeth to be 17.

Once we able to get the dimension of the sprocket, from this we can calculate the speed of our robot by using Eqn. (2), where N is the rotation per minute of a sprocket and for this robot we use it to be 200 rpm:

$$v = \frac{\pi d N}{60} \tag{2}$$

we can use Eqn. 3 to calculate the required drive power transmitted of all four motors in a robot, each with a stall current of 10A. By using the dimension and velocity which we have calculated and maximum torque to be generalized to be 5 Nm as mentioned on each motor.

$$P = \frac{2\pi N T}{60} \tag{3}$$

The block diagram consists of an Arduino microcontroller, which is the main controller of the entire system along with DC motors, sensors and all helping components as shown in Fig 4. Solar panels connected to the battery for energy storage, as well as a charging circuit that supplies + 5 V to the Arduino board. Our battery system provides + 12 V power supply to control the DC motor using the L298 motor driver module.

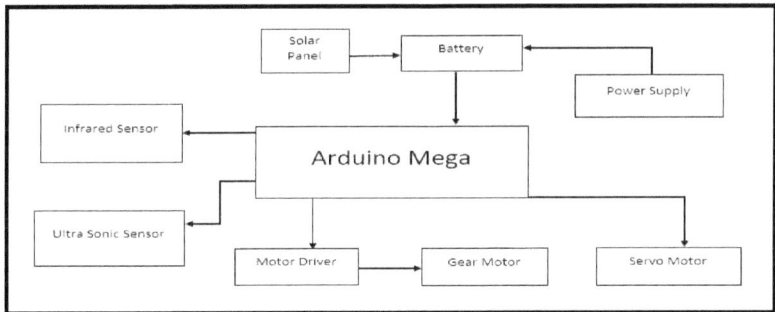

Figure 4. Blok diagram of complete electronic components

4.2. Electronic Components

The electronic system of robot consists of a microcontroller and drive circuits required to connect a robot to various mechanical parts as shown in Fig 4. The microcontroller (MCU) is an Arduino Mega 2560, which is the main controller and brain of a complete electronic system. This was chosen because of its high processing power and wide range of peripherals attaching capability, which makes it easy to apply control algorithms. When the robot is placed on the surface of the field, it intelligently orients around the field to cover all areas. The edge of the field was detected using the ultrasonic proximity sensor HC-SR04. This component is used to identify areas that reside on field boundaries [8]. The MCU uses a motor driver connected to a DC motor to control the movement of the robot. Each motor driver consists of a H bridge power MOSFET connected to the MCU.

4.2.1. Arduino Mega 2560

Arduino is an open source physical computing platform based on a simple I / O board and development environment. It is an Atmega 2560 based microcontroller board, which is used to interact with all sensors and actuators. It keeps all the components well synchronized for the proper functioning of the robot.

4.2.2. DC Motor Driver L298N

The L298N is a H-bridge driver with high current load. The L298N can supply up to 1A at voltages between 4.5 and 36. The L298N is designed to drive inductive loads such as solenoids, relays, DC motors, bipolar stepper motors, and other powerful motors. application of current or high voltage.

4.2.3. Ultrasonic Sensor HC-SR04

HC-SR04 ultrasonic sensor is used to calculate the distance to an object using sonar signal. It uses a non-contact ultrasonic sounder to measure the distance to an object and consists of two ultrasonic transmitters (mainly loudspeakers), a receiver and a control circuit [9]. The transmitter emits high-frequency ultrasonic waves that are reflected off nearby solids, and the receiver tracks the reflected echo. This echo is then processed by the control circuit to calculate the time difference between the transmit and receive signals [10].

4.2.4. Servo Motor

Servo motors are half-turn actuators used in applications that require a high degree of swing angle accuracy. This is because the servo motor has a feedback sensor. The feedback sensor measures the difference between the set and the required angle to ensure that you reach the end position more accurately.

4.2.5. Power Source

A battery is a source of energy that provides a surge (voltage) of energy to circulate current in a circuit. It powers the entire circuitry including all the components attached to the system. Our entire system is powered by a pair of 18650 Li-ion batteries through the battery management system(BMS) [11]. Sonar sensors have a built-in voltage regulator; hence it is powered from any DC voltage from 5 volts. Modules are a group of cells that are electrically connected and packaged in a frame (most commonly known as a solar panel).

5. Functioning of The Proposed Robot

The working principle of the developed system is described as a flowchart in Fig. 5. The code is developed to enable the robot to move in all directions which is being controlled intelligent by the microcontroller Arduino Mega 2560 to automate the whole process of seeding.

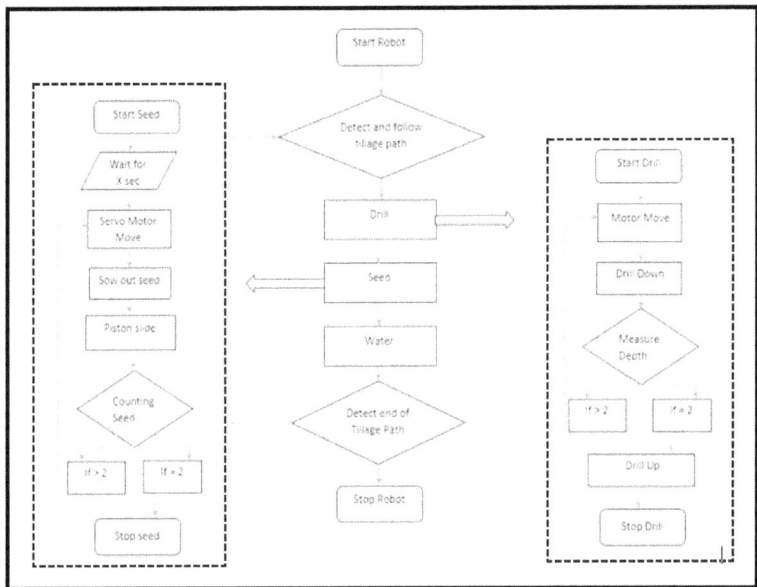

Figure 5. Flowchart of functioning of proposed robot

The robot starts to move forward and the seeds are planted evenly at an equal distance. The seeds are stored in a tapered seed bottle, so the seeds can easily enter the funnel through the attached hose. The seed placement mechanism runs smoothly and uses servo motors to complete the seeding tasks. It consists of two-stroke sliders used to control the opening of the hole from which the seeds are poured. After each drop of a seed, there is an infrared sensor which is used to count the number of seeds passed through the funnel to the hole. After pouring of pair of seeds, water sprinkle is placed, its duty is to sprinkle water on that sowed seed. After watering task is to flat the surface and for that a flap which is attached behind the robot to cover the soil as shown in Fig 6.

| (a) | (b) | (c) | (d) |

Figure 6. (a) Seeding drop mechanism, (b) Water sprinkler, (c) Flapper to flap surface & (d) Seed container.

When the robot deviates from the intended direction of motion due to undulations in the tillage path, the input provided by the ultrasonic sensor varies. This aberration is reduced by using the microcontroller which alters the power of the individual motors as described in Table 1. Thus the controller allows the device to traverse in the correct direction.

Table 1. Motor direction controlling algorithm

Movement Type	Motor A(Left Front)	Motor B(Left Rear)	Motor C(Right Front)	Motor D(Right Rear)
Straight	ON	ON	ON	ON
Left	OFF	OFF	ON	ON
Right	ON	ON	OFF	OFF

Motor A and B are on left side of a robot and Motor C and D are on right hand side of a robot. Ultrasonic sensors are used to determine the end of the field and, when detected, rotate the robot in different directions of travel. This allows the robot to plant seeds in subsequent rows that are zigzag known as tillage rows as shown in Fig 7.

     (a)   (b)   (c)

Figure 7. (a) Top view of robot, (b) Front bottom view of robot & (c) Side view of robot.

6. Results and Discussion

In the first series of experiments which was performed to validate the possibility to drill the accurate required depth to pour a pair of pea seeds to germinate. In this experiment, the planting testbed replicated the field environment which comprises of soil. The drill was prepared so that the precisely hole can be achieved at precise depth and pouring of 2 seeds as a pair in that hole. The prepared drill was prepared and dig into soil and waited for 30 days after successfully sowing of seeds. The moisture in the soil would dissolve the seeds entirely. All seeds inside a hole successfully germinated into pea cotyledon which is shown in Fig. 8.

     (a)   (b)   (c)   (d)

Figure 8. Seed germination after drilling hole from a robot.

In the second series of experiments, we have tried to sow different types of seeds and analyze required depth for each seed and the required spacing between sowing of seeds in field, we also analyze the size of different seeds and come to know that each seeds with different size required different size of digging hole for sowing, but our robot at current state is not able to dig a customized hole according to the seed, it ca be a future work to implement. Table 2 shows the comparison of the types of seeds, their dimension, required spacing of sowing and the required depth.

Table 2. Seed Analysis and Comparison

Type of seed	Length(mm)	Width (mm)	Diameter(mm)	Spacing(inch)	Depth(inch)
Pea	5.91	5.01	5.18	5-7.5	5
Tomato	6.3	6.9	4.23	6.5-8	5
Sun Flower	9.52	5.12	3.27	6-7	3
Chia	2.11	1.3	0.8	3.5-5	4
Flax	4.64	2.37	1.00	4.5-6	5

Figure 9 shows the graph of approximation of dropping seeds inside the hole dig by the drill in the different experiments we did before installing that seeding part on the robot. The graph explains that number of times the dropping of seeds and its accuracy to drop exactly inside the hole or scattering around the hole.

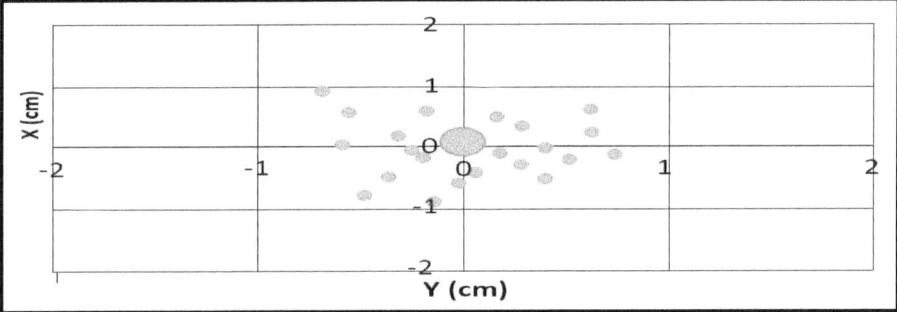

Figure 9. Experiment result of conducting different times seed dropping in hole.

7. Conclusion and Future Work

The design and development of a robot along with the Working procedure of the autonomous seed sowing mechanism have been proposed for the sole purpose of which is to automate the seed sowing process with flattening of the surface and watering over newly sowed seeds is discussed in this paper. All components except the sensors are individually designed, and the entire seed mechanism workflow and sequence of executions are illustrated in the block diagram above. Different models of seed collectors have also been developed, depending on the type of seed.

For the future purpose, the robot can be equipped with Global Positioning System (GPS) for more accuracy of the movement of robot and also for future perspective, robot can be equipped with more powerful processing board which can be used to calculate intelligent algorithm including machine learning algorithm for better performance of robot in context of accuracy in navigation around the field.

8. Acknowledgment

We would like to thank our respected Country Head Ms Raana Sarmad, for the inspiration and motivation towards creating a research culture in undergraduate students.

References

[1] PJ, S. K., "A brief survey of classification techniques applied Sheela to soil fertility prediction", International Conference on Emerging Trends, 2015.

[2] Pratik Hore. a, Goluprasad Gupta. a, Shrikrishn Deshmukh, "Modern seed sowing techniques and developing technologies," International Journal of Innovative and Emerging Research in Engineering, Volume 4, Issue 3, 2017.

[3] D. Ramesh, H.P. Girishkumar, Agricultural Seed Sowing Equipments: A Review International Journal of Science, Engineering and Technology Research (IJSETR), Volume 3, Issue 7, July 2014

[4] D. Ramesh and H.P. Girishkumar, "Agriculture Seed Sowing Equipments : A Review", International Journal of Science, Engineering and Technology Research (IJSETR), vol. 3, Issue 7, July 2014, pp. 1987 - 1992.

[5] H. Pota, R. Eaton, J. Katapriya and S. D. Pathirana, "Agricultural robotics: A streamlined approach to realization autonomous farming," in IEEE conference on industrial and information systems, 2007, pp. 85-90.

[6] Gollakota, A. and Srinivas, M.B., (2011), December. "Agribot —a multipurpose agricultural robot", In India Conference (INDICON), 2011 Annual IEEE (pp. 1-4). IEEE

[7] Lalwani, A., et al., 2015, December, "A Review: Autonomous Agribot For Smart Farming", Proceedings of 46th IRF International Conference, 27th December 2015, Pune, India

[8] J. Borenstein and Y. Koram, "Obstacle Avoidance with Ultrasonic Sensors", IEEE Journal of Robotics and Automation, vol. 4, no. 2, April 1988, pp. 213-218.

[9] S. Barai, D. Biswas and B. Sau, "Estimate distance measurement using NodeMCU ESP8266 based on RSSI technique," 2017 IEEE Conference on Antenna Measurements & Applications (CAMA), Tsukuba, 2017, pp. 170-1

[10] F. Noor, M. Swaied, M. AlMesned and N. AlMuzini, "A Method to Detect Object's Width with Ultrasonic Sensor," 2018 International Conference on Computing, Electronics & Communications Engineering (iCCECE), Southend, United Kingdom, 2018, pp. 266-271

[11] Khanna,A;Ranjan,"Solar-powred Android based Speed Control of DC motors through Secure Bluetooth," Communication systems and network technologies CSNT 2015 international conference (IEEE Publication), pp 1244-1249.

[12] P. Amiribavandpour, A. Kapoor, W. Shen and J. Shearer, "Thevmathematical model of 18650 lithium-ion battery in electric vehicles," 2013 IEEE 8th Conference on Industrial Electronics and Applications (ICIEA), Melbourne, VIC, Australia, 2013, pp. 1264-1269

Intelligent Environments 2021
E. Bashir and M. Luštrek (Eds.)
© 2021 The authors and IOS Press.
This article is published online with Open Access by IOS Press and distributed under the terms
of the Creative Commons Attribution Non-Commercial License 4.0 (CC BY-NC 4.0).

Subject Index

Intelligent Environments 2021
E. Bashir and M. Luštrek (Eds.)
© *2021 The authors and IOS Press.*

267

Author Index